Mechanics of Magnetostrictive Materials and Structures

Mechanics of Magnetostrictive Materials and Structures demonstrates the practical applications and uses for cutting-edge smart magnetostrictive materials. Exploring the analytical and numerical solution procedures and characteristics of these materials more generally, the book details how these materials respond to external factors. Exceptionally adjustable and adaptable, magnetostrictive materials are artificial structures that offer distinctive physical properties. Providing clear illustrations throughout, this book includes a comprehensive guide to the theory of magnorestrictive materials and its applications. Comprehensively assessing the practicalities of these smart materials, it also discusses vibration and buckling under different loads, alongside dynamic behavior.

Features:

- Presents vibration analysis of magnetostrictive materials and structures
- Demonstrates and analyzes the effect of implementing boundary conditions on the mechanical responses of magnetostrictive structures
- Examines the use of smart materials in engineering structures

Aimed at students and professionals working in the field of mechanics, materials and dynamics, the book is an essential guide to this rapidly developing area.

Mechanics of Magnetostrictive Materials and Structures

Farzad Ebrahimi and
Mehrdad Farajzadeh Ahari

CRC Press
Taylor & Francis Group
Boca Raton London New York

CRC Press is an imprint of the
Taylor & Francis Group, an **informa** business

First edition published 2024
by CRC Press
2385 Executive Center Drive, Suite 320, Boca Raton, FL 33431

and by CRC Press
4 Park Square, Milton Park, Abingdon, Oxon, OX14 4RN

CRC Press is an imprint of Taylor & Francis Group, LLC

© 2024 Farzad Ebrahimi and Mehrdad Farajzadeh Ahari

ISBN: 978-1-032-40926-9 (hbk)
ISBN: 978-1-032-40934-4 (pbk)
ISBN: 978-1-003-35542-7 (ebk)

DOI: 10.1201/9781003355427

Typeset in Times
by Apex CoVantage, LLC

Dedication

To my parents...
Farzad Ebrahimi,

To my family, friends and reasons...
Mehrdad Farajzadeh Ahari

Contents

3.2.4   Results 164
3.2.5   Concluding Remarks 172
References 181

Chapter 4   Wave Propagation of Magnetostrictive Materials
and Structures 183

4.1   Wave Propagation of Magnetostrictive Plates 184
4.1.1   Formulation 184
4.1.2   Solution 189
4.1.3   Results 190
4.1.4   Concluding Remarks 215
References 216

Chapter 5   Dynamic Analysis of Magnetostrictive Materials
and Structures 217

5.1   Dynamic Behavior of a Magnetostrictive Nanoplate
Subjected to a Nano-Mass-Spring-Damper
Stimulator 218
5.1.1   Laplace Transforms 218
5.1.2   Formulation 218
5.1.3   Solution 222
5.1.4   Results 227
5.1.5   Concluding Remarks 238
References 246

Index 249

Preface

One of the primary objectives of the book is to familiarize readers with the basic ideas and problems that might arise from the presence of magnetostrictive materials. It is essential to emphasize the fact that anybody may circumvent the challenge of analyzing various structures and situations that magnetostrictive material can withstand by reading this book.

This is a book that can be read on two separate ways, however equally enjoyable, levels. To begin, it is not necessary for the reader to have any prior knowledge of science in order to understand it. This readership is always kept in mind when writing any chapter of this work. The information provided may, at times, seem to be too much to take in: complicated diagrams with many structures.

I really hope that you won't be someone who gives up easy. Even if the chemical information or physical relations in a particular chapter is difficult to grasp, the scientific evidence that is provided, the citations from the original sources that are used, the conclusions that are derived, and the suggestions that are made may all be clearly understood.

The second set of readers will be made up of people who work in the academic world. It is not realistic for me to assume that members of the scientific community would see the thoughts and information presented in this book the same way I do. On the other hand, it is my sincere wish that the insights and data provided here will serve as a wake-up call for both the general public and the scientific community.

There are five different chapters in this book. Magnetostrictive materials and structures are presented with a high-level overview in Chapter 1. The static behavior of these materials is discussed in Chapter 2. The vibrational behavior of magnetostrictive materials and structures is covered in Chapter 3. The phenomenon of wave propagation in magnetostrictive materials is discussed in Chapter 4. Understanding the dynamic behavior of smart magnetostrictive materials and structures requires a fundamental grasp of the information presented in Chapter 5.

Farzad Ebrahimi
Associate Professor
Department of Mechanical Engineering
Faculty of Engineering
Imam Khomeini International University
Qazvin, Iran

Mehrdad Farajzadeh Ahari
Graduate Student
Department of Mechanical Engineering
Faculty of Engineering
Imam Khomeini International University
Qazvin, Iran

Acknowledgment

The authors are eager to extend their gratitude to those who played significant roles for bringing this book to life. The authors would also like to express their gratitude to Taylor & Francis Group/CRC Press for their assistance with the proposed book. The writers significantly depended upon their wisdom and friendliness. The authors' special gratitude goes to Jonathan Plant and Bhavna Saxena, executive editor and editorial assistant, respectively, at CRC Press, for their indispensable efforts during the publication procedure.

Authors

Farzad Ebrahimi is an associate professor at the Department of Mechanical Engineering, IKIU, Qazvin, Iran. His research interests include mechanical behaviors of nano-engineered systems, mechanics of composites and nanocomposites, functionally graded materials, viscoelasticity, and smart materials and structures. Dr. Ebrahimi has authored more than 400 high-quality peer-reviewed research articles in his fields of interest. He has also edited and authored multiple books for well-known publishers. He is an associate editor of the journal *Shock and Vibration*, an editorial board member of the *Journal of Computational Applied Mechanics*, and a distinguished reviewer whose expertise helps the editors of prestigious journals judge research articles.

Mehrdad Farajzadeh Ahari received his Master's Degree from Imam Khomeini International University as one of the top five students. His research interests are nanocomposites, the dynamic behavior of smart materials, and functionally graded materials. He has authored articles in his research field in well-known journals. As an undergraduate student, he also published a paper on his research in the *Journal of Energy*. He is a dedicated and experienced mechanical engineer with extensive knowledge of engineering principles, theories, specifications, and standards.

Abbreviations

Acronym	Definition
AC	Alternating Current
AVC	Active Vibration Control
CPT	Classical Plate Theory
DC	Direct Current
FGM	Functionally Graded Material
FSDT	First-Order Shear Deformation Theory
GDQT	Generalized Differential Quadrature Technique
GMA	Giant Magnetostrictive Actuator
GMI	Giant Magnetostrictive Injector
GMM	Giant Magnetostrictive Material
hOB	Human Osteoblast Cell
HSDT	Higher-Order Shear Deformation Theory
MA	Magnetostrictive Actuator
MCA	Magnetocrystalline Anisotropy
MCG	Magnetocardiography
MD	Molecular Dynamics
MEA	Magnetoelastic Anisotropy
MEG	Magnetoencephalography
MSNP	Magnetostrictive Sandwich Nanoplate
NDT	Nondestructive Test
NSGT	Nonlocal Strain Gradient Theory
XMCD	X-Ray Magnetic Circular Dichroism

Symbols

Symbol	Definition	Symbol	Definition	Symbol	Definition
U	Displacement vector in the x direction	k	Gradient index	∇^2	Laplacian operator
V	Displacement vector in the y direction	ε_{ij}	Strain vector	X_m	Admissible Glerkin functions along the x direction
W	Displacement vector in the z direction	σ_{ij}	Stress vector	Y_n	Admissible Glerkin functions along the y direction
ϕ_x	Rotation in the x direction	μ	Non-local parameter	∂T	Variation of strain energy
ϕ_y	Rotation in the y direction	λ	Length scale parameter	∂U	Variation of potential energy
u_0	Initial displacement in the x direction	$[C]$	Damping Matrix	∂W	Variation of external work
v_0	Initial displacement in the x direction	$[K]$	Stiffness Matrix	z_0	Neutral location
w_0	Initial displacement in the z direction	$[M]$	Mass Matrix	k_s	Shear correction factor
a	length	I_i	Moment of inertia	δ	Variation
b	width	E_f	Young's modulus of FG materials	ε	Strain
h_c	The thickness of the magnetostrictive layer	E_m	Young's modulus of magnetostrictive materials	σ	Stress
h_f	The thickness of the FG layer	ρ_f	Density of FG materials	K_w	Winkler coefficient
K_c	Coil constant	ρ_m	Density of magnetostrictive materials	K_p	Pasternak coefficient
e_{ij}	Magnetic modules	ω	Frequency	I_i	Mass moment of inertia
N_{xx}	Normal forces	Ω	Non-dimensional frequency	C_{ijk}	Elastic coefficient
M_{xx}	Bending moment	P	Critical buckling load	t	Time
ξ	The amount of porosity	H_{zz}	Magnetic field		

1 An Introduction to Magnetostrictive Materials and Structures

Background

In today's world, there is a significant focus on the use of intelligent materials to dampen the vibrations. Because of their rapid reaction to stimulation, low weight and volume, and simple transformation of electrical and magnetic energy into mechanical energy and vice versa, these materials are held in very high esteem. The use of composite materials is unavoidable because of the numerous benefits they offer, including high levels of strength and low overall weight, as well as the ability to control the structural characteristics of a material by altering the orientation of its fibers and the number of layers it contains. As a result, it is essential to have knowledge of the static and buckling behavior of composite structures and to have an accurate estimation of the change in locations in the direction of better design. In layered composites, an abrupt change in the material characteristics of one layer may induce undesired stresses to develop at the interface between the layers. These stresses can lead to plastic deformation or even the separation of the layers altogether.

An Introduction to Magnetostrictive Materials and Structures

In this chapter, a thorough analysis of magnetostrictive materials is obtained, covering both the big picture and the finer points of the topic. Several distinct effects and occurrences are broken down in extensive detail. Additionally, a wide range of magnetostrictive applications are covered in this chapter. This chapter may include information that is essential for understanding the subsequent chapters.

1.1 GENERAL CONCEPTS

It is not an exaggeration to say that the study of magnetostrictive materials is one of the areas of interest related to the exploration of smart materials. In point of fact, these materials are capable of transforming magnetic stimulations into elastic reactions, which may take the form of either stretching or contracting. It is important to take note of the fact that the magnetoelastic interaction in question

DOI: 10.1201/9781003355427-1

manifests itself when the structure in question is subjected to a magnetic field. Structures that have a magnetic effect are great candidates for use as ultrasonic generators, ultrasonic receivers, and echo detectors due to the qualities listed earlier.

Active metallic compounds known as magnetostrictive materials change shape in response to magnetic fields. Magnetoelastic coupling and the associated dependence of magnetic moment orientation on interatomic distance are responsible for these distortions. The linear or Joule magnetostriction is the most prevalent kind of magnetoelastic coupling, and it refers to the situation in which strains are recorded parallel to the direction of the magnetic field (see Figure 1.1 for an illustration of this). It should be observed that irrespective of the direction of rotation of magnetic moments, the material will elongate if the magnetostriction is positive. Moreover, the transverse dimension will decrease, which will result in the volume remaining the same. When the magnetostriction occurs in the negative direction, the sample diameter grows while sample length becomes shorter. After this, a magnetostriction curve that is symmetric is formed as a result of cycling the magnetic field. Although though Joule magnetostriction is present in the majority of magnetic materials, only a select few compounds containing rare-earth elements may produce stresses that are more than $1,000 \times 10^{-6}$. Magnetostrictive materials alter their magnetic state in response to stresses because of the symmetric magnetoelastic coupling. The Villari effect allows for the measurement methods of force and displacement.

1.1.1 JOULE PHENOMENA

The Joule effect, also known as magnetostriction, is the transformation of a ferromagnetic material that takes place in response to an applied magnetic field. The

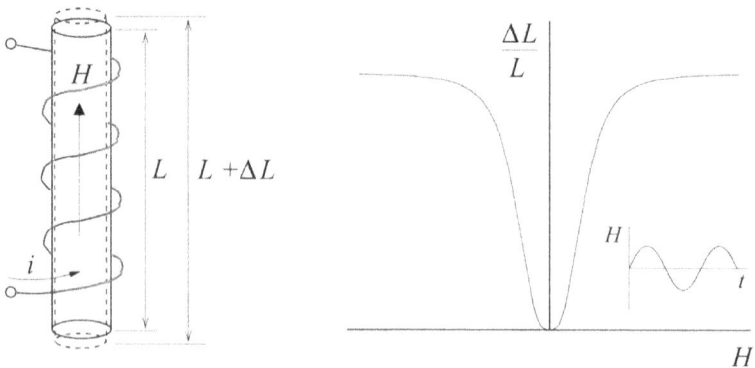

FIGURE 1.1 H-field magnetostriction, induced by a Joule force. It can be seen that (a) Sinusoidal field variation results in a field H that is roughly equivalent to the current in the direction of the solenoid's current when a voltage is supplied to the solenoid, and (b) the field H is proportional to the ratio of the diameter of the solenoid to the length of the field (inset).

implications of this phenomenon may be seen in a variety of applications. In the year 1842, an English scientist by the name of James Joule was the first person to describe the phenomenon that is known as magnetostriction. He made the discovery that a length change is shown by a control sample made of a ferromagnetic material, such as iron, when it is placed in the proximity of a magnetic field. In addition, Joule identified a chemical with a negative magnetostriction, although up to that point, scientists had only found compounds with a positive magnetostriction [1].

Materials may be broken down into one of two broad groups when discussing magnetostriction and the Joule phenomena.

- Positive magnetostriction
- Negative magnetostriction

When a material has positive magnetostriction, the material will expand when it is magnetic, while a substance with negative magnetostriction would cause the material to shrink when it is magnetized [2]. This variation in length may be rationalized as a consequence of the spinning of a number of relatively tiny magnetic spheres. This rotation and re-rotation generates internal strain in the structure of the material, which causes the material to deform. The strain in the structure causes the material to stretch (a process known as positive magnetostriction) in the direction of the applied magnetic field (Figure 1.2). The application of a higher powerful field results in a re-rotation of the pools that is both more pronounced and more pronounced in the direction of the magnetic field [1]. The incorrect point is achieved when all of the magnetic basins are oriented corresponding to the direction of the magnetic field. During this process of longitudinal expansion, there is a reduction in the cross-sectional area, which allows the volume to stay virtually the same. Due to the negligible nature of the volume change, it is possible to disregard its effects under typical circumstances of operation.

Figure 1.2 illustrates how the length change should behave in an ideal situation in response to an externally supplied magnetic field. The direction of a magnetic field is said to be "negative" when it is applied in the opposite direction. Yet, the

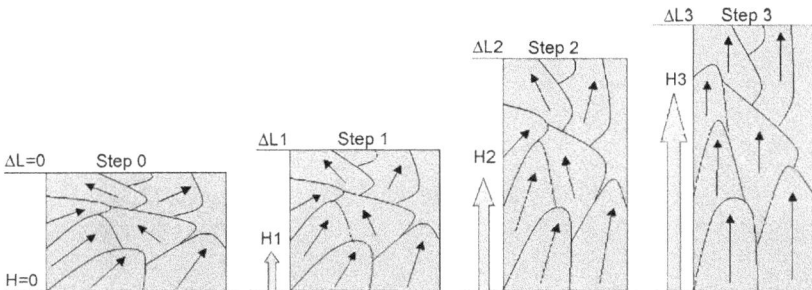

FIGURE 1.2 Schematic of magnetostriction phenomenon.

length change that is caused in magnetostrictive materials by a negative magnetic field is the same as that is caused by a positive magnetic field. Since the contour of the curve is comparable to that of a butterfly, the diagram in question has been given the name of a butterfly diagram. Figure 1.3 provides a simplified and schematic representation of the underlying physical processes that are responsible for the recycling of magnetic basins [3].

According to Figure 1.3, in the region between points 0 and 1 where the intensity of the externally supplied magnetic field is quite low, the magnetic basins exhibit a pattern of very little rotation. There is a minor rotating pattern that might appear in the permanent magnet depending on how the material was created. This pattern is known as the bias of the magnet. The degree of homogeneity of the base structure of the magnetostrictive material and the formulation of the material both have a significant impact on the strain that is produced by the material. Idealistically speaking, between points 1 and 2, the strain and magnetic field should have a linear relationship. Because of how easy it is to communicate, it is much simpler to anticipate how the material will behave, and a great number of devices are created specifically to operate in this field. Following point 2, the behavior changes to become nonlinear, and the cause for this change is viewed as being due to the fact that a large number of magnetic basins are aligned with the magnetic field. The occurrence of saturation takes place at point 3, which precludes an increase in strain from occurring beyond that point [3]. It is possible to achieve optimal strain via the use of pre-tension as well as magnetic bias. Because of the constantly shifting circumstances that occur during operation, magnetostrictive materials provide significant difficulties due to their behavior in a wide range of contexts. This is because the features of the material are altered. When it comes to defining the mechanical output of magnetostrictive actuators (Mas),

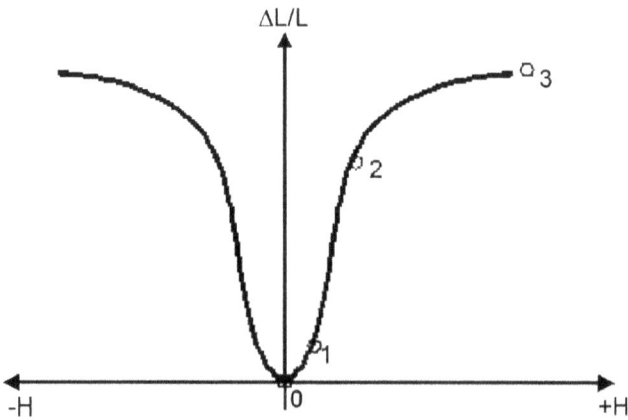

FIGURE 1.3   Strain in terms of magnetic field.

one of the essential factors that play a role is the maximum usable magnetoelastic strain [3]. When the magnetic field is removed from the vicinity of the magnetostrictive material, the form of the material goes back to how it was before. Hence, the process of magnetoattraction may work in both directions. Actuators may be built using the one-of-a-kind quality that magnetostrictive materials possess. Technology based on magnetostrictive forces provides controllability in addition to great power density [3].

Magnetostriction is a property that is shared by all ferromagnetic materials; nevertheless, its value is often rather low for the majority of these materials. When combined with alloys that raise their Curie temperature above the surrounding temperature, certain transition metals and rare elements exhibit a greater degree of magnetostriction than others do. This is especially true when the Curie temperature of the alloys is raised above the surrounding temperature. The Terfenol-D alloy, which is a mixture of Terbion, iron, and Dispersium, is the magnetostrictive material that has the best circumstances in terms of the amount of strain and the Curie temperature [3, 4]. This material is the winner among the magnetostrictive materials. As the field strength is increased, the positive value of the magnetostrictive coefficient in iron flips to a negative value [2]. Magnetostriction is a phenomenon that only happens in materials when temperatures drop below the Curie point; nevertheless, the temperature of the surrounding environment is almost always lower than the temperature of the furnace; this allows magnetostriction to have a practical use [3]. Currently, magnetostrictive materials that have the best possible characteristics are continuing to increase and improve.

Table 1.1 is a compilation of the nominal longitudinal strain for a variety of materials. It is interesting to note that some materials, like nickel, have a negative magnetostrictive coefficient, meaning that their length shortens when exposed to a magnetic field. In contrast, other materials, like Terfenol-D, have a positive magnetostrictive coefficient and their length lengthens when exposed to a magnetic field [3, 4].

1.1.2 Villari Phenomena

In addition to magnetostriction, there is also a phenomenon known as the Villari phenomenon [1], which is also often referred to as the piezomagnetic phenomenon. This phenomenon was initially described by Villari in 1864 [5] and has since gained widespread usage. It has been shown that a ferromagnet sample would undergo this phenomenon when subjected to mechanical stress, there is a corresponding change in the sample's magnetic permeability, which results in a density of magnetic flux and how it changes (magnetization) that is transmitted through the sample. In other words, the form of the magnetic material will change throughout the magnetization process (a phenomenon known as magnetostriction), and this will cause the material's magnetic characteristics to alter as a result of the strain. It may be reversed, and it is used in applications that include

TABLE 1.1

Nominal Saturation Strain and Curie Temperature of Magnetostrictive Materials

Material	Saturation strain (ppm)	Curie Temperature (K)
$SmFe_2$	−1,560	--
$CoFe_2O_4$	−110	793
Cobalt	−62	1,388
Nickel	−40	630
Iron	−14	1,040
$Co_{72}Fe_3B_6A_{13}$ (Amorphous)	0	--
82%Ni-18%Fe, Permalloy	0	--
Fe-3.2%Si	+9	1,015
45%Ni-55%Fe, Permalloy45	+27	--
87%Fe-13%Al	+30	673
$Fe_{66}Co_{18}B_{15}Si$	+35	--
Fe_3O_4, Magnetite	+40	860
49%Co-49%Fe-2%V, permendur	+70	1,253
$Fe_{100-x}Ga_x$, $15 \leq x \leq 28$ (Amorphous)	+250	--
Nickel-Cobalt	+186	--
$TbFe_2$	+1,753	--
$Tb_{0.3}D_{0.7}Fe_{1.93}$, Terfenol-D	+2,000	650
$Tb_{0.5}Dy_xZn$	+5,000	200
$Tb_{0.5}Zn_{0.5}$	+5,500	180

sensing [1]. In addition, increasing the stress above the elastic limit of the material (also known as hard labor) has the effect of lowering the magnetic permeability of annealed materials. The presence of internal strains, which are the result of excessive mechanical strains or hard sedimentation, makes the material harder not only from the mechanical point of view but also from the magnetic point of view. This is because internal strains are the result of hard sedimentation or excessive mechanical strains. Materials may be broken down into one of two broad groups when talking about the Villari phenomena, which is similar to the magnetostriction phenomenon are divided [2].

- Positive magnetostriction
- Negative magnetostriction

When a material has positive magnetostriction, an increase in the magnetization of the material occurs when it is stretched. On the other hand, when a material has negative magnetostriction, the degree to which it softens as it is stretched decreases. The sign of the strain caused by the stress (an increase for positive strain and a decrease for negative strain) and the type of material both have an effect on the magnitude of the change in magnetic flux that is caused by the change in magnetic permeability. This change can be either increased or decreased. It is possible to describe the Villari phenomena in terms of thermodynamic connections.

$$\frac{1}{l}\frac{\partial l}{\partial H} = \frac{1}{4\pi}\frac{\partial B}{\partial \sigma} \tag{1.1}$$

In response to the intensity of an applied magnetic field (H), the connection between changes in induction (B), tension (σ), and length (l) is described in Eq. (1.1). These three variables are all affected by the magnetic field. As can be seen from Eq. (1.1), magnetization rises with the application of tensile stress (provided that magnetostriction is positive), rises continuously with the field, and falls if Al is negative. All of these changes take place with the assumption that magnetostriction is positive. Moreover, the dependence of magnetization on stress may be calculated via energy terms linked to stress and the direction of self-excited magnetization in a pond. This can be done in a same manner as described before. The magnetic strain energy density is calculated as follows when seen in this light:

$$E_\sigma = \frac{3}{2}\lambda_s sin^2\theta. \tag{1.2}$$

This is seen in the connection between magnetostrictive expansion at Eq. (1.2), in saturation (λ_s) and θ is the angle that exists between saturation magnetization and tensile stress (σ). The above formula demonstrates that when λ_s and σ are positive and when stress is applied to the iron, the amount of energy required to maintain $\theta = 0$ is at its lowest possible value. As a result, the pools are stable when e is parallel to sigma. As a consequence of this, for materials of this kind, it often rises with elasticity. The pools have a tendency to spin with y perpendicular to the axis of tension when sigma is negative (as it is when nickel is under strain), and the magnetization diminishes as the tension increases. Figure 1.4 is a collection of diagrams that illustrate how the effects of stress and pressure on a variety of materials may be expressed graphically. The preceding illustrations illustrate, in a straightforward manner, the structure of the pools of materials, which helps in predicting the behavior of such materials [2].

Figure 1.5 depicts the impact that stretching has on the magnetization diagram of samples of Permalloy 68, which has positive magnetostriction, and nickel, which has negative magnetostriction. Some of the physical occurrences that are associated with the magnetostrictive phenomenon are shown in Figure 1.6, along with the theoretical connections that exist between the Joule and Villari phenomena.

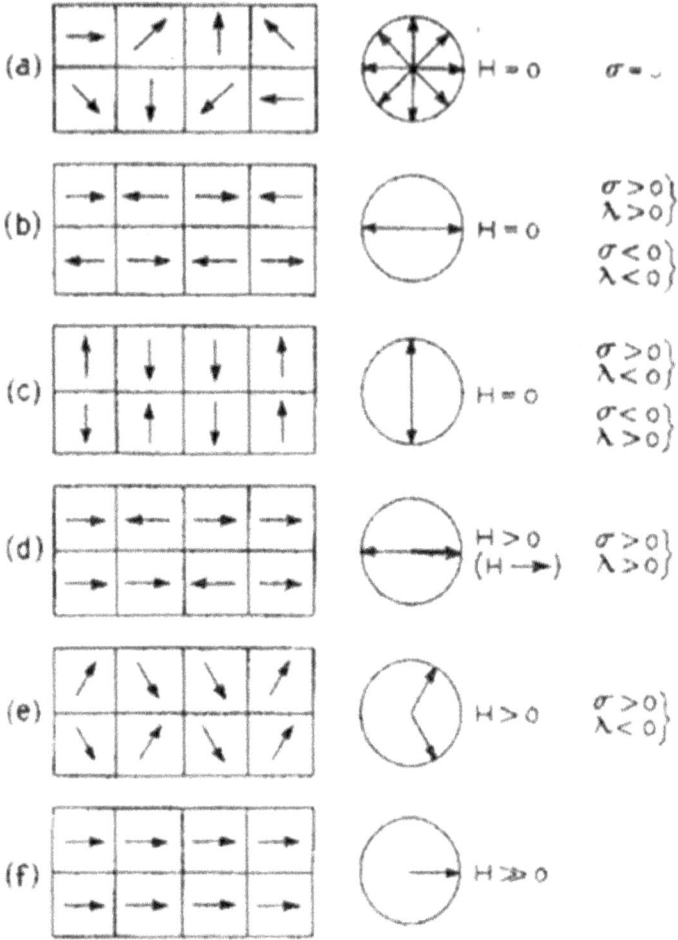

FIGURE 1.4 The effect of stress on the structure of pools of materials with positive or negative magnetostriction.

This section and the amount of magnetostriction of the materials in Table 1.2 lead one to the conclusion that among the available magnetostrictive materials, the Terfenol-D material has many capabilities, including magnetostriction properties and a very suitable Curie temperature. This can be deduced from the fact that the Terfenol-D material is included in this section. It has also been said, in credible sources, that the greater magnetostriction coefficient of the material is a marker of the prevalence of Villari memories in the material. This was found to be the case when the material was tested. Terfenol-D seems to be a promising contender for inclusion in the magnetostrictive force sensor construction for this reason. In order to accomplish this goal, the magnetostrictive material corresponding to Terfenol-D will be precisely examined in the next portion.

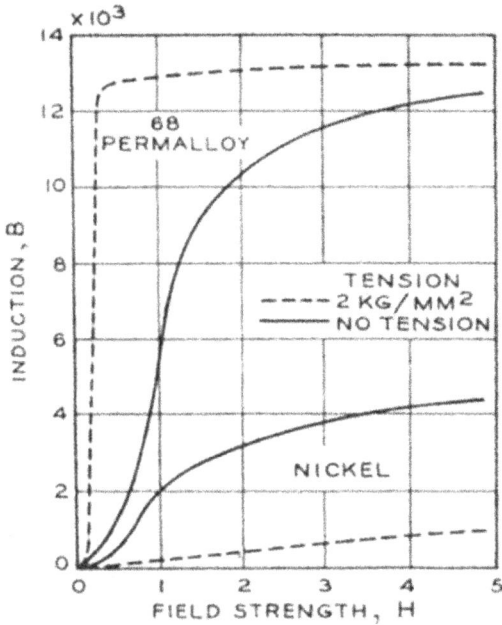

FIGURE 1.5 The effect of tension on the magnetization curve of Permalloy 68 and nickel.

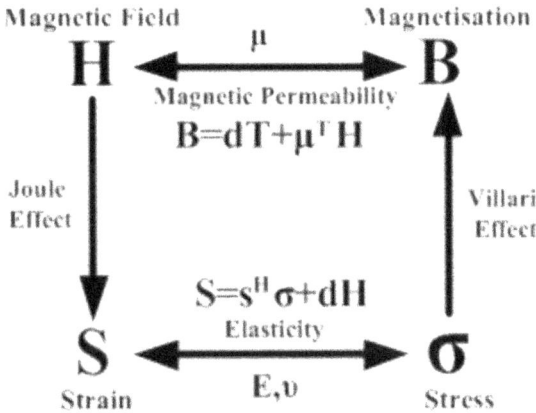

FIGURE 1.6 Magnetostrictive phenomena.

1.1.3 ΔE EFFECT

Magnetostrictive materials are elastic in two ways: the traditional stress-strain elasticity that results from interatomic forces, and the magnetoelastic component that is caused by the rotation of magnetic moments and the strain that results from applying a stress to the material. As the material is stretched or compressed, both of these

TABLE 1.2
Certain Magnetostrictive Materials Having Magnetoelastic Characteristics

Material	$\frac{3}{2}\lambda_s$ (10^{-6})	$\rho (g/cm^3)$	$B_s (T)$	$T_c (C)$	$E (GPa)$	k
Fe	−14 (Clark 1980)	7.88 (Bozorth 1968)	2.15 (Bozorth 1968)	770 (Bozorth 1968)	285 (Bozorth 1968)	—
Ni	−50 (Bozorth 1968)	8.9 (Bozorth 1968)	0.61 (Bozorth 1968)	358 (Bozorth 1968)	210 (Engdahl 2000)	0.31 (Clark 1980)
Co	−93 (Bozorth 1968)	8.9 (Bozorth 1968)	1.79 (Bozorth 1968)	1120 (Bozorth 1968)	210 (Engdahl 2000)	—
50%Co-50%Fe	87 (E. du Tremolet de Lacheisserie 1993)	8.25 (Clark 1980)	2.45 (Jiles 1994)	500 (Bozorth 1968)	—	0.35 (Clark 1980)
50%Ni-50%Fe	19 (E. du Tremolet de Lacheisserie 1993)	—	1.60 (Jiles 1994)	500 (Bozorth 1968)	—	—
TbFe$_2$	2630 (Clark 1980)	9.1 (Bozorth 1968)	1.1 (E. du Tremolet de Lacheisserie 1993)	423 (Clark 1980)	—	0.35 (Clark 1980)
Tb	3000(−196C) (Restroff 1994)	8.33 (Bozorth 1968)	—	−48 (Cullity 1972)	55.7 (Engdahl 2000)	—
Dy	6000(−196C) (Restroff 1994)	8.56 (Bozorth 1968)	—	−184 (Engdahl 2000)	61.4 (Engdahl 2000)	—
Terfenol-D	1620 (Clark 1980)	9.25	1.0	380 (Jiles 1994)	110 (Kellogg and Flatau 1999)	0.77 (Clark et al. 1990)
Tb$_{0.6}$Dy$_{0.4}$	6000(−196C) (Restroff 1994)	—	—	—	—	—
Metglas 2605SC	60 (Restroff 1994)	7.32 (E. du Tremolet de Lacheisserie 1993)	1.65 (Jiles 1994)	370 (E. du Tremolet de Lacheisserie 1993)	25–200 (E. du Tremolet de Lacheisserie 1993)	0.92 (Engdahl 2000)

Note: Unless otherwise noted, all measurements were conducted at room temperature.

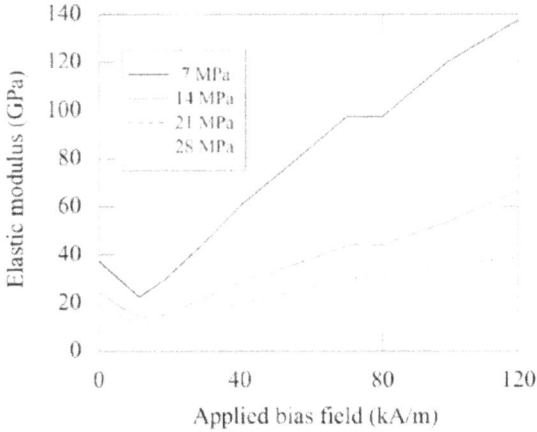

FIGURE 1.7 $Tb_{0.3} Dy_{0.7} Fe_2$ magnetoelastic modulus under different stress conditions.

qualities become apparent. The ΔE effect, or the difference between the elastic modulus at magnetic saturation and the lowest elastic modulus, may be expressed as $\Delta E = (E_s - E_0) / E_0$, where the minimum elastic modulus is denoted by E_0 and saturation magnetic modulus of elasticity is denoted by E_s. When the moments are allowed to freely spin, the material becomes more flexible because the strain created by the rotation of the magnetic moments adds to the strain induced by the non-magnetic moments. This idea is well-depicted in Figure 1.7. Keep in mind that when the saturation point is approached, the material will become increasingly more stiff while simultaneously suffering a decrease in its magnetic moment mobility. The ΔE effect is negligible in nickel ($\Delta E = 0.06$), but it may become important in Terfenol-D (ΔE up to 5) and in particular transverse-field annealed $Fe_{81} B_{13.5} Si_{3.5} C_2$ (Metglas 2605SC) amorphous ribbons ($\Delta E = 10$). The ΔE of Terfenol-D may be very useful in broadband sonar systems and adjustable vibration absorbers [6].

1.1.4 WIEDEMANN EFFECT

In a plane perpendicular to the wire, a current-carrying ferromagnetic or amorphous wire will generate a circular magnetic field, with the moments aligned mostly in the circumferential direction. Some of the moments will line up in a helical pattern when an axial magnetic field is applied. The Wiedemann effect describes the wire's twist. The axial magnetization of a wire that conducts electricity changes as it is twisted. The Matteuci effect, sometimes known as the inverse Wiedemann effect, describes this phenomenon [7] has further details.

1.1.5 MAGNETOVOLUME EFFECT

It is possible for the volume of a magnetostrictive material to vary in reaction to magnetic fields under some severe operating regimes. Nevertheless, during regular operation, the volume of a magnetostrictive material almost never changes. The term "volume magnetostriction" or "Barret effect" is used to refer to this abnormal change in volume. The effect has very limited relevance in systems that use smart structures. For example, whereas a graph of nickel's magnetorestriction properties quickly approaches 35×10^{-6} at just 10 kA/m, the volume change is only 0.1×10^{-6} with a considerably greater field of 80 kA/m. This indicates that the magnetostriction curve of nickel is quite sensitive to the field strength. To account for Invar's natural thermal expansion, engineers calculated a volume fractional change at the Curie temperature, which is somewhat higher than room temperature (36% nickel and 64% iron). This produces a substance with almost little thermal expansion at normal temperature. The Nagaoka-Honda effect is the name given to the shift in magnetic state that results from a volumetric change [2, 7]. The Barret effect has the opposite effect here. Figure 1.8 explains these effects.

Magnetostriction does not diminish over time, unlike other piezoelectric compounds, since it is a fundamental characteristic of magnetic materials. This is in stark contrast to the situation with piezoelectric materials. In addition, the strains, forces, energy densities, and coupling coefficients provided by current magnetostrictive materials are competitive with those of more established transducer technologies like those based on piezoelectric materials. This is due to the fact that magnetostrictive materials may change the direction of magnetic fields. However, the use of magnetostrictive materials is made more difficult by a number of design and modeling issues that must be addressed. Magnetostrictive transducers, for example, are typically larger and more cumbersome than their piezoelectric or electrostrictive counterparts. This is because magnetostrictive transducers require a solenoid and other related magnetic circuit components. Thus, the vibration control of heavy constructions is one of the key applications for magnetostrictive materials. An additional factor to take into account is that the production of the most technologically advanced magnetostrictive compounds is very expensive. This is due to the fact that the production of advanced crystalline transducer drivers necessitates via crystal growth methods that cause solidification to occur mostly along the driving axis, in addition, all ends of the cut-to-length pieces are machined to be perfectly parallel and the laminations are machined to be the exact diameter. In contrast to other smart materials like electrostrictives,

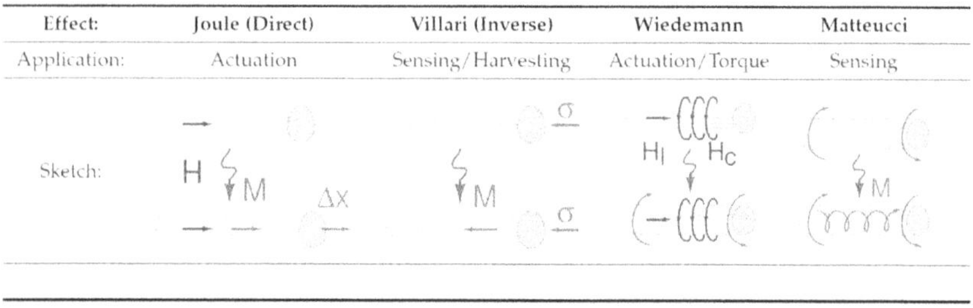

Effect:	Joule (Direct)	Villari (Inverse)	Wiedemann	Matteucci
Application:	Actuation	Sensing/Harvesting	Actuation/Torque	Sensing
Sketch:				

FIGURE 1.8   Magnetic effects.

magnetostrictive materials exhibit significant nonlinearities and hysteresis. This is crucial from the perspective of the implementation of devices using these materials. By the use of feedback control strategies, it is possible to avoid the negative impacts that are brought on by these actions. Nevertheless, in many magnetostrictive systems, it has remained challenging to create broadband feedback control approaches that maintain performance throughout a wide performance range. This is due to the fact that hysteresis, thermal creep, and significant material property fluctuations in these systems reduce the effectiveness of real-time monitoring and feedback control. In addition, the development of broadband feedback control solutions has been elusive in many magnetostrictive systems. The creation of feedforward loops that make use of constitutive laws that describe the behavior of a material in terms of its intrinsic physical qualities is one successful method for accounting for and mitigating the detrimental consequences of these concerns. Recent years have seen significant progress made in the modeling of magnetostrictive materials and structures. Model thoroughness and accuracy will likely increase as transducer designers see opportunities to develop novel applications. Furthermore, recent advancements in materials science research have allowed the fabrication of more competent for a wide range of magnetostrictive materials. Sintered powder compacts are great for mass production of small irregular forms, and magnetostrictive particle-aligned polymer matrix composite structures are another examples, and amorphous or crystalline thin films.

1.2   THE ORIGIN OF MAGNETISM IN MATERIALS

Magnetic materials are made up of various different domains that are together referred to as magnetic domains (see Figure 1.9). Magnetic dipoles in a region of matter are said to be parallel and aligned in the same direction to form a domain. The total of all vectors from each domain is considered to be zero when the material is non-magnetic; however, when the material is magnetic, this value is considered to be something other than zero. When an external magnetic field is applied to magnetic materials, they get magnetized in a manner that is specific to their structure. Likewise, when the exposure to the magnetic field is removed, the magnetic materials demagnetize in a manner that is specific to their structure. The magnetic and non-magnetic properties of this process allow for the classification of materials into the following four groups: paramagnetic, ferromagnetic, antiferromagnetic, and ferrimagnetic, each of which will be discussed in further detail in the following paragraphs. When it comes to paramagnetic materials, the magnetic vectors are dispersed randomly throughout the item and are aligned in the direction of the external field; however, once they exit the field, they revert back to their initial configuration. If there is a magnetic field available, the magnetic vectors in ferromagnetic materials have a tendency to line up in a parallel fashion with one another. On the other hand, in contrast to materials that are paramagnetic, they do not change position when an external field is removed. Because the magnetic vectors in antiferromagnetic materials are parallel to each other but pointing in opposite directions, their magnitudes are typically the same, and the magnetic

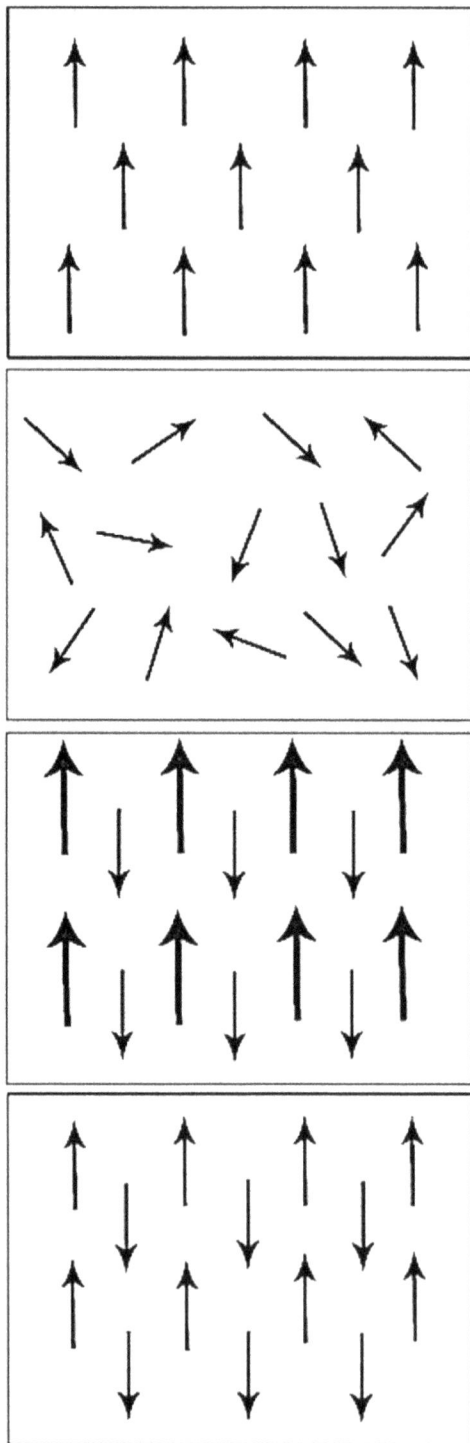

FIGURE 1.9 Vectors in materials (a) paramagnetic, (b) ferromagnetic, (c) antiferromagnetic, and (d) ferrimagnetic.

intensity is therefore zero throughout the entire object when an external field is not present, antiferromagnetic materials have a magnetic intensity of zero. The third and final classification is known as ferromagnetic materials. Once the vectors in these materials are aligned parallel to one another, as indicated in the figure, the amplitude and strength of these vectors are not equal, which results in the production of a magnetic field that permeates the whole item. In most cases, the presence of more than one ion in the chemical is responsible for causing this quality.

The magnetic dipole moments of individual atoms are where magnetism is born on the subatomic level. Atoms have a net magnetic dipole moment because their electron shells are only partly filled, which results in unpaired electrons. These unpaired electrons contribute to the atom's orbital and spin angular moments. The spin angular momentum is an inherent feature of an electron that originates from quantum mechanics and does not have a classical equivalent. The orbital angular momentum is caused by electrons circling atomic nuclei, but the spin angular momentum has no classical think of an electron orbiting the nucleus of an atom in a Bohr orbit, which has a radius of r, a velocity of v, a mass of m_e, and a charge of e^-, as illustrated in Figure 1.10. This will help put this phenomenon into better perspective. An electric current, denoted by the equation $i = ev / 2\pi r$, is created when an electron, which is a charged particle, orbits the nucleus of an atom. Because of the current loop that was established, it resulted in the creation of a magnetic dipole that is perpendicular to the path of the electron. The magnetic moment of the dipole that is produced is equal to $\mu_l = iA$, where A is the area that is covered by the orbit of the electron and i is the current that is produced by the movement of the charged electron. The motion of an electron along a circular path around the nucleus (which is distinct from the motion of a charged particle) also provides angular momentum $\vec{L}$ with the value $L = vrm_e$. The ratio of L to μ_l may be calculated using the basic constants, as shown by Eq. (1.3).

$$\frac{\mu_l}{L} = \frac{\mu_B}{h} \tag{1.3}$$

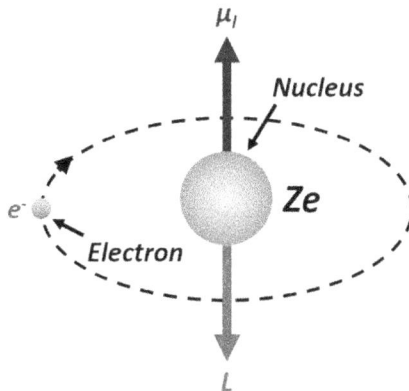

FIGURE 1.10   An electron circling the atomic nucleus of a simplified model of an atom.

The Bohr magneton is denoted by μ_B in this context

$$\mu_B = \frac{e\hbar}{2m_e} = 0.927 \times 10^{-23} \left(Am^2\right) \text{or} \left(J/T\right) \tag{1.4}$$

and $\hbar = \frac{h}{2\pi} = 1.054571800(13) \times 10^{-23}\, Js$ is the reduced Planck's constant.

In order to determine the magnitude and direction of μ_l in relation to $\vec{L}$, one may rewrite Eq. (1.3) such that it is in the form of a vector.

$$\vec{\mu_l} = -\frac{\mu_B}{\hbar}\vec{L} \tag{1.5}$$

Hence, according to the laws of quantum mechanics, the value of $\vec{L}$ may be expressed as:

$$L = \hbar\sqrt{l(l+1)} \tag{1.6}$$

where l is the quantum number of the orbital angular momentum, which may take on any positive integer value between 0 and $n-1$. In this context, n refers to the fundamental quantum number that controls both the length of an electron's orbit and the amount of energy it has. As a consequence of this, the magnetic dipole moment may be expressed as:

$$\mu_l = \mu_B\sqrt{l(l+1)} \tag{1.7}$$

and the value of its z component may be written as:

$$\mu_{lz} = -\mu_B m_l \tag{1.8}$$

where m_l denotes the magnetic quantum number and may take on the integer values of either $+l$ or $-l$ depending on the circumstances. Since electrons have negative charges, the sign of the moment and the orbital angular momentum in Eq. (1.8) is negative, which implies that these two quantities are antiparallel to one another. In 1922, O. Stern and W. Geralch came to the conclusion that electrons have an inherent quantum mechanical spin angular momentum denoted by the symbol $\vec{S}$. The size of this momentum is characterized by the formula [6]:

$$S = \hbar\sqrt{s(s+1)} \tag{1.9}$$

where s denotes the rotational angular momentum quantitative value, which takes on the value $1/2$, and its subsequent z component is provided by:

$$S_z = h m_s \qquad (1.10)$$

where m_s is the spin projection quantum number, which may have values of $+s$ and $-s$. The magnetic moment in vector form that is created from the spin angular momentum has a relationship to the direction of the spin angular momentum that may be expressed as the following:

$$\vec{\mu}_s = -g_e \mu_B \vec{S} \qquad (1.11)$$

putting g_e equal to 2.002290716(10)2 for a free electron, and the splitting factor, which is often referred to as the gyromagnetic factor (g-factor), is equal to 2 [6]. The equation that describes the z component of the spin moment is as follows:

$$\mu_{sz} = -g_e \mu_B m_s \qquad (1.12)$$

Keep in mind that μ_{sz} can only take on two possible values because, as shown experimentally, an electron can only deflect in two different ways when driven in a non-uniform magnetic field. It is important to keep this point in mind. This is consistent with the quantum numbers, as an example, it is known that the "spin-up" and "spin-down" states of an electron with s equal to 1/2 respectively correspond to the equation $m_s = \pm s$. Via a spin-orbit contact known as Russel-Saunder's coupling, the orbital angular momentum and the spin angular momentum are linked together. By adding the angular momenta of the electron's spin and orbit, we may calculate the electron's total angular momenta. This value is denoted by the letter J. When $\vec{L}$ and $\vec{S}$ have revolved around $\vec{J}$ in the manner seen in Figure 1.11, the total angular momentum of an electron may be expressed as a vector by using the equation $\vec{J} = \vec{J} + \vec{S}$ The magnitude of the total angular momentum is determined in a way similar to that of the orbital and spin angular momenta:

$$J = h\sqrt{j(j+1)} \qquad (1.13)$$

where j is the total angular momentum quantum number, and the equations that are utilized to figure out j are as follows:

$$\left|\vec{J}\right| = \left|\vec{L} + \vec{S}\right| \geq \left\|\vec{L}\right| - \left|\vec{S}\right\| \qquad (1.14)$$

This, when written in terms of the quantum numbers, corresponds to the quantities:

$$\left|\sqrt{j(j+1)}h\right| \geq \left|\sqrt{l(l+1)}h - \sqrt{s(s+1)}h\right| \qquad (1.15)$$

and for an electron with the property $s=1/2$, the conditions $j=l+1/2$ and $j=l-1/2$ are met for the preceding. The expression gives the z component of the value of j.

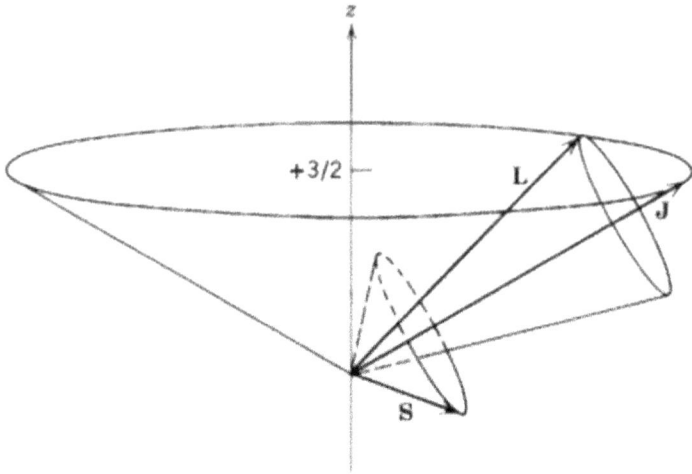

FIGURE 1.11 $\vec{L}$, $\vec{S}$, and $\vec{J}$ are the angular momentum vectors. With the value of the vector sum of $\vec{J} = \vec{L} + \vec{S}$, $\vec{L}$ and $\vec{S}$ revolve around $\vec{J}$, which is the origin of the spin $\left(\vec{S}\right)$ -orbit $\left(\vec{S}\right)$ coupling.

$$J_z = m_j h \qquad (1.16)$$

where m_j takes on values between $+j$ and $-j$. Hence, the total magnetic moment may be expressed as the product of the spin magnetic moment and the orbital magnetic moment using the formula:

$$\vec{\mu} = \mu_B \left(\vec{L} + 2\vec{S}\right) \qquad (1.17)$$

The values of $\vec{L}$, $\vec{S}$, and $\vec{J}$ may be calculated for a free atom while it is in the ground state by using Hund's laws, which can be summarized as follows [4]:

1) Electrons are allowed to take on the maximum value of S without going against the Puali exclusion principle.
2) To the greatest extent possible, increase the orbital angular momentum L, provided that this does not go against rule 1.
3) The ground state of an electron has the form $J = L - S$ if the shell is less than half full, but it takes the form $J = L + S$ if the shell is more than half full. If the shell is less than half full, J takes its least feasible value as the ground state.

It is essential to keep in mind that the Russel-Saunders coupling that was discussed earlier is applicable to lighter atoms with a Z value of less than 30, in which the total spin $\vec{S}$ of the group of electrons is coupled with the total angular momentum $\vec{L}$, which results in the total angular momentum $\vec{J}$. In contrast, the individual spin and orbital angular momenta of the electrons in heavier atoms, such as rare-earth elements, are firmly related to the individual total angular momentum for that electron alone. This is because heavier atoms have more electrons. The L-S coupling, in which the angular momentum of the individual electrons is summed together to produce the angular momentum J for the atom, is more appropriate for describing the electron interactions here than the Russel-Saunders coupling. The calculation for this can be found as follows:

$$\vec{J} = \sum_{i=1} \vec{j_i} = \vec{l_i} + \vec{s_i} \tag{1.18}$$

1.2.1 Ordering by Magnetism

Individual attractive or repulsive forces governed by magnetic dipole interactions of the atoms of the material generate the magnetic ordering. According to the orientation of their magnetic dipoles in the absence or presence of an applied magnetic field, magnetic materials may be classified as diamagnetic, paramagnetic, ferromagnetic, antiferromagnetic, or ferrimagnetic (Figure 1.12). While there have been many additional categories of magnetic materials designated to this day, just the handful that have been stated earlier are required to offer the appropriate context for the material that will be covered in subsequent chapters.

1.2.1.1 Diamagnetism

When electrons in an atom have a closed shell, the spin and orbital moments of those electrons are often orientated in such a way that the atom as a whole does not have a net dipole moment. Elements without any unpaired electrons in their outer orbital shell exhibit a feature called diamagnetism. Thus, in the periodic table, diamagnetism is a property that is experienced by monoatomic gases (i.e., H_2, N_2). Ionic solids, such as NaCl, operate according to similar principles. In ionic solids, the formation of the bonding ionic bonding occurs when an electron is transferred from the Na atoms to the Cl atoms. Ionic compounds are characterized by their general diamagnetic nature due to the fact that the ions that are produced as a by-product of the reaction-in this case, Na^+ and Cl^- both have entirely filled electron shells. Because of the filled shell concept, compounds that are covalently bound and in which electrons are shared evenly between atoms are likewise diamagnetic. Examples of such compounds include carbon (diamond), silicon, germanium, and others.

1.2.1.2 Paramagnetism

Both conventional physics and quantum mechanics may be used to describe the behavior of paramagnetic materials. The Curie Law, which illustrates that the mass

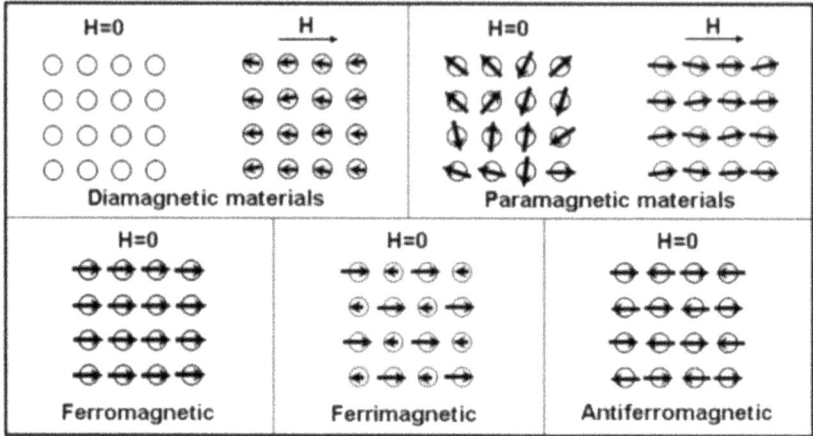

FIGURE 1.12 The configuration of the magnetic dipoles in diamagnetic.

susceptibility x_m changes in an inversely proportionate manner with the absolute temperature, is the best way to express the classical theory of paramagnetism:

$$x_m = C / T \tag{1.19}$$

using the Curie constant C on a per-gram scale. A more accurate reflection of the paramagnetic behavior of materials is given by the Curie-Weiss law, which was shown to be a special case of the Curie Law in later years as:

$$x_m = \frac{C}{T - \theta}, (T > \theta > 0) \tag{1.20}$$

where θ is a constant expressed in terms of the units used to measure absolute temperature T. Langevin was the first person to propose the concept of paramagnetism. Langevin said that materials exhibiting this property include atoms that all share the same overall magnetic moment μ and that the spin and orbital moments contained inside individual atoms do not cancel each other out [6]. But, if there is no H magnetic field that is being supplied from the outside, the atomic moments will point in various directions, which will result in the sample having no overall magnetization. If there is no force that may counterbalance the effect of H being applied, it seems to reason that when H is applied, the atoms' magnetic moments will gravitate in that direction. To align the magnetic moments in the direction of H, however, quite large fields are necessary. Because of this, there must be an existent thermal agitation force that has a tendency to align the individual moments in a random fashion. This leads to the moments having a partial alignment with H as a consequence of a positive magnetic susceptibility that is despite its tiny magnitude. Increasing the temperature has the consequence of increasing the randomizing effects, which, in turn, results in a decreased susceptibility. In order to put this into perspective, the normal values for paramagnetically active compounds are in the range of $x_m = 10^{-5}$ to 10^0.

Quantum mechanics provides an approachable framework for comprehending paramagnetism. In quantum physics, *quanta*, or discrete energy levels, are required since the energy of a system is not continuously variable. In the classical model, an individual atom's moment may be arbitrarily oriented with respect to space and time, but in the quantum mechanical model, there are only certain angles θ that are allowed (i.e., θ_1, θ_2). As a consequence of this, the states that have distinct J_z are degenerate when H is not present. When H is introduced, a magnetic moment in the field appears with the energy necessary to nullify the degeneracy (Zeeman energy):

$$\epsilon_z = J_z H \tag{1.21}$$

1.2.1.3   Ferromagnetism

Ferromagnets, on the other hand, have magnetic susceptibilities that are much greater than those of paramagnets, which range between $x_m = 10^0 - 10^7$. Because of this, the magnetic behavior of such materials is quite strong, and they have a significant amount of induced magnetization. Without an external magnetic field, the material's significant magnetic behavior may be traced back to the existence of unpaired electrons, which creates strong atomic moments, and the presence of parallel aligned magnetic domains, which causes those atomic moments to line up with one another. It is not always necessary for the magnetic domains to align entirely parallel with one another, which results in the formation of what are termed domain barriers. The change in volume of the individual magnetic domains causes the domains to line up in the same direction as the externally applied field.

Weiss was the first person to suggest the idea of spontaneous magnetization, and he did so by defining an effective molecular field [7]. Since then, several researchers have built on his work. It was hypothesized that the molecular field would be proportional to the bulk magnetic saturation; however, this hypothesis did not hold true for ferromagnetic materials because of the inherent domain structure of such materials. Nonetheless, he was not completely incorrect since in this scenario, the molecular field was proportional to the saturation magnetization when the temperature was set to 0 degrees Celsius. The exchange contact between nearby spins, denoted by the symbols $\vec{s_i}$ and $\vec{s_j}$, was first proposed by W. K. Heisenberg in 1928. This interaction helps to explain the creation of enormous molecular fields in ferromagnets. The eigenvalue of the Hamiltonian is equal to the amount of energy contained in the exchange interaction ϵ_{ex} denoted by:

$$\mathcal{H}_{ex} = \mathcal{H}_{ij} = -2J\vec{s_i}\vec{s_j}\cos(\varnothing) \tag{1.22}$$

where J represents the exchange integral and $\varnothing$ angle between the two spins (not to be confused with the total angular momentum). It is important to distinguish J from the total angular momentum. Conventionally speaking, a value of J that is positive indicates a parallel spin (also known as a ferromagnetic state), while a value of J that is negative indicates an antiparallel spin (i.e., antiferromagnetic state).

It is possible to characterize the temperature dependency of the magnetic behavior of ferromagnets by referring to their magnetic susceptibility. Because of their more powerful magnetic moments and comparatively longer range ordering, these materials have much bigger magnetic susceptibilities in compared to paramagnets. Nevertheless, as the temperature is raised over a particular point, which is referred to as the Curie temperature T_c, these materials begin to act in a paramagnetic manner. Because of this, the magnetic susceptibility of a ferromagnet may be characterized as:

$$x_m = \frac{C}{T - T_c} \tag{1.23}$$

Ferromagnetism can only be maintained up to an absolute maximum temperature known as the Curie temperature. As the temperature reaches T_c, the magnetic susceptibility starts to diverge, which indicates that there is some magnetization present even if there is no external magnetic field.

1.2.1.4 Antiferromagnetism

In the 1930s, Louis Neel won the prestigious Nobel Prize [8] for his description of the fundamental physics of antiferromagnetism. This work was published in the journal *Physical Review*. Antiferromagnets may be conceptualized as two ferromagnetic sublattices that are oriented in such a way as to be in direct opposition to one another, with the result that the magnetic moments are totally cancelled out by one another. The ordering of antiferromagnetic materials is controlled in a way similar to that of ferromagnetic materials by the exchange interaction for negative values of J. As a consequence of this, the spins in these materials are antiparallel, which leads to an overall net zero magnetization since the moments fully cancel each other out. Nevertheless, in canted or disordered antiferromagnets or polycrystalline antiferromagnetic specimens, which are constituted of pinned moments owing to grain boundaries, a very modest magnetism could develop. This is because pinned moments are a kind of pinned moment. These materials have a tiny magnetic susceptibility on the order of $x_m = 10^5 - 10^2$, which is comparable to that of paramagnetism. The difference between the two phenomena is that antiferromagnets have ordered magnetism. On the other hand, antiferromagnet susceptibility is maximum at the Neel temperature T_N and rises with temperature. In a manner similar to that of ferromagnets, the antiferromagnetic ordering disappears above T_N, and the Curie-Weiss equation dictates that the paramagnetic ordering takes its place.

1.2.1.5 Ferrimagnetism

Intriguing magnetic materials known as ferrimagnets have two or more antiparallel sublattices that do not completely cancel each other out. This arises as a result of the magnetic moments of the sublattices not being comparable in magnitude to one another but rather being opposed to one another. These magnetic materials exhibit certain characteristics and qualities that are derived from ferromagnetic as well as antiferromagnetic substances. In order to put this into perspective, we may say that the value of J in ferrimagnets must be positive because, unlike antiferromagnets, ferrimagnets have a spontaneous net moment. Yet, the values of J that are found in ferrimagnets are a significant amount lower than the values that are found in ferromagnets. In addition, ferrimagnets experience a rise in susceptibility that is qualitatively analogous to that of antiferrimagnets. Ferrimagnets have a magnetic susceptibility that rises with temperature and become paramagnetic when the temperature reaches T_N or above.

1.3 GENERAL INTRODUCTION TO HISTORY

The first major advancement in magnetostrictive materials occurred in the early 1960s, when the largest-known magnetostriction was discovered in the rare-earth elements terbium and dysprosium. This was the beginning of the field of magnetostrictive materials. These elements are experiencing strains on the order of $10,000 \times 10^{-6}$, which is three orders of magnitude bigger than the strains in nickel. However, in order to attain these strains, the elements must be cooled to cryogenic temperatures. An entirely new category of transducer materials that could endure high strains at room temperature was sought for in the early 1970s. The development of magnetostrictive materials was hampered by the temperature limitation, as well as the fact that the field of piezoelectricity was reaching a level of technical maturity at the same time. Specific chemical synthesis and fast sputtering were employed to mix the extremely magnetostrictive rare earths (R), principally samarium (Sm), terbium (Tb), and dysprosium (Dy), with the magnetic transition metals nickel, cobalt, and iron to generate amorphous alloys. These processes resulted in the formation of amorphous alloys. The Curie temperature of R-Fe compounds rises as the concentration of rare-earth elements rises, which is the opposite of what often happens with R-Ni and R-Co compounds. This is the case because the R-Fe compounds include more rare earths [8]. Specifically in the $TbFe_2$ combination, this peculiar feature makes it possible for enormous magnetostrictions to exist at ambient temperature, with values reaching as high as $3,000 \times 10^{-6}$. However, due to the fact that magnetostriction arises from the strain dependency of magnetic anisotropy, the large magnetostriction that can be achieved in these compounds comes at the expense of large anisotropies. Since it would need impractically enormous fields of more than 2 MA/m to get these chemicals to saturation, there is a roadblock to technological progress. The magnetostriction and anisotropy characteristics of the $TbFe_2$ system were significantly enhanced as a consequence of a partial replacement of dysprosium for terbium.

The resulting pseudobinary compound, designated $Tb_{0.3}Dy_{0.7}Fe_{1.9-1.95}$, has been sold under the trade name Terfenol-D since the 1980s (Ter = terbium, Fe = iron, N = Naval, O = Ordnance, L = Laboratory, D = dysprosium). Terfenol-D reaches its maximum magnetostriction of $1,600\times10^{-6}$ at a moderate saturation field of 0.16 MA/m when the material is measured at room temperature. However, when utilized in transducers driven at mechanical resonance, this material is capable of producing even larger magnetostrictions of up to $3,600\times10^{-6}$. Terfenol-D may be acquired in a variety of various forms, including monolithic rods [7, 9], particle-aligned polymer matrix composites, and thin films. Since they have a high magnetostriction anisotropy and a significant magnetoelasticity, Terfenol-D and other pseudobinary rare-earth-iron compounds may be produced to demonstrate a broad range of characteristics. This makes it possible for these compounds to be synthesized [8]. A second novel magnetostrictive material based on amorphous metal was introduced in 1978. This substance was created by rapidly cooling alloys containing iron, nickel, cobalt, and maybe silicon, boron, and phosphorus. These alloys are primarily produced in thin-ribbon geometries and are sold under the brand name Metglas, which stands for "metallic glass." Because of its extraordinarily high coupling coefficients (k > 0.92), Metglas is a promising material for use in sensors [7]. These characteristics make Metglas an excellent choice for devices that must transform mechanical energy into an electric current or voltage.

1.3.1 MAGNETIC ANISOTROPY

The only one that might be regarded as a material characteristic is crystal anisotropy. The magnetic moments in many crystalline materials tend to align themselves along preferred crystallographic orientations rather than freely spinning in response to applied fields. This is due to the material's crystalline lattice structure. The Curie temperature of R-Fe compounds rises as the concentration of rare-earth elements rises, which is the opposite of what often happens with R-Ni and R-Co compounds, in which they are most firmly aligned is referred to as the anisotropy energy, and this phenomenon is known as magnetocrystalline (or crystal) anisotropy. Magnetostriction would not exist if crystal anisotropy energy were unconnected to strain state [2], but the two phenomena are inextricably linked. Models for complex crystal structures typically rely on simplifying assumptions that limit the analysis to the simplest of situations, despite the fact that precise models for crystal anisotropy and its link with the magnetization process exist for the simple examples of cubic and hexagonal crystals [10]. Consider operating regimes where the stress anisotropy is more prominent than the crystal anisotropy, for example, when designing a transducer. Depending on whether the magnetostrictive material exhibits a positive or negative magnetostrictive, these regimes may be produced in structural design by subjecting the material to a compressive or tensile prestress. In the actuator seen in Figure 1.13, a precompression spring is employed. Anisotropy of crystals is discussed in further detail.

The term "anisotropy" comes from the Greek terms "anisos," which means "unequal," and tropos, which means "turn." As was mentioned before, the magnetic dipoles that are responsible for magnetism are inherently oriented in a certain

Prestress Bolt   Permanent Magnet   Solenoid   Compression Spring

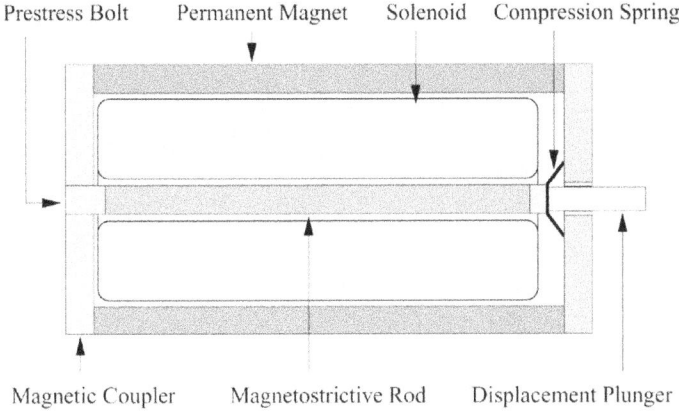

Magnetic Coupler   Magnetostrictive Rod   Displacement Plunger

FIGURE 1.13 Magnetostrictive transducer cross section.

direction. Since different magnetic materials might have different arrangements of magnetic dipoles, it is important to understand how these dipoles behave, there must be a preferred orientation in certain materials, and this preferred orientation is determined by the phenomenon that is referred to as magnetic anisotropy. The role of magnetic anisotropy in the design of numerous magnetic commercial devices utilizing ferro and ferrimagnetic materials is of significant importance as reflected by the coercive field of magnetic materials. These devices may be found in a variety of different industries. This is because the intrinsic magnetic anisotropy of magnetic materials results in magnetic characteristics that exhibit distinct features in different directions. Thus, a comprehensive comprehension of magnetic anisotropy is essential in order to thoroughly comprehend the magnetic behavior in ferro and ferrimagnetic materials. This is because magnetic anisotropy is a basic property of these materials. The magnetocrystalline anisotropy (MCA), magneto-elastic anisotropy (MEA), elastic anisotropy, and the shape anisotropy, also known as the demagnetizing field (demag for short), the Zeeman energy, and finally the energy resulting from the exchange interaction, will be discussed here. As a result, the formula for calculating the total energy of a magnetic system, E_T, is as follows:

$$E_T = E_{MCA} + E_{MEA} + E_{ex} + E_{demag} + E_{zeeman} \left(J / m^3 \right) \qquad (1.24)$$

where MCA energy (EMCA), MEA energy (EMA), elastic anisotropy energy (E_{el}), dipole moment energy (E_{Demag}), zeeman interaction energy (E_{zeeman}), and exchange interaction energy (E_{ex}) are all represented. Unlike other contacts, the exchange contact does not add to the overall anisotropy since it is completely isotropic. Because of this, the formula for calculating the total anisotropic energy present in a magnetic system may be recast as:

$$E_{an} = E_{MCA} + E_{MEA} + E_{el} + E_{demag} + E_{zeeman} \left(J / m^3 \right) \qquad (1.25)$$

Saturation magnetization M_s and the coercive field H_c are related to the overall anisotropy E_{an} of a magnetic material:

$$E_{an} = M_s H_c \tag{1.26}$$

The findings that are reported in the subsequent sections will place a significant emphasis on Eqs. (1.25) and (1.26). The individual contributors to the overall magnetic anisotropy will be explored in more detail further in the text.

1.3.2 MAGNETOCRYSTALLINE ANISOTROPY

The spin-orbit coupling is the sole source of anisotropy that is inherent to a material, and it is known as the MCA [11]. One would think that the exchange contact between two nearby spins, which controls whether the spins are aligned in a parallel or antiparallel fashion, would be the cause of the MCA. As Eq. (1.22) shows that the exchange contact depends solely on the angle between them, it is strong yet completely isotropic. The interaction energies between the spin, orbit, and lattice must be compared to one another, as illustrated in Figure 1.14, in order to comprehend how the MCA emerges from spin-orbit coupling. The spin-orbit coupling is high, but in comparison to the strength of the orbit with the lattice, it is comparatively modest. This is because to the quenching of orbital moments. Since their orbits are quenched, their orientations are locked to the lattice and cannot be changed, not even by the injection of extremely powerful fields.

Now take into consideration the interaction that the spin and orbital motion of the electrons have with one another. When a magnetic field is applied, there is a tendency for the electron to get reoriented, and the orbit of the electron also has a tendency to become reoriented. Yet, the orbit is tightly connected with the lattice, so any efforts to get reoriented will be met with resistance. Consequently,

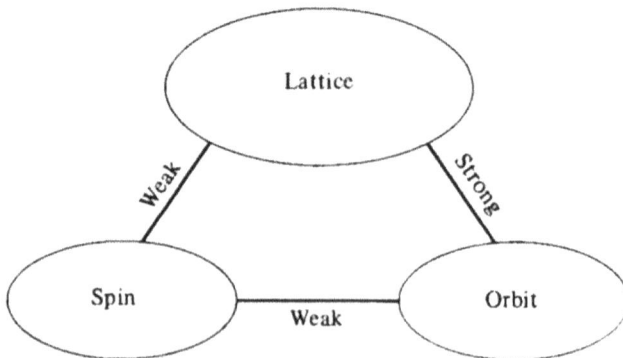

FIGURE 1.14   The strength of the interactions between the spin, orbit, and lattice.

the maximum cyclic acceleration, or MCA, refers to the amount of energy that is necessary to overcome the spin-orbit coupling, which, in turn, will rotate the complete spin-system. When temperatures rise, the magnitude of the MCA energy declines more quickly than the magnetization, and it disappears entirely at the Curie point. The MCA's vitality The crystal structure of the magnetic substance naturally affects EMCA. Nonetheless, two significant equations for EMCA have been found that properly explain the energy in all crystal systems. These equations may be broken down into the cubic anisotropy and the uniaxial anisotropy.

1.3.2.1 MCA in Cubic Crystals

E_{MCA} for cubic crystals, approximately corresponding to the 6th order, given by:

$$E_{MCA}^{Cubic} = K_0 + K_1\left(\alpha_1^2 + \alpha_2^2 + \alpha_3^2\right) + K_2\left(\alpha_1^2\alpha_2^2\alpha_3^2\right) \tag{1.27}$$

where K_1 and K_2 are the magnetocrystalline anisotropy constants, which are different for each material system, and where α_1, α_2, and α_3 are the directional cosine terms with regard to the direction of the magnetization vector $\overline{M_s}$.

There is an easy direction and a hard direction associated with every cubic crystal structure. As can be seen in Figure 1.15, the easy axis of BCC iron is in

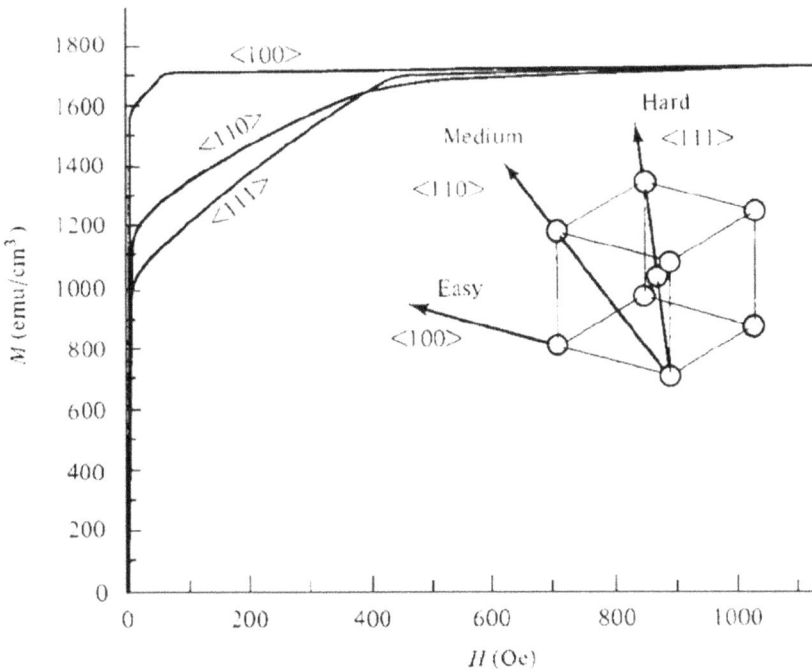

FIGURE 1.15 Magnetization curve for single crystal BCC iron, where the easy axis of magnetization falls along the <100> direction of the crystallographic direction.

the direction of <100>, while the hard axis is in the direction of <111>. This helps to put things into perspective. The preferred crystallographic orientation of the MCA may be largely attributed to the values of the magnetocystalline anisotropy constants, which play a crucial part in the process. Nonetheless, nickel crystallizes in FCC, which, as shown in Figure 1.16, has the favored easy axis running along the <111> direction and the hard axis running along the <100> direction. As can be seen in Figure 1.17a, for iron, K_1 is greater than K_2, and because the dominating K_1 approaching the fourth term has a positive value, this causes <100> to be the easy axis in BCC systems. As can be seen in Figure 1.17b, for nickel, the relationship between K_1 and K_2 is such that the dominant K_1 approaching the fourth term has a negative value, which causes <111> to be the easy axis in FCC systems.

1.3.2.2 MCA in Uniaxial Crystals

Uniaxial crystals have one crystallographic axis that is significantly longer in length than the other crystallographic axes in the crystal. These crystalline systems include hexagonal, tetragonal, and trigonal crystal structures, and the c-axis is seen to be the considerably longer axis; as a result, the phrase "uniaxial crystal"

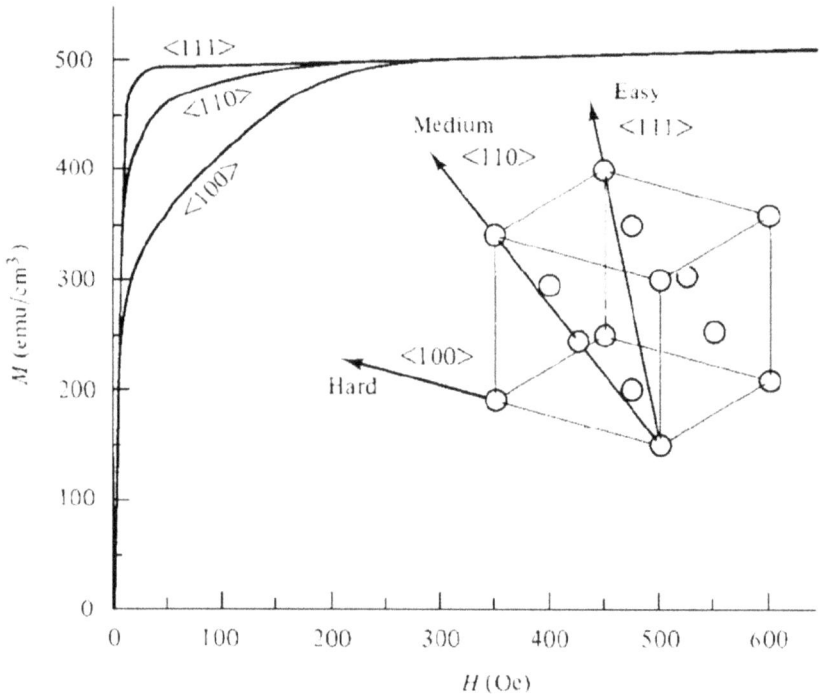

FIGURE 1.16 The magnetization curve for a single crystal of FCC nickel, where the easy axis of magnetization falls along the <111> direction of the crystallographic plane.

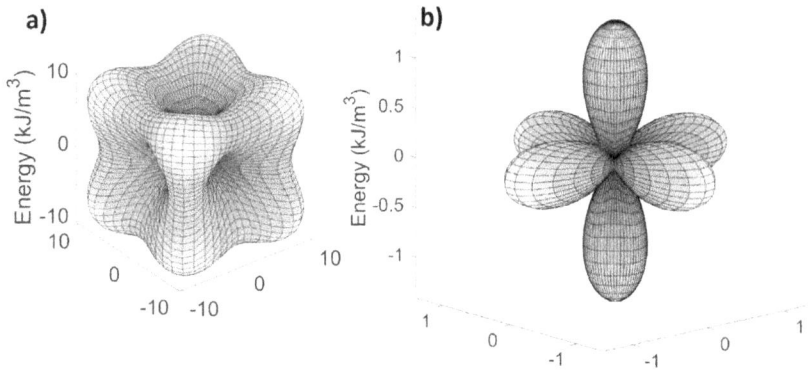

FIGURE 1.17 Surface energy maps showing the MCA energy in (a) BCC iron with K_1 greater than K_2 and (b) FCC nickel with K_1 less than K_2.

describes these types of crystals. As a consequence of this, the E_{MCA} for uniaxial crystals, approximated to the fourth, is given by:

$$E_{MCA} = K_0 + K_1 sin^2\left(\theta\right) + K_2 sin^4\left(\theta\right) \tag{1.28}$$

where θ is the angle between the crystallographic c-axis and the magnetization vector $\overrightarrow{M_s}$ (the easy axis of magnetization) and K_1 and K_2 are the MCA constants that are specific to each magnetic material. The K_1 and K_2 terms determine the direction of the magnetization in a manner similar to that which is accomplished by the cubic MCA constants.

If both K_1 and K_2 are positive, then the $\overrightarrow{M_s}$ value will be found at the c-axis, which is where the smallest amount of energy may be obtained when $\theta = 0°$. If both K_1 and K_2 are negative, then $\overrightarrow{M_s}$ will be located along a basal plane that is perpendicular to the c-axis. This is the plane in which the smallest amount of energy will be found when is equal to $\theta = 90$ degrees. If K_1 and K_2 both have signs that are opposite to one another, the scenario becomes far more complicated, as seen in Figure 1.18. In the event when K_1 is positive and K_2 is negative, the line $K_2 = -K_1$ serves as the boundary between the magnetization that is in the basal plane and the magnetization that lies along the c-axis. In this particular scenario, if $K_1 = -K_2$, there are two straightforward directions that lie at 0 degrees and 90 degrees. Now, if K_1 is negative and K_2 is positive, the limit of the easy-plane behavior follows $K_2 < \dfrac{-1}{2}K_1$, and within this range, the lowest E_{MCA} falls between the angles of 0–90 degrees, and there is an *easy cone* of magnetization in this condition. This is the case if K_1 is negative and K_2 is positive.

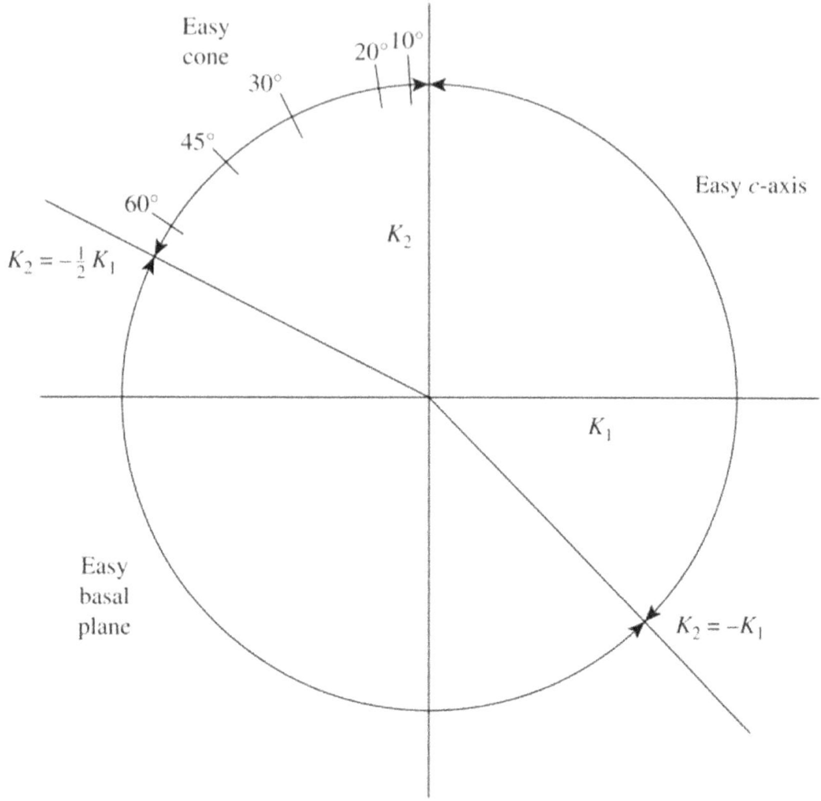

FIGURE 1.18 Directions and planes in uniaxial crystals that are simple and straightforward for any feasible value of K_1 and K_2.

1.3.3 MAGNETOELASTIC ANISOTROPY

The MEA is caused by the influence of stresses (also known as strains), which change the structure of the magnetic domains. In a manner similar to that of the MCA, the genesis of the MEA may be traced back to the modulation of the spin-orbit coupling, which governs the magnetic domains [12]. To put this into perspective, any kind of mechanical strain has an immediate and direct impact on the crystal lattice, which is the location where a change to the spin-orbit coupling takes place [13]. In addition, the MEA is defined for cubic as well as uniaxial crystals, much as the MCA, which will be covered in more detail in the following paragraphs.

1.3.3.1 MEA in Cubic Crystals

This equation represents the cubic MCA energy term to the 4th order.

$$E_{MCA}^{Cubic} = B_1 \left(\epsilon_1 \alpha_1^2 + \epsilon_2 \alpha_2^2 + \epsilon_3 \alpha_2^2 \right) + 2B_2 \left(\epsilon_4 \alpha_2 \alpha_3 + \epsilon_5 \alpha_1 \alpha_3 + \epsilon_6 \alpha_1 \alpha_2 \right) \quad (1.29)$$

where α_i represents the directional cosine terms with respect to the magnetization vector $\overrightarrow{M_s}$, ϵ_i represents the strains measured along the crystallographic directions, and B_1 and B_2 represent the magnetoelastic coupling coefficients, respectively. Both B_1 and B_2 may be derived from:

$$B_1 = -\frac{3}{2}(c_{11} - c_{12})\lambda_{100}, \, B_2 = -3c_{44}\lambda_{111} \tag{1.30}$$

where c_{11} and c_{12} are the elastic constants and λ_{100} and λ_{111} are the magnetostriction constants, respectively. The magnetic energy asymmetry and magnetostriction (MEA and magnetostriction) are the same thing, with the former referring to the energy and the latter describing the physical consequence. Consequently, it is evident that without having a grasp of the physical origin of MEA, it is almost difficult to properly explain the physical origin of magnetostriction. So, the important relevance of magnetic anisotropy can be clearly demonstrated here because magnetostriction (MEA), which drives many of today's devices in industry, could not exist without magnetic anisotropy. Here is an excellent example.

It is possible to rewrite Eq. (1.31) in terms of the magnetostriction constants as follows:

$$E_{MCA}^{Cubic} = -\frac{3}{2}\lambda_{100}\sigma\left(\alpha_1^2\gamma_1^2 + \alpha_2^2\gamma_2^2 + \alpha_3^2\gamma_3^2\right) - 3\lambda_{111}\sigma\left(\alpha_1\alpha_2\gamma_1\gamma_2 + \alpha_2\alpha_3\gamma_2\gamma_3 + \alpha_1\alpha_3\gamma_1\gamma_3\right) \tag{1.31}$$

where γ_i is the directional cosine terms with respect to the stress vector $\vec{\sigma}$, and σ is the magnitude of the stress vector ($\sigma = |\vec{\sigma}|$). The MEA for a magnetic sample consisting of a single crystal is represented by Eq. (1.31). Nonetheless, the majority of magnetic materials used nowadays are polycrystalline. In light of this, it is possible to reduce the complexity of Eq. (1.31) for polycrystals by assuming that $\lambda_{100} = \lambda_{111} = \lambda_{si}$.

$$E_{MCA,si}^{Cubic} = \frac{3}{2}\lambda_{si}\sigma\sin^2(\theta) \tag{1.32}$$

where θ denotes the angle that exists between the stress vector $\vec{\sigma}$ and the magnetization vector $\overrightarrow{M_s}$.

1.3.3.2 MEA in Uniaxial Crystals

It can be shown in Eq. (1.32) that the MEA of uniaxial crystals is comparable to that of polycrystalline cubic materials. On the other hand, it is possible to rewrite it for uniaxial crystals.

$$E_{MCA}^{uniaxial} = K_\sigma \sin^2(\theta) \tag{1.33}$$

where K_σ is the stress anisotropy constant, which may be calculated using the formula $K_\sigma = \frac{3}{2}\lambda_{si}\sigma$. When the equation is written in this fashion, the axis of stress is the same as the easy axis when K_σ is positive, but the easy axis is in a plane that is perpendicular to the axis of stress when K_σ is negative.

1.3.4 ELASTIC ANISOTROPY

The elastic energy is another significant component that contributes to the magnetic anisotropy of the material. It is possible to think of the elastic energy as the potential energy that is stored in the material while it undergoes plastic deformation as a result of a force or work being applied to it. Stretching, shearing, bending, twisting, and other deformations all contribute in their own unique ways to the elastic energy of the deformed material. The elastic energy that is contained in bulk materials may be transferred in a variety of different ways. As a consequence of this, the elastic energy of every crystal may be expressed as:

$$E_{el} = \frac{1}{2} C_{ijkl} \epsilon_{ij} \epsilon_{kl} \tag{1.34}$$

where C_{ijkl} denotes the elastic/stiffness tensor of the fourth order and ϵ_{ij} and ϵ_{kl} denote the strain tensors, respectively. The equation for the elastic energy of a material with cubic symmetry looks like this:

$$E_{el} = \frac{1}{2} c_{11} \left(\epsilon_1^2 + \epsilon_2^2 + \epsilon_3^2 \right) + \frac{1}{2} c_{12} \left(\epsilon_1 \epsilon_2 + \epsilon_2 \epsilon_3 + \epsilon_1 \epsilon_3 \right) + \frac{1}{2} c_{44} \left(\epsilon_4^2 + \epsilon_5^2 + \epsilon_6^2 \right) \tag{1.35}$$

1.3.5 SHAPE ANISOTROPY

Both terms, shape anisotropy and shape anisotropy, are used synonymously to express the same underlying physical phenomenon. The presence of magnetic "free poles" on the surface of a magnetic material is caused by a discontinuous shift in the magnetization when seen perpendicular to the surface of the material. This is because magnetic domains in magnetic materials are aligned in the same direction. Because of this, a sizable demagnetizing field and a sizeable amount of magnetostatic energy are created as a consequence. In addition, this results in the production of magnetic domains in the materials themselves. Domains are generated in order to reduce the demagnetizing forces that are present by striking a balance between the exchange energy and the magnetostatic energy. This allows us to write the energy term for the demagnetization energy as:

$$E_{demag} = \frac{1}{2} \mu_0 \left(N.\vec{M}_s \right).\vec{M}_s \tag{1.36}$$

where it is possible to write it for an ellipsoid in general as:

$$E_{demag}^{ellip} = \frac{1}{2} \mu_0 M_s^2 \left(N_1 \alpha_1^2 + N_2 \alpha_2^2 + N_3 \alpha_1^2 \right) \tag{1.37}$$

where N is the demagnetization tensor of the second order and α_i are the directional cosine terms with respect to the magnetization vector $\overrightarrow{M_s}$.

The inherent dependence of the form anisotropy on the geometric shape of the magnetic substance may be stated as follows. The shape anisotropy makes consistent efforts to be reduced to the barest minimum. Figure 1.19 depicts a top view as well as a side view of two circular disks with comparable radii but distinguishable thicknesses. The brown lines in the figure reflect the demagnetization field of each disk. The demagnetization field that is shown in the thinner disk on the left may be made to be as little as possible by choosing to travel over and under the disk rather than passing across the edges of the disk. On the other hand, the demagnetization field wants to travel around the thicker disk that is illustrated on the left so that it may save as much energy as possible. While domains are inherently a technique of reducing the demagnetization field, the shape anisotropy also affects the domain structure. This is because domains are naturally organized in a hierarchical fashion. Hence, it is possible to produce single domain or multi-domain states in magnetic structures by the process of micropatterning variously shaped magnetic structures (such as disks).

1.3.6 ZEEMAN ENERGY

The energy that is brought into the system as a result of the application of a magnetic field from the outside is referred to as Zeeman energy. When continuous magnetic moments are exposed to an external magnetic field, it is a kind of potential energy that may be described by the formula:

$$E_{zeeman} = \mu_0 \int_V M.HdV \tag{1.38}$$

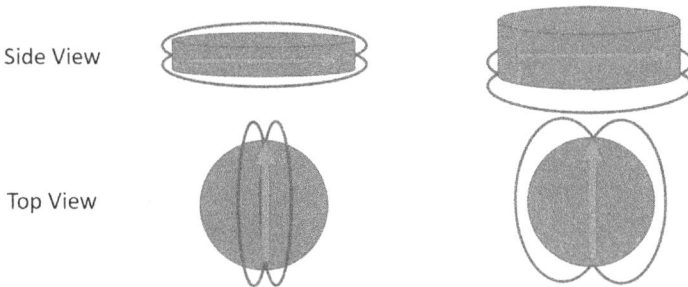

FIGURE 1.19   A field capable of demagnetizing two circular disks that are magnetized in the same plane. The disk on the left is far smaller than the disk on the right, and it has a demagnetizing field that runs through both the top and the bottom of the disk. The disk on the right is more substantial, and the demagnetizing field wraps around it in order to reduce the amount of energy it has.

where V stands for the volume of the body, μ_0 stands for the permeability of free space, M stands for the magnetization, and H stands for the magnetic field that is applied from the outside.

1.3.7 MAGNETOSTRICTION

When an external magnetic field is applied to a magnetic material, a phenomenon known as magnetostriction occurs. Joule made the discovery in 1842 when he observed a change in the dimensions of an iron rod caused by the application of a small magnetic field down the length of an iron rod [14, 15]. The fractional change in length $\lambda = \dfrac{\Delta l}{l}$ is inherently a strain, and it is possible to differentiate it from an imposed strain as a consequence of stress by defining it as the difference between the original length and the new length.

$$\lambda = \frac{\Delta l}{l} \tag{1.39}$$

where is the unitless magnetostriction term, l is the length, and Δl is the change in length, respectively. As illustrated in Figure 1.20, the value of that is measured at the saturation magnetization M_s is referred to as the saturation magnetostriction and is indicated by the symbol λ_s.

Since the magnetostrictive effect is derived from MEA, it follows that the magnetostrictive effect also comes from spin-orbit coupling. As a result, domain wall motion causes the magnetostrictive effect [16]. In many magnetic materials, where λ_s normally has values on the order of 10^{-5}, magnetostriction is not a notable phenomenon. Nevertheless, magnetic properties like permeability and coercive field H_c are modified as a result of inverse magnetostrictive actions like the application of various external stressors. This, in turn, affects the size and form of the magnetic hysteresis loop, which is especially beneficial for magnetic devices that make use of materials that are very magnetostrictive.

There are two different forms of magnetostriction that may happen: spontaneous and forced.

When a substance is heated beyond its Curie temperature T_c and subsequently cooled below that temperature, a phenomenon known as spontaneous magnetostriction may take place. When a sample is exposed to an externally applied magnetic field that is substantial enough to enhance the magnetization of the domain over its spontaneous value, a phenomenon known as forced magnetostriction takes place. The spin order of the system is increased by both forms of magnetostriction.

1.3.8 SINGLE-CRYSTAL MAGNETOSTRICTION

The scientific community places a significant amount of emphasis on the measurement of the magnetostriction coefficients λ_{100} and λ_{111} for a variety of

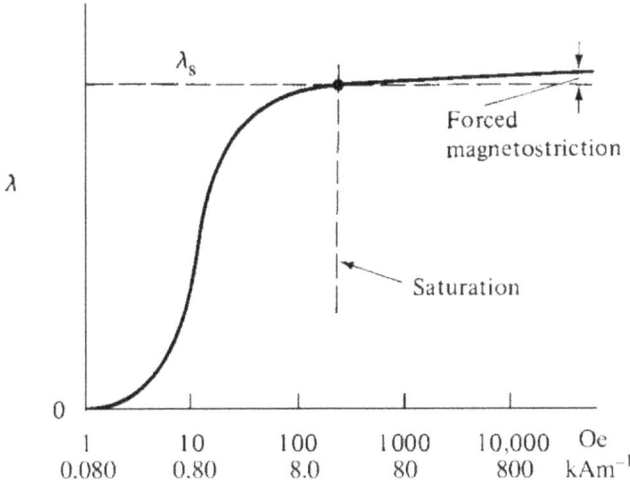

FIGURE 1.20 Magnetostriction as a function of the applied field.

magnetostrictive materials, in especially for those whose micromagnetic models rely on these numbers. The measurement of magnetostriction in single crystal iron in the [100] direction is shown in Figure 1.21. In the example of a single crystal of iron shown in Figure 1.21a, the multidomain structure may be observed rather well. While the magnetostrictive effect is exaggerated in Figure 1.21b, it is still abundantly visible that the vast majority of the domains have spun in the same direction as the external magnetic field. Since domain wall motion causes the domains that were initially aligned along the [10] direction to rotate, the domain must expand in the [100] direction and contract in the [10] direction, which ultimately results in the lengthening of the crystal Δl. As a result, the measured difference in length represented by Δl is utilized to compute λ_{100} for iron in this case. The value of λ_{111} may likewise be established using a method similar to that used in the [111] direction.

1.3.9 SATURATION MAGNETOSTRICTION IN CUBIC CRYSTALS

There is a widespread misunderstanding that cubic crystals that go through magnetostriction are entirely cubic. In reality, this only happens when the material is heated to a temperature that is greater than Tc. On the other hand, the tetragonality is not very high, and it is reasonable to believe that the crystals are pseudocubic. As a consequence of this, for these crystals, the saturation magnetostriction term may be written as:

$$\lambda_s = \lambda_{100} + 3(\lambda_{111} - \lambda_{100})(\alpha_1^2 \alpha_2^2 + \alpha_2^2 \alpha_3^2 + \alpha_1^2 \alpha_3^2) \tag{1.40}$$

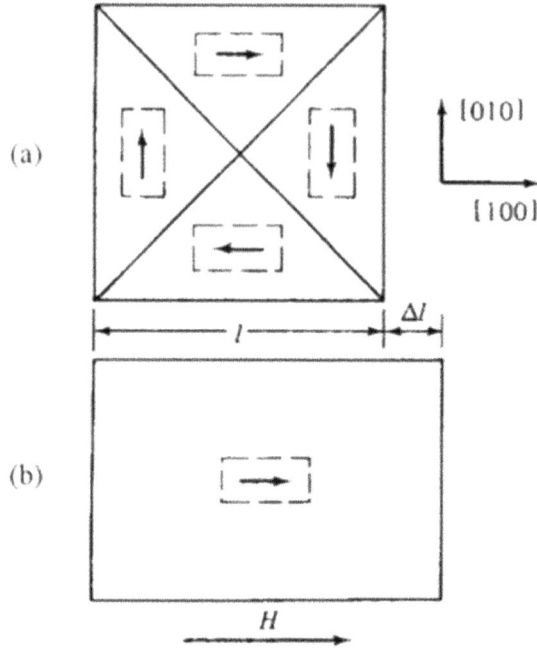

FIGURE 1.21 (a) Measurement of the magnetostriction in single crystal iron in the [100] direction (b) after saturation in the presence of magnetic field.

1.3.10 EXAMINING TERFENOL-MAGNETIC D'S ANISOTROPY

To control the magnetic moment using external fields or mechanical strain, magnetic anisotropy must be overcome or controlled. This is an essential process for devices used in spintronics. It is challenging to develop and manufacture magnetic anisotropies customized to particular applications for rare-earth materials because of a lack of understanding of the effect of elemental spin-orbital moments on the overall magnetic response [1–3]. This is crucial because the orientation of the magnetic moments and magnetic hardness are determined by the lattice crystal field, which interacts with the spin-orbit coupling between the element's spins and their orbits [4, 5]. Changes in the spin-orbital moments of individual elements within a material are known to have a profound effect on the material's magnetic anisotropy [6–9]. With this discovery, we may get closer to creating materials optimized for certain uses. Nevertheless, there is a lack of in-depth research concentrating on rare-earth magnetostrictive materials, where the magnetoelastic and magnetocrystalline anisotropic responses are quite large and complex and hence impact the overall magnetic response.

Due to its huge magnetostriction at ambient temperature and its tiny MCA, the rare-earth ferrimagnetic $Tb_{0.3} Dy_{0.7} Fe_2$ (Terfenol-D) is one of the gigantic magnetoelastic materials that have received the most research attention [10]. In order to reduce the MCA while maintaining a high magnetostriction, Terfenol-magnetic D's characteristics were modified by alloying ferrimagnetic $TbFe_2$ with $DyFe_2$ [11, 12]. Terfenol-D is the name given to the winning composition (i.e., $Tb_{0.3}Dy_{0.7}Fe_2$) after previous studies explored several compositions to maximize the magnetostriction while reducing the MCA for sonar applications [13]. Rare-earth thin films, such as Terfenol-D, have lately been extensively employed in a variety of device applications [14, 15]. As a result, there is an inherent necessity for the research of diverse contributions toward the magnetic anisotropy in addition to the magnetocrystalline ones. Due to the residual stresses in thin films, it is necessary to investigate the magnetoelastic contributions to magnetic anisotropy [16].

Several measurement techniques, such as soft X-ray absorption (XAS) and X-ray magnetic circular dichroism (XMCD), along with the XMCD sum rule calculations, have been utilized in the process of researching elemental spin-orbit moments and the influence that these moments have on the magnetic anisotropy of a variety of materials [17–19]. These experiments are carried out by reducing the temperatures, which leads to an increase in magnetic anisotropy as a result of a higher spin-orbit coupling, which can be detected precisely by XMCD [4]. An XMCD investigation that was conducted on the ferrimagnetic magnetostrictive alloy $DyCo_3$ indicated that increases in the magnetic moment of the Dy element were the source of the augmentation of the thermally induced magnetic anisotropy at lower temperatures [20, 21]. XMCD spectroscopy of ferrimagnetic multiferroic $TbMnO_3$ [22, 23] also revealed MCA alterations thermally driven by large increases in the Tb orbital moment. We perform XMCD on Terfenol-D thin films from 100 to 300 K (using liquid nitrogen cooling) and use the XMCD sum rules to directly determine the relationship between the elemental spin and orbital moments and the coercive field, and to then formulate these contributions in terms of the magnetocrystalline and magnetoelastic anisotropies.

1.3.11 MAGNETIZATION AND MAGNETIC HYSTERESIS CURVES

Changes in magnetization brought on by an external magnetic field may be either reversible or irreversible, depending on which category they fall into. When a tiny magnetic field is applied to a material, reversible magnetization changes may take place. These changes are energy-efficient because the magnetization of the material can be restored to its original state once the field is removed. Dissipative magnetizations are caused by irreversible magnetizations because the magnetism can only be returned to its initial condition with the application of external restoring forces, such as when significant fields are used. During the magnetization process of applications, two sorts of processes might be found taking place. It is possible to explain the magnetization, regardless of whether it is reversible or irreversible, by taking into consideration the rotation of moments and the shifting of domain barriers are two related phenomena. Figure 1.22 depicts the typical

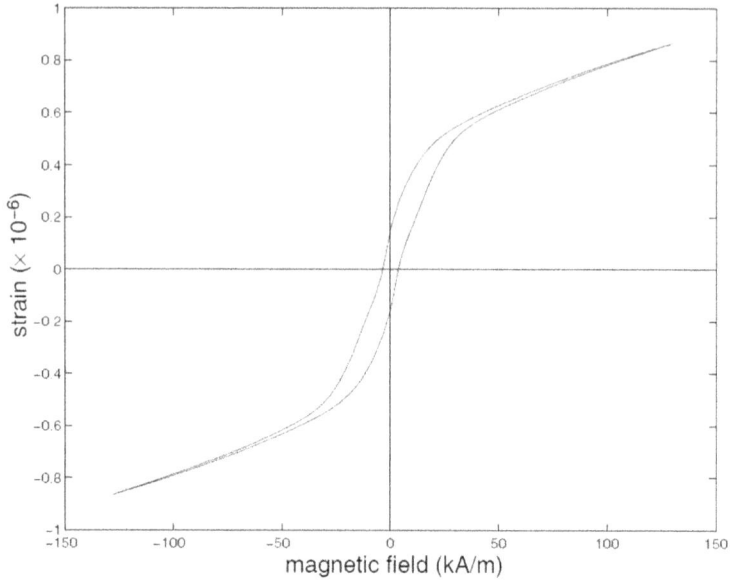

(a)

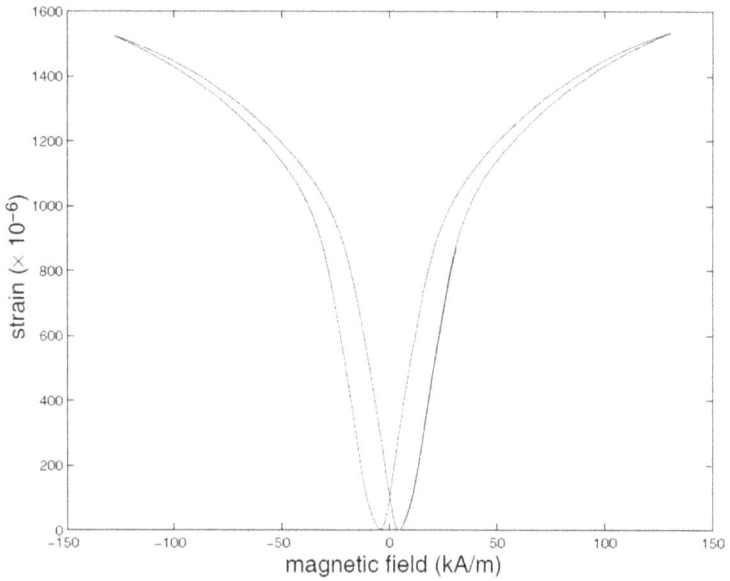

(b)

FIGURE 1.22 Total strain and magnetic induction measured experimentally from 6.9 MPa Terfenol-D loading.

magnetization and strain loops for Terfenol-D. These loops indicate the magnetic hysteresis and saturation effects that occur in the compound. Partial excursions in the M-H or μ-H curve will resemble a linear relationship when the magnetic field strength is low. Nonetheless, hysteresis is always present to some degree in certain situations. As domain walls travel over twin boundaries in Terfenol-D, for example, this might cause hysteresis because there is an irreversible obstacle to domain motion caused by pinning sites. This can be linked to the phenomenon. At the moment, one of the most important aspects of the planning and management of magnetostrictive materials is the modeling of hysteresis and nonlinear behavior. The book provides further information on the subject of ferromagnetic hysteresis [4, 24, 25].

The investigation of the magnetic hysteresis loops present in all categories of magnetic materials provides valuable information on the characteristics of such materials. Hysteresis loops are often researched for ferro and ferrimagnetic materials. As a consequence of this, the first thing that will be covered in this part is the hysteresis loop for a hypothetical ferro/ferrimagnet, which may be seen in Figure 1.23. The magnetic flux density of a material that has previously been magnetized is described by the hysteresis loop as a function of the applied magnetic field H, which is represented by the symbol B and is equal to:

$$B(H) = \mu_0 \left[H + M(H) \right] \tag{1.41}$$

where $\mu_0 = 1.25663706 \ 10^{-7} \text{(H/m)}$ is the permeability of empty space and $M(H)$ is the magnetization of the material as a function of H. The saturation induction, denoted by B_m on the B curve, and the value of B when the applied field H is removed are respectively denoted by B_r. The coercive field H_c, which describes the strength of the applied field to return the material to zero induction, is another critical quantity (i.e., demagnetize the material). The link between B and H may now be examined since the B hysteresis loop has been fully understood. The slope of the BH curve, as such, determines the permeability of a magnetic sample.

$$\mu = \frac{B}{H} \tag{1.42}$$

This may be phrased in terms of the relative permeability of the material as

$$\mu_r = \frac{\mu}{\mu_0} = \frac{B}{\mu_0 H} \tag{1.43}$$

comparable to in that it has no dimensions. The inner curve in Figure 1.23 depicts the magnetization curve of the material in its virgin state. This is a description of the magnetization of a ferro/ferrimagnetic substance that has never been magnetized before. In this context, the term "M_s" refers to the saturation magnetization of the material, which is measured to its highest value and is equivalent to "B_m". It is essential to keep in mind that when the material has been magnetized for

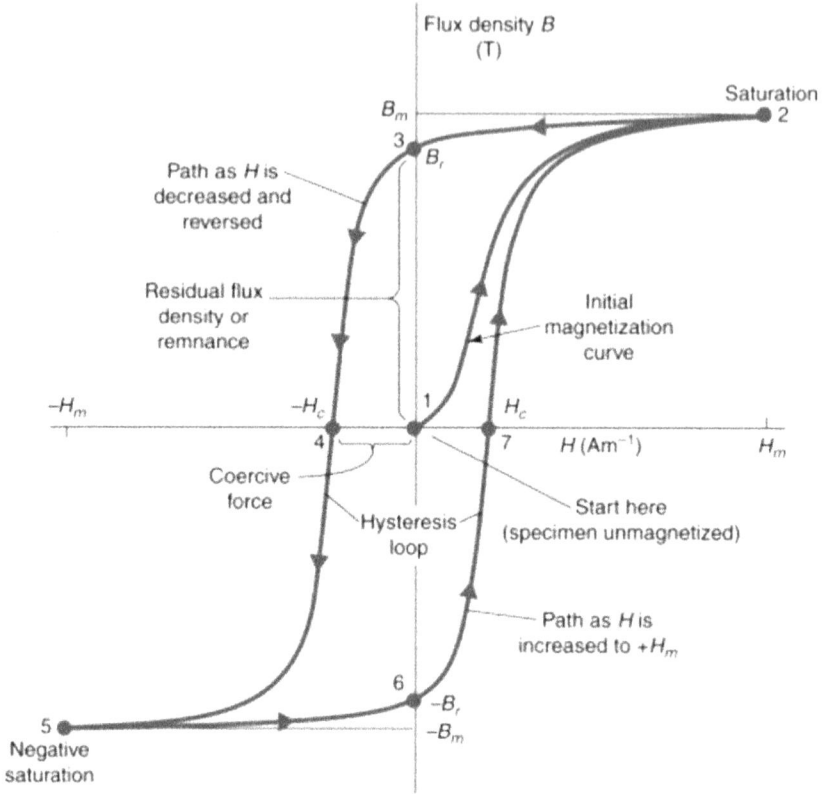

FIGURE 1.23 A perfect ferro/ferri magnet's magnetization and hysteresis loop.

the first time, the magnetic hysteresis curve will begin to behave in a manner that is analogous to the outer loop. Notably, the inner curve that is supposed to describe the virgin magnetization starts at a value that is close to zero in both the magnetization and the applied field (origin of the *BH* axis). Yet, once the material has been subjected to a magnetic field H_m that is strong enough to saturate it, the induced magnetization *B* will never fully return to that state. As a result, the magnetization curve, denoted by *MH*, is capable of providing useful information on the susceptibility of the material.

$$x = \frac{M}{H} \tag{1.44}$$

where *x* represents the susceptibility of the material and *M* represents the magnetization of the material as a function of the applied field *H*. There are several different types of susceptibilities that are used, including mass, atomic, volume, and molar susceptibilities. The kind of susceptibility that is utilized is determined by the units in which the magnetization *M* was measured. For example, because the mass susceptibility was discussed at length in the preceding section, the mass susceptibility may be expressed as x_m.

$$x_m\left(\frac{emu}{Oe.g}\right) = \frac{M\left(\frac{emu}{cm^3}\right)}{H(Oe)}\rho\left(g/cm^3\right) \tag{1.45}$$

where ρ is the density of the magnetic substance, and it is possible to establish a relationship between the mass susceptibility, denoted by x_m, and the relative permeability, denoted by μ_r, by combining Eqs. (1.43) and (1.44) as follows:

$$\mu_r = \mu_0\left(1+x_m\right) \tag{1.46}$$

It is possible to generalize the permeability and susceptibility of the different kinds of magnetic materials discussed in the preceding section using the information shown in Table 1.3.

As was said previously, paramagnets and antiferromagnets exhibit hysteresis curves that are quite similar to one another. Moreover, both types of materials have very low yet positive magnetic permeability μ and susceptibility x.

1.4 MATERIALS WITH MAGNETOSTRICTIVE PROPERTY

1.4.1 MAGNETOSTRICTIVE FE-GA ALLOYS AND FE-GA-X ALLOYS

As a result of its high magnetostrictive strain at low saturating field, low brittleness, high strength, and remarkable machinability, Fe-Ga alloys have emerged as the most promising new magnetostrictive material [11]. In addition, their application potential in the ultrasonic sector, transducer, micro-displacement devices, and MAs was revealed. In low magnetic fields, Fe-Ga alloys have a high tensile strength (~500 MPa), strong thermal-mechanical characteristics, and a moderate magnetostriction saturation (~350 ppm) under low magnetic fields (~100 Oe). Fe-Ga alloys bridge the gap between conventional magnetostrictive materials and rare-earth giant magnetostrictive materials (GMMs). Hence, this section will

---

TABLE 1.3
Permeability and Susceptibility Defined for a Wide Variety of Materials

Material type	μ	x
Free Space	1	0
Diamagnets	$\mu < 1$	$x < 0$
Ferro and Ferrimagnets	$\mu \gg 1$	$x \gg 1$
Para and Antiferromagnets	$\mu > 1$	$x > 0$

concentrate primarily on the magneto-mechanical behavior of binary and ternary Fe-Ga alloys.

1.4.2 Fe-Ga Alloys

Substitutional solid solution is produced when Ga is dissolved in bcc-Fe. The magnetostrictive properties of Fe-Ga alloys are directly influenced by the formation of different phase structures during heat treatment, as shown in Figure 1.24a, when the atomic fraction of Ga in the substitutional solid solution is continuously increased from 15% to 30% [12]. These structures include A2, DO_3, Ll_2, B2, DO_{19}, and Modified-DO_3. The corresponding phase symbols may be used to identify these phase structures. Of these phase configurations, the disordered A2 phase at high temperatures contributes the most to the magnetostrictive property. Nevertheless, the development of the ordered phase would greatly reduce the magnetostrictive behavior, hence the Modified-DO_3 phase may also improve the alloys' magnetoelastic coupling density. The high magnetostriction of Fe-Ga alloys is believed to occur mostly from the Ga-Ga atom pairs and the asymmetrical Ga clusters created when nearby Fe atoms are replaced by Ga atoms in the direction in the bcc-Fe structure. This is the prevailing opinion among scientists. Clark et al. explored the relationship between the gadolinium content and the magnetostrictive properties of single crystal orientated Fe-Ga alloys [13]. When the proportion of Ga in the alloys was relatively low, the phases disordered A2 and ordered DO_3 emerged. Over this time span, the number of isolated Ga atom pairs increased, leading to a rise in the magnetostrictive strain at saturation. When the quantity of Ga in the alloy rose, subclusters of asymmetric Ga formed. These sub-clusters then produced short-range order phases in the alloy, resulting in a decrease in magnetostriction. The best magnetostrictive properties were shown by Fe-Ga alloys with Ga atomic percentages between 19% and 27.5%, according to subsequent study. As shown in Figure 1.24b, the magnetostrictive strain reached its maximum value of around 400 ppm when the atomic fraction of Ga was increased to 19%. The A2 phase, which was only stable at high temperatures, was retained at ambient temperature after the quenching treatment. This was because the A2 phase was only stable at elevated temperatures. The second peak value came at around 27.5% atomic fraction for the element Ga, which was mostly due to the very low shear modulus. In addition, the DO_3 structure was almost entirely present in the alloy in this case.

1.4.3 Fe-Ga-X Alloys

In recent years, researchers have been investigating various approaches that may be beneficial in further enhancing the magnetostrictive characteristics of Fe-Ga alloys. The incorporation of third elements into Fe-Ga alloys has been shown to be a workable process, and it is now one of the primary focuses of study in this

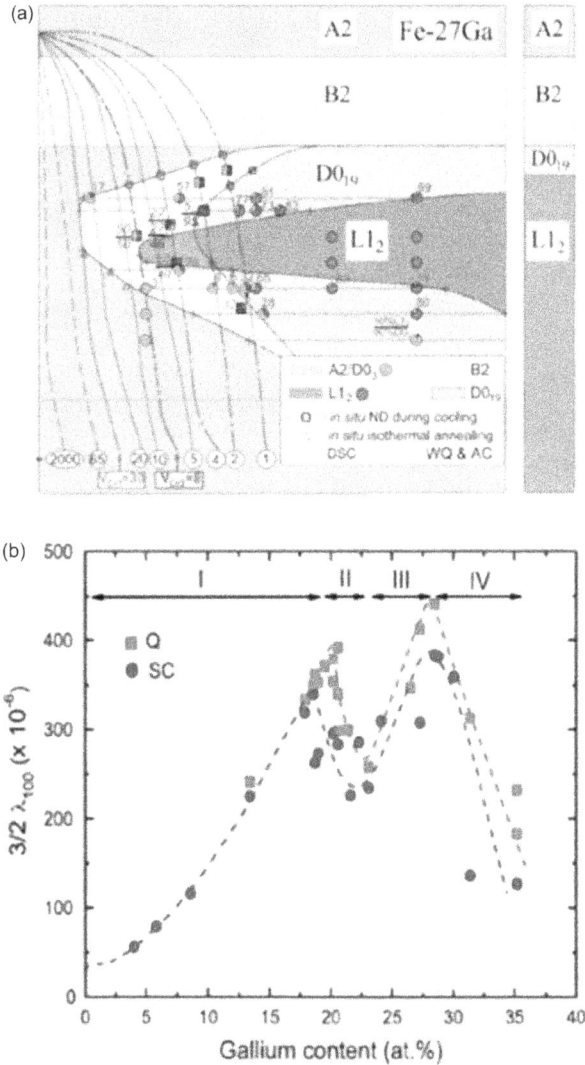

FIGURE 1.24 (a) depicts the time-temperature-transformation curve for the Fe-27Ga alloy. A thick green line represents the first crucial cooling rate, while a thick red line represents the second essential cooling rate. The X-ray diffraction (XRD) experiment results are shown as a pie chart with horizontal lines that demonstrate the lever rule. The lines for the different cooling speeds are shown in kilometers per minute. The plus-shaped spots are determined using either in situ neutron diffraction isothermal annealing or XRD analysis. (b) Saturated magnetostriction of Fe-Ga alloys as a function of the ga concentration, with arrows indicating the four regimes $I \sim IV$. Q: water-cooled at 10 degrees Celsius per minute from 1,000 degrees Celsius; SC: water-cooled at 10 degrees Celsius per minute from 1,000 degrees Celsius.

area. The effects of adding a third element can be broadly divided into the following categories:

- Enhancing Fe-Ga alloy plasticity and machinability boosting the stability of the A2 phase and the solubility limit of the Ga atom in solid solutions based on Fe.
- Improving the stability of $D0_3$ phase, reduces the solubility limit of Ga in α -Fe-based solid solution, and promoting the decomposition of A2 phase.
- Improving plasticity and machinability of Fe-Ga atoms.
- At the A2 phase, encouraging the development of certain oriented grains and giving rise to texture.
- Enhancing various A2 phase physical qualities, and so forth.

This is a list of the four different kinds of third elements that have been successfully introduced to Fe-Ga alloys so far as a result of extensive study.

1.4.3.1   3d and 4d Transition Components (Ni, Mo, V, Cr, Mn, Rh, and Co)

The influence that the incorporation of transition elements has on the magnetostrictive characteristics of Fe-Ga alloys has been the subject of a great number of research projects. According to the findings, the incorporation of even a trace quantity of Ni or Mo brought about a reduction in the values of λ_{111} and λ_{100} [14]. In addition, the magnetostrictive characteristics of Fe-Ga alloys were negatively affected by the incorporation of elements such as V, Cr, Mn, or Rh (Figure 1.25a). The reason for this was due to the fact that these elements had a consistent influence on the $D0_3$ structure and would encourage the creation of ordered $D0_3$ phase, neither of which were favorable to the magnetostrictive characteristics of Fe-Ga alloys [87/53]. Nonetheless, the incorporation of Co was shown to have some beneficial consequences. Magnetostriction, magnetism, and T_c of Fe-Ga alloys were found to rise when a minor amount (less than 10%) of cobalt (Co) atoms were substituted for iron (Fe). This was shown to be a realistic method for increasing the service temperature.

1.4.3.2   Si, Ge, Sn, and Al Elements

In the table of the elements, Si can be found at the period that comes before Ga and has an unoccupied d electronic shell, while Ge can be found at the same period as Ga and possesses an occupied d electronic shell. Sn possesses complete 3d and 4d electronic shells and locates in the next period after Ga in the periodic table. The magnetostrictive characteristics of Fe-Ga alloys are affected in a variety of distinct ways by the presence of these three elements, which are located at three consecutive periods of the same group. The magnetostrictive characteristics of the alloys may be marginally improved by exchanging a tiny percentage of Sn for another element [15]. Nevertheless, the reduction in magnetostrictive characteristics of the alloys that results from replacing Ga atoms with Si or Ge atoms may be found in references. Since Si and Ge atoms may occupy the site of Ga atoms, substantial stresses can be formed, which hinders the production of Ga-Ga atomic pairs. The

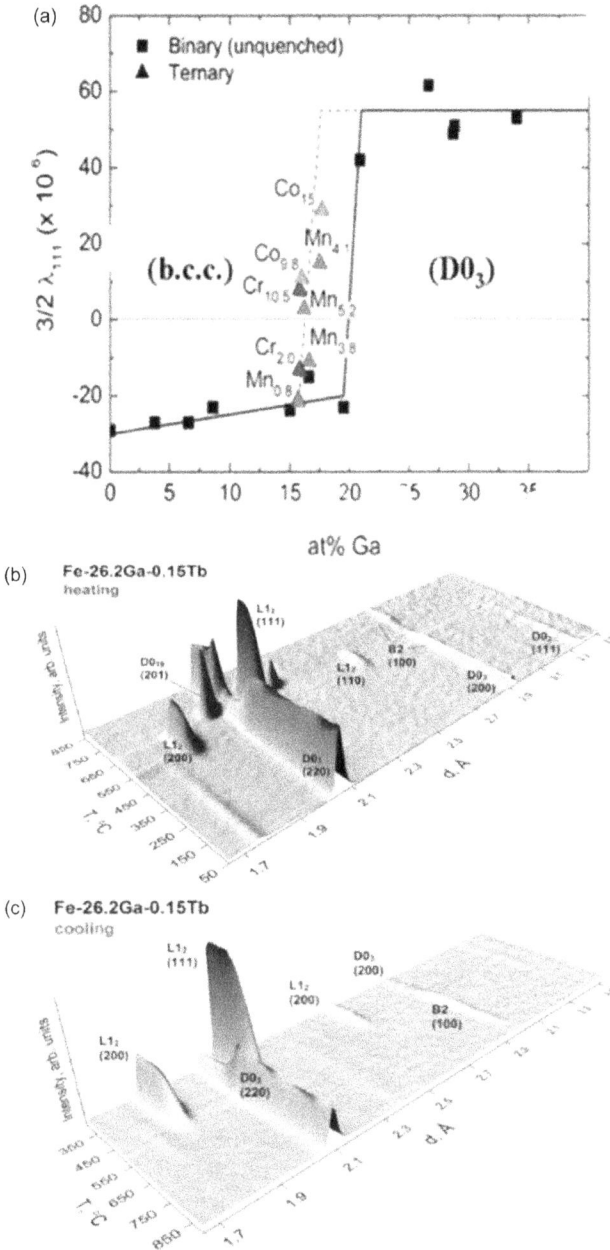

FIGURE 1.25 (a) Saturated magnetostriction equals $3/2\ \lambda_{111}$ for the Fe-Ga binary system and when Cr, Mn, or Co are added to the Fe-Ga ternary system. The visualization of the diffraction pattern in three dimensions (3D) as a consequence of (b) heating and (c) subsequent cooling of direct solidified Fe-26.2 Ga-0.15 Tb alloys.

reason for this is as follows: Al and Ga are both members of group IIIA, occupying neighboring places, and have atomic sizes and electrical configurations that are comparable to one another. As a result, the expansion of the lattice constant of Fe in Fe-Ga alloys is affected similarly by the presence of Ga atoms as well as Al atoms. The grain size of alloys with a composition of $Fe_{80} Ga_{20-x} Al_x$ ($x = 0$, 6, 9, 14) rose, while the magnetostrictive strain reduced as the amount of Al in the alloy grew. In addition, the incorporation of Al into Fe-Ga alloys has been shown to greatly enhance the material's malleability. For instance, the elongation rate of the alloys improved from 1.3% to 16.5% following the addition of 9% Al.

1.4.3.3   Interstitial Elements (C, B, N)

It is believed that the presence of Ga atoms is good for boosting the magnetostriction because it generates distortion of the cubic lattice in pure Fe. It has been shown that in single crystal Fe-Ga alloys, tiny interstitial atoms such as C, B, and N may enter the gap sites. This not only improves the magnetostriction of single crystal Fe-Ga alloys but also produces cubic lattice distortion and prevents the creation of stable $D0_3$ structures in the alloys. Dong et al. developed a highly sensitive sensor by taking advantage of the low anisotropy and high saturation magnetostriction of FeGaB alloy. With a low saturation field of 12 Oe, the magnetically annealed FeGaB films may achieve a high saturation magnetostriction of 75 ppm. Nevertheless, the magnetostrictive properties of polycrystalline Fe-Ga alloys are significantly dampened by the incorporation of C and B as small interstitial atoms.

1.4.3.4   Rare-Earth Elements

Due to the fact that rare-earth elements have an empty 4f electronic layer, they have the ability to increase the magnetic characteristics of most materials. The alloy $Fe_{83}Ga_{17}$ that had a trace quantity of Tb doped into it was the first gigantic magnetostrictive material that was identified. Doping with Tb resulted in the stabilization of bcc-born phases (A2, B2, and $D0_3$) and the prevention of the appearance of closed-packed phases (fcc ordered $L1_2$ and hcp ordered $D0_{19}$) (Figure 1.25b,c). This resulted in an increased number of grains oriented in the <100> direction, which led to an improvement in the magnetostrictive properties. In addition, research has shown that the incorporation of the rare-earth Dy into Fe-Ga alloys will not result in a change to the alloy's initial bcc-Fe structure. Dy was primarily concentrated near the grain boundary, where it formed a precipitated phase, while Ga in the matrix was skewed towards the rare-earth phase that included Dy. In addition to this, the incorporation of Dy resulted in a significant rise in the <100> preferred orientation of Fe-Ga polycrystalline alloys, which led to an increase in the magnetostrictive strain. Subsequent investigation revealed that light rare-earth elements such as Pr, La, and others performed the best when doped, and the vertical magnetostriction could reach up to −800 ppm, which was five times higher than the magnetostrictive qualities that existed before doping. The findings of these experiments point researchers in a new path toward the creation of novel GMMs.

1.4.4   FURTHER MAGNETOSTRICTIVE ALLOYS FE-BASED

1.4.4.1   Alloys Composed of Fe and Al

Since 1928, researchers have been trying to perfect Fe-Al alloys as a magnetostrictive material. It was found that Fe-Al alloys were comparable to Fe-Ga alloys in many respects. At 19 at.% Al content, Fe-Al alloys showed saturation magnetostriction λ_{100}, which was strikingly comparable to that of Fe-Ga alloys (Figure 1.26a). The similarities to the performance of Fe-Ga alloys were striking. Binary Fe-Al alloys showed a phase diagram that was extremely similar to that of Fe-Ga alloys in the iron-rich region. This was shown in Figure (1.26b). Hence, it is envisaged that magnetostrictive Fe-Al materials with higher overall properties would be produced. Pigott was the first person to identify significant magnetostriction in Fe-Al alloys with 13 at.% Al in 1956. Due to their high magnetomechanical coupling strain and low eddy current losses, these alloys, which were later given the name Alfenol, started to be exploited in commercial energy harvesters. Furthermore, compared to other Fe-based alloys, they exhibited greater levels of hardness and strength, a lower density, and a cheaper price.

1.4.4.2   Alloys Composed of Fe and Co

In addition to possessing significant magnetostrictive capabilities at room temperature, Fe-Co alloys are distinguished by their excellent ductility, exceptionally soft magnetic properties, and low magnetic field at saturation. Fe-Co alloys have abundant resources and a lower cost than Fe-Ga alloys, making them attractive candidates for a variety of actuator and power-generating device applications. Figure (1.26c) shows the relationship between the atomic percentage of cobalt and the saturated magnetostriction of iron and cobalt alloys fabricated by a variety of methods. Observably, there are two peaks in the s of the alloys, each corresponding to a distinct cobalt content percentage: 45% and 70%. In an alloy consisting of $Fe_{30}Co_{70}$, the maximum magnetostrictive strain is around 92 ppm, while the maximum magnetostrictive strain in an alloy composed of $Fe_{55}Co_{45}$ is 80 ppm. The alloy's thermal conductivity is the lowest. The magnetostriction of Fe-Co alloy is also associated with the cooling process, and the phase boundary with the highest (fcc + bcc)/bcc magnetostriction is where the maximum magnetostriction may be found (Figure 1.26d). The results show that the magnetostrictive increase is caused by the precipitation of fcc Co-rich grains into the bcc -Fe matrix, which is identical to the precipitation of $D0_3$ in the Fe-Ga alloy. It was also discovered that the maximum strain in Fe-Co films might approach 130 ppm, which was associated with the softening of the elastic modulus in Co-rich alloys. The CoFeC alloy sheet created by Wang and colleagues has a high-voltage electromagnetic coefficient (10.3 ppm/Oe), a high saturation magnetostrictive constant (75 ppm), and a low coercivity (less than 2 Oe). High saturation magnetostrictive constant and piezoelectric coefficient were traced back to the presence of a cubic phase and an amorphous phase at the core of the nanocrystalline structure.

1.4.4.3 Alloys Composed of Fe and Pd

Because to its unique martensitic transformation and shape memory features, Fe-Pd alloys have attracted a great deal of attention in the areas of micromechanics and magnetic field-controlled intelligent material system. Because of the extraordinary magnetostrictive properties they display when exposed to low magnetic fields, they have just recently been brought back into the limelight. Ren et al. reported a low-field-triggered massive magnetostriction in Fe-Pd alloys, which displayed a magnetostrictive strain of 800 ppm at a saturation field of 0.8 kOe and a Pd content of 32.3 at.%. This magnetostrictive strain was observed at 0.8 kOe of saturation field. In addition, Kubota and colleagues [16] found that a rapid-solidified Fe-29.6% Pd alloy exhibited a magnetostriction of around 650 ppm. The considerable magnetostriction of Fe-Pd alloys is mostly attributable to two factors: first, the presence of a magnetic field influences the reorientation of martensite twins, and second, martensitic transformation occurs in Fe-Pd alloys.

1.4.5 ALLOYS COMPOSED OF RARE-EARTH MAGNETS WITH MAGNETOSTRICTIVE PROPERTIES

1.4.5.1 Tb-Fe Alloys and Dy-Fe Alloys

In the 1960s, it was noticed that the bulk of rare-earth metals showed negligible magnetostriction at room temperature, but exhibited strong magnetostriction below their critical temperatures (T_c). This was due to the fact that their T_c were much lower than room temperature. As a consequence, a vast number of studies have been conducted in an attempt to create alloys with a higher T_c and a greater magnetostriction. Clark et al. [17] were able to increase the T_c of heavy rare-earth metals such as Tb and Dy by alloying magnetic transition metals like Fe, Co, and Ni. In addition, the generated $TbFe_2$ and $DyFe_2$, together with the other binary rare-earth compounds, have a cubic Laves structure [18]. At room temperature, the magnetostriction value of a $TbFe_2$ single crystal in the direction of easy magnetization <111> was more than 2,600 ppm. Nevertheless, $TbFe_2$ and $DyFe_2$ alloys in cubic Laves structures showed a considerable anisotropy, indicating that they were only capable of creating a substantial magnetostrictive strain when exposed to a magnetic field of 800 kA/m or larger. Hence, their practical uses were restricted.

1.4.5.2 Tb-Dy-Fe Alloys

The typical formula for magnetostrictive Tb-Dy-Fe alloys is $Tb_{1-x} Dy_x Fe_2$, where x represents the Dy content. And when x equals 0.70, the alloy $Tb_{0.3} Dy_{0.7} Fe_2$, also known as Terfenol-D, has the lowest anisotropy at room temperature [19]. Terfenol-D has a substantially lower saturation magnetization field than $TbFe_2$ and $DyFe_2$ alloys, with a value significantly less than 80 kA/m. This is due to the fact that at ambient temperature, the magnetic anisotropy of Dy and Tb are similar in magnitude but point in opposite directions. Hence, the inclusion of Dy and Tb into a matrix of Fe leads in Terfenol-D with dramatically decreased anisotropy and higher magnetostrictive stresses at ambient temperature. In addition, recent study has shown that heat-treating Terfenol-D in a strong magnetic field produces magnetostrictive stresses that are more than 2,000 (ppm). And this change in magnetostriction is associated with the changing of the light-to-dark contrast generated by strong magnetic fields in the domain image. Terfenol-D is currently the magnetostrictive material with the broadest use across several sectors, including but not limited to sonar systems, high-power ultra-large ultrasonic devices, precision control systems, various valves, and so forth. Terfenol-D may have a far larger range of applications; nevertheless, as the frequency exceeds a few kHz, its eddy current loss increases significantly. Under alternating current, Wan et al. investigated the magnetostrictive characteristics of Terfenol-D/epoxy and piezoelectric lead-zirconate-titanate [Pb $(Zr_{0.52} Ti_{0.48})O_3$]/epoxy composites (AC). Because of the efficient ME coupling between the two components, they determined that the eddy current loss of this composite was insignificant at frequencies as high as ~200 kHz. In another work, the process of oblique field annealing was utilized to raise the number of domain walls in magnetostrictive material to 180°. This technique reduced the magnetostrictive material's high-frequency eddy current loss (especially effective in the frequency range of approximately 50 Hz to 20 kHz) [19]. This approach was shown to be successful between around 50 Hz and 20 kHz in frequency range. In addition, there are concerns regarding the usage of Terfenol-D including its high price, strong magnetic field, brittleness, and difficulty in machining.

1.4.6 MATERIALS THAT EXHIBIT MAGNETOSTRICTIVE FERRITE BEHAVIOR

Materials known as ferrite are composite oxides that are made up of iron oxides as well as one or more other types of metal oxides. The majority of ferrite materials exhibit robust magnetism in addition to high electrical resistance and minimal eddy current losses. In addition, further research revealed that some ferrite materials displayed high T_c together with magnetostrictive stresses that were of a high quality. As a result of these remarkable qualities, ferrite nanoparticles have found widespread use in a variety of fields, including high-density information storage, medication delivery, microwave devices, and permanent magnets. In order to investigate the various ferrite material components, a great number of tests have been carried out. The majority of ferrite materials were found to have a modest MCA constant and magnetostrictive strain. Cobalt ferrite, on the other hand, exhibited distinct properties, including a greater magnetostrictive strain and a tunable MCA constant. In 2007, it was discovered that the magnetostrictive strain of sintered cobalt ferrite could reach up to −197 ppm [20], while following study indicated that the magnetostrictive strain of polycrystalline cobalt ferrite could reach up to −400 ppm. Researchers have high hopes that cobalt ferrite

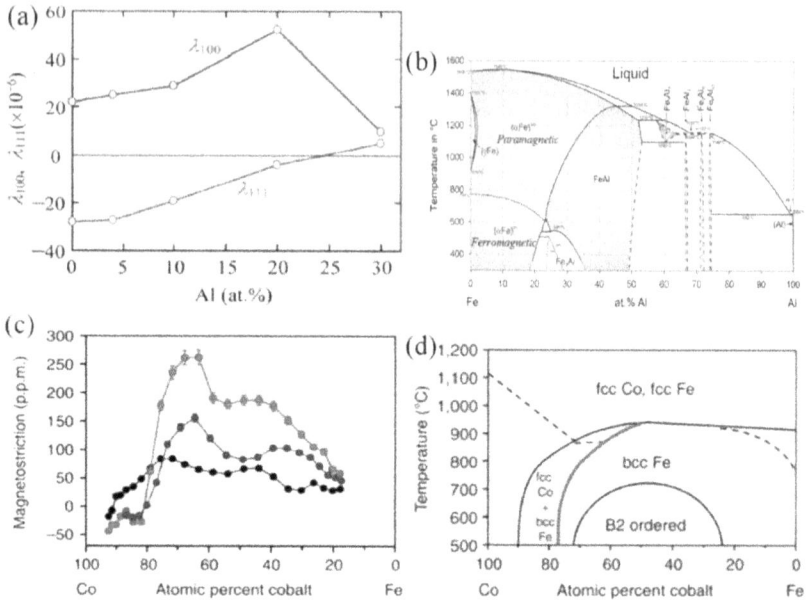

FIGURE 1.26 (a) Magnetostriction constants, λ_{100} and λ_{111}, of (b) Fe-Al (001) single-crystal films as a function of Al concentration. (c) Saturated magnetostriction versus the atomic percentage of Co for Fe-Co alloys made using three distinct techniques: as-deposited (black dots), slow-cooled (blue dots), and quenched (red dots). (d) Fe-Co-phase diagram. The approximate phase boundary between (fcc Co + bcc Fe) and bcc Fe is shown by the red curve.

magnetostrictive materials may one day be able to replace rare-earth magnetostrictive materials. These hopes are based on the features described earlier.

1.5 TERFENOL-D

The Terfenol-D material is the magnetostrictive material that has the best conditions in terms of strain and Curie temperature. The study and development of magnetostrictive supermaterials was initiated in the 1960s by researchers such as Clark and others. Terfenol-D (see Figure 1.27) was found in the 1970s at the Naval Ordinance Laboratory by a research group that Keller was leading. Keller was responsible for the laboratory's name. When magnetostrictive materials were employed competitively in transducer technology in the 1960s, before magnetostrictive superalloys were found, magnetostrictive materials were the only option. MAs, sensors, and dampers have seen a significant uptick in use over the last two decades due to the availability of super magnetostrictive materials on the market, such as Terfenol-D, which has improved levels of dependability, strain,

(a)   (b)

FIGURE 1.27   Commercial products of Terfenol-D (a) integrated (b) in powder form.

and force. Terfenol-D, a formulation of super magnetostrictive materials, was first made available on the commercial market in the middle of the 1980s. This is a significant advancement in electromagnetic devices, since they now have a better energy density and accuracy, in addition to a quicker reaction time than in the past. Sound and vibration sources, sonar systems, underwater information exchange, mechanical impact actuators and experimental acoustics, structure mechanics, active vibration control (AVC), small movement control, magnetostrictive motors, hydraulics, ultrasonics, mechanical operations, process chemicals, sensors and electrical generators, and GDI injectors are just a few examples of the many fields that benefit from these materials.

Terfenol-D was first used in high-performance sonar transducers, which was one of its earliest uses. Terpinol-D has the potential to generate positive magnetostriction at levels greater than 1,000 to 2,000 ppm when exposed to a magnetic field in the range of 50 to 200 KA/m. The coefficient k_{33} is what decides whether or not the electrical energy can be converted into elastic energy. It is assembled in Table 1.4, which even in higher projections and fields has adequate values, according to the experimental findings that were obtained in the instance of Terfenal-D, which are in accordance with other ways by employing a two-mass vibrator. In addition, in order to generate dynamic overstrain and high output power, it is necessary to have a combination of a high coupling, a high pre-tension, and a high magnetic resistance. In general, actuators that include Terfenol-D have the capability of producing 220 N to 880 N of force, and the time response of this material to an applied electric current is around a few microseconds in monolithic materials.

In the presence of strong magnetic fields, this material is able to exhibit a positive magnetostriction of around 4,000 ppm in its mechanical resonance frequency. In the real world, super magnetostrictive material is used in applications such as diesel fuel injection systems for rail cars, vibration dampers for turboprop

TABLE 1.4
Longitudinal Magnetoelastic Coefficients of Terfenol-D at 90 KA/m Degree Bias in Terms of Prestress T_0

T_0 (MPa)	30	35	40	50
Y^H (GPa)	29	21	23	40
S_{33}^H (GPa^{-1})	0.034	0.048	0.043	0.025
Q^H	4.6	3.5	4.3	8.3
μ_{33}^T / μ_0	3.7	4.2	3.8	3.0
Q^T	2.0	1.9	2.2	2.8
d_{33} (nm A^{-1})	8.0	11.0	9.7	5.0
k_{33} (%)	63.1	69.3	67.4	52.0

aircraft, micro-feeding devices, and many more. The magnetostriction strain of Terfenol-D material is 5–25 times greater than that of piezoelectric materials, and it is 50 times higher than that of pure nickel. Magnetostrictive stresses create higher compressive force, and their energy conversion efficiency may reach up to 70%, while the efficiency of energy conversion in nickel alloy is just 16%, and it ranges from 40% to 60% in piezoelectric material.

It is possible to draw the conclusion that in addition to its use in actuators, Terfenol-D will exhibit a behavior that is suitable for use in sensors due to the reasons discussed in previous sections, as well as the direct connection that exists between the magnetostriction of the material and the clarity of its Villari property. Yet, one of the most evident drawbacks of Terfenol-D is its hysteresis, which both restricts its production and makes it more difficult to manage. Also, the wetness of the material is regarded to be one of its drawbacks. Additionally, the surface quality of the support conditions and the contact surfaces of the material must be adequate, flat tolerance must be maintained, and parallelism must be maintained.

1.6   AN INTRODUCTION TO GIANT MAGNETOSTRICTIVE MATERIAL

The term "giant magnetostrictive material" (GMM) (see Figure 1.28) refers to a kind of smart material that is extensively used and has various complicated coupling physical phenomena, such as the Joule effect, the Villari effect, the e effect, and so on [21]. Because of the inverse magnetostrictive effect, several recent investigations have shown that GMM can detect displacement, velocity, stress and force without the need for additional sensors. These capabilities have been attributed

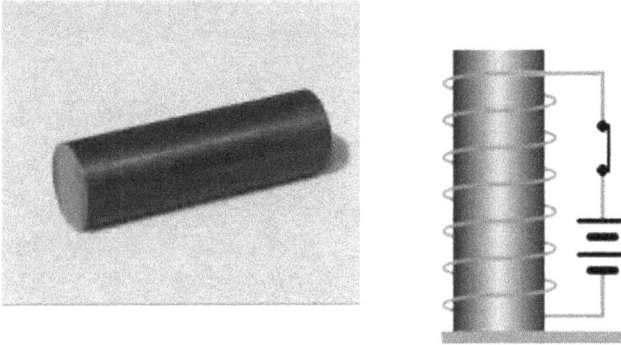

FIGURE 1.28 A giant magnetostrictive material.

to the inverse magnetostrictive effect. The GMM has been created as a combined sensor and actuator device, also known as a self-sensing actuator. This gives the GMM a greater energy density as well as an inbuilt resilience. Traditional methods for applications such as valve mechanisms for diesel injection systems may be simplified because of the self-sensing actuator's small size, which decreases the complexity of these mechanisms [19]. GMM is a kind of functional material that has a variety of outstanding qualities such as a high Curie temperature, a huge output force and displacement, a quick reacting speed, and a lot more.

A smart magnetic material with great performance is designated as 61 GMM. In comparison to piezoelectric materials and other classic magnetostrictive materials, it has a significant magnetostrictive strain, a big output force, and a quick reaction speed. As a result, it is used in a variety of applications, including precision actuation, active control, and others.

The following table provides a summary of the benefits and drawbacks associated with gigantic magnetostrictive materials.

Advantages

- Massive magnetostriction
- The choice between positive and negative magnetostriction, which may be manipulated by the composition of the substance
- Temperature much over the Curie point
- Very low vulnerability to failure due to tiredness
- Non-contact drive through magnetic field
- Low voltage drive (due to low impedance)
- Exceptional energy density
- Small hysteresis
- Quick response
- Temperature properties that can be controlled
- High levels of both elasticity and rigidity
- Significantly high coupling coefficient

TABLE 1.5
Physical Properties of Terfenol-D

Standard composition	$Tb_{0.3} Dy_{0.7} Fe_{1.92}$
Mechanical Properties	
Density	9,200–9,300 Kg/m^3
Young's Modulus	50–90 GPa
Bulk Modulus	90 GPa
Speed of sound	1,395–2,444 m/s
Tensile strength	28–40 MPa
Compressive strength	300–880 MPa
Vicker's Hardness	650 HV
Thermal Properties	
CTE	11 ppm/$°C$ @ 25$°C$
Specific heat	0.33 KJ/(kg-K)
Thermal conductivity	13.5 W/(m-k) @ 25$°C$
Melting point	1,240 @ 25$°C$
Electrical Properties	
Resistivity	60×10^{-8} Ohm-meters
Curie temperature	380$°C$
Magnetostrictive Properties	
Strain (estimated linear)	800–1,200 ppm
Energy density	4.9–25 KJ/m^3
Piezomagnetic constant, d_{33}	6–10 nm/A
Magnetomechanical Properties	
Coupling factor	0.7–0.8
Magnetic Properties	
Relative permeability	2–10
Saturation flux density	1 Tesla

Disadvantages

- The need for a driving mechanism in order to provide a magnetic field
- The driving coil's contribution to the joule heat
- Loss of eddy current in conditions of high frequency driving range
- Low corrosion resistance
- Expensive in cost

In Table 1.5, physical properties of Terfenol-D presented.

1.7 APPLICATIONS

This section will focus on the many uses of magnetostrictive materials, including those in the fields of industry, medicine, and other related fields. Smart materials may be conveniently classified as either sensors or actuators, depending on their roles in engineering. To complete its intended function, a smart structure integrates sensors, actuators, and controls into its composite construction.

1.7.1 ACTUATORS

The effectiveness of smart materials and devices is greatly influenced by Mas. This is because they are a practical example of magnetostriction, one of the first examples of a smart behavior in materials to be identified. One of the first identified intelligent characteristics of materials is called magnetostriction.

Actuators based on Tb-Dy-Fe materials have been created to serve a variety of purposes, as shown by the current studies or applications that will be presented further in the chapter. These actuators are rather distinct from one another.

As a direct result of the high energy density, high force, broad frequency bandwidth, and quick response that magnetostrictive materials are capable of providing, the number of actuator applications that are based on magnetostrictive materials, most prominently Terfenol-D, is continuously increasing. Even though the price of Terfenol-D is somewhat expensive right now, it is very probable that the number of uses for this compound will continue to broaden even as production methods get further refined and costs fall.

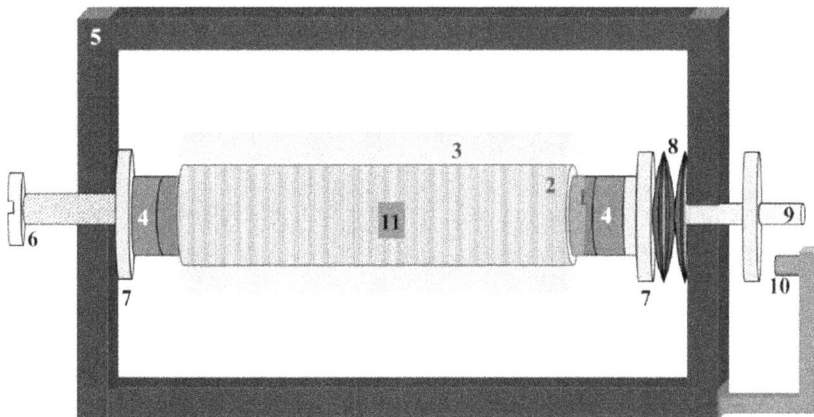

FIGURE 1.29 Not shown to scale, this illustration of a linear magnetostrictive actuator.

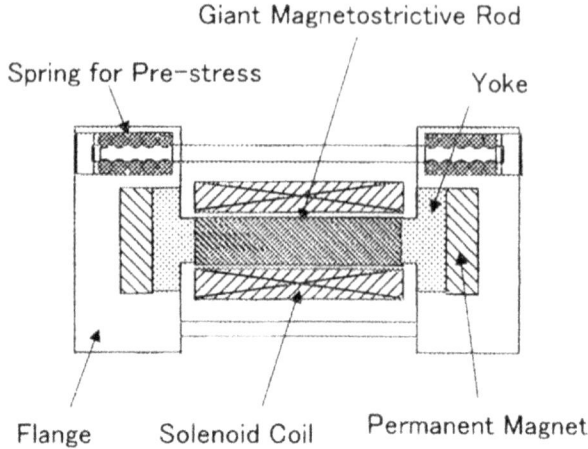

FIGURE 1.30   Terfenol-D test unit.

1.7.1.1 Actuators with a High Degree of Accuracy

Smart material actuators, such as those based on massive magnetostrictive materials like Terfenol-D and Galfenol, have a response speed orders of magnitude faster than that of conventional materials. This new actuation method has various advantages, including a straightforward design, a high-frequency drive, and a high-resolution displacement output. That's a big boost for the progress of developing actuator parts for future planes [22]. GMMs don't fast grow brittle with age, unlike piezoelectric materials, which are prone to drift, aging, overvoltage breakdown, and other issues. If the magnetostrictive effect abruptly disappears at the Curie temperature, the material will return to normal once the temperature is lowered below the Curie point, and thermal failure will not be an issue. This is very reliable and safe for use in complex aeronautical applications. The functional requirements of actuators may be broken down into categories such as the strain and forces needed at various frequencies, the need for amplifiers, the amount of space and weight allowed, and the amount of loss allowed.

The use of magnetostrictives for actuation may be broken down into the following categories, each of which is explained in more depth below.

1.7.1.2 Micropositioning

Terfenol and Galfenol alloys, with their micrometric strokes and high exerted pressures, are used for micropositioning tasks, often at low or extremely low operating frequencies. In most cases, actuations on the micrometric or sub-micrometric scale may be achieved. The rate-independent memory effects, also known as hysteresis, are primarily shown by rare-earth-based ternary compounds. These

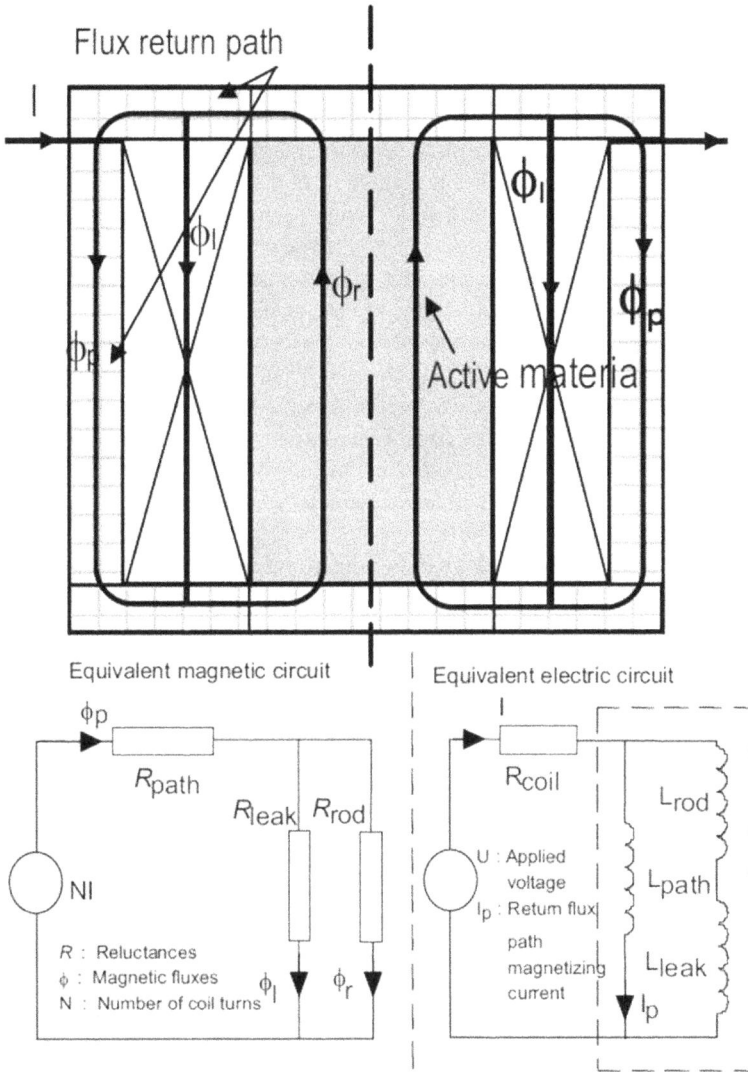

FIGURE 1.31 A magnetostrictive actuator's electrical and magnetic equivalent circuits for the driving coil and flux return routes.

effects have the potential to compromise the actuation precision, which is the fundamental performance parameter of such an application.

1.7.1.3 Motors

Since magnetostrictive materials give relatively high strokes, it is possible to conceive of micro- or inchworm-motors that have promising properties in terms of forces/torques and resolution at low speed.

1.7.1.3.1 Linear Motors

Because of the direct connection that exists between the load and the magnetostrictive element in Figure 1.58, it can be deduced that the magnetostriction will set a cap on the total net load displacement. As an example, an 11.4-centimeter Terfenol-D actuator will provide maximum displacements of around 0.2 millimeters. When it comes to vibration control applications, this sort of displacement is adequate in most cases; however, some systems, such as flow control valves or aviation flap positioners, often need significantly bigger strokes. With the Kiesewetter motor, the fact that Joule magnetostriction takes place at a constant volume is used in order to displacing loads above the maximum strain that is typically attainable with a Terfenol-D rod. This motor is made up of a cylindrical rod made of Terfenol-D, which, in the absence of a magnetic field, is able to fit snuggly inside of a rigid stator tube. In order to generate a magnetic field profile that travels down the Terfenol-D rod, the stator is surrounded on all sides by a number of shorter coils. When one of the coils is activated, for example coil No. 1 in Figure 1.32(a), the portion of the rod that is immediately exposed to the magnetic field elongates and then contracts. When the field is taken away, the rod will re-clamp itself within the stator, but it will do so at a distance of d to the left of where it was initially located. The movement of the rod is in the direction that is counter to the direction that the magnetic field profile is being swept in as the other coils in the assembly are successively powered. Inverting the order in which the coils get their electrical current results in a change to the motion's direction. From a design point of view, the only thing that may restrict the total displacement is the length of the Terfenol-D rod; nevertheless, the speed of motion is related to the sweeping frequency as well as the magnetostriction of the rod. Additional aspects, such as the number of traveling pulses, the distance between excitation coils, the stiffness of the Terfenol-D material, and the skin effect degradation caused by eddy currents, influence the smoothness and speed of the motor. The fact that the Kiesewetter motor may automatically lock itself in the off position is a feature that is useful in a wide variety of robotic applications. The proof-of-concept Kiesewetter motor that was presented in has a force of 1000 N, a useful stroke of 200 mm, and a speed of 20 mm/s. It is designed for applications such as the control of coat weight and fiber distribution in the paper industry or the operation of valves and precision positioners in the machine tool industry. Several of the technical problems that are associated with the Kiesewetter motor are addressed by an improved design that was given in, namely the deterioration of fit between the stator and the rod that is caused by wear and thermal expansion. Moreover, in comparison to the original Kiesewetter design, this updated model makes it feasible to generate rotational motion in a manner that was previously thought to be impossible.

See Figure 1.32 for yet another use of the "inchworm" idea (b). These motor parts include a load shaft, pusher transducers, translating clamps, fixed clamps, and a load shaft. By synchronizing the clamping and unclamping of the clamps with the action of the pushing transducers, it is possible to generate motion in both directions along the load shaft. Maximum allowable loads are determined by the frictional force between the clamps and the load shaft. It's worth noting that the inchworm idea may be applied to the usage of other kinds of smart materials,

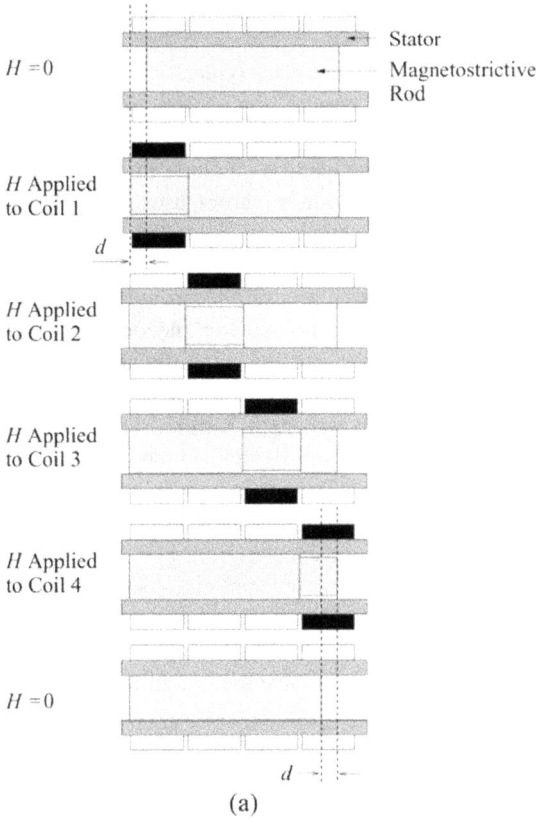

(a)

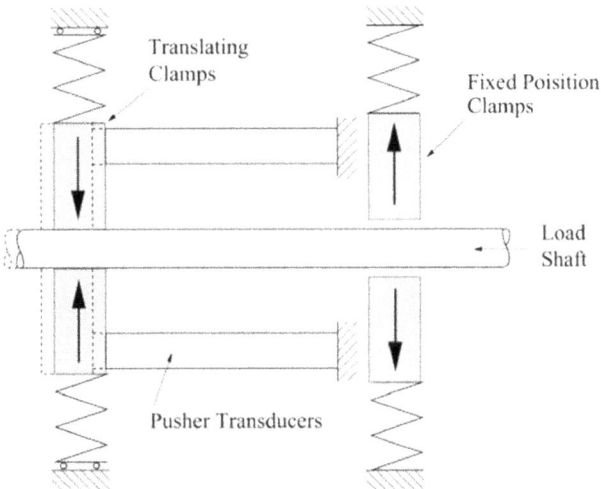

(b)

FIGURE 1.32 (a) Kiesewetter inchworm motor. In this diagram, the black rectangles represent active coils, while the white ones represent those that aren't receiving power. (b) A linear motor with an inchworm.

including piezoelectric stacks or a mix of piezoelectric and magnetostrictive components.

Although piezoelectric transducers are typically favored for ultrasonic power generation in the MHz range, certain applications in the low-ultrasonic range benefit from the magnetostrictive materials' ruggedness as well as their lack of depoling mechanisms. These applications include those that involve a magnetic field. For example, nickel has a wide range of uses, including the degassing of liquids at 20–50 kHz and the cleaning of dental or jewelry components (over 50 kHz). Recently, a surgical ultrasonic instrument that is based on Terfenol-D has been developed. This instrument is said to provide enhanced power and displacement outputs in comparison to existing piezoelectric instruments, while also being lighter, more compact, and featuring the ability to deliver a 600 V, 1 MHz signal to cauterize bleeds without interfering with the instrument's surgical function. A laminated quarter wavelength Terfenol-D rod is linked to a quarter wavelength titanium waveguide in this device, which supplies the resonant subassembly to which a half wavelength acoustic horn is attached. This device is seen in Figure 1.33. An amplification factor of between 15 and 30 is provided by the acoustic horn, which results in high accelerations and a concentration of energy at the very tip of the instrument. Additional applications for such a transducer design include industrial cleaning, sonic cell disruption and sterilization, friction welding, and the treatment of a variety of chemical and biological processes. These applications either are now in use or have the potential to be in use.

1.7.1.3.2 Rotational Motors

Not only are they technically feasible, but smart material motors that operate on the magnetostrictive principle have the potential to be less complicated and more dependable than traditional hydraulic or electromagnetic systems. A rotary motor that

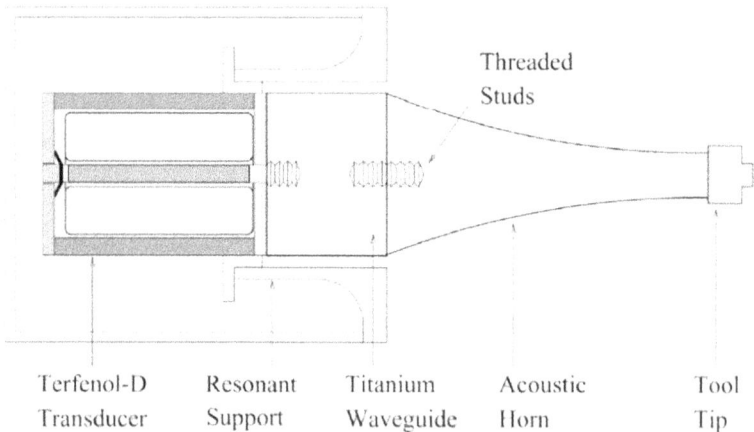

| Terfenol-D | Resonant | Titanium | Acoustic | Tool |
| Transducer | Support | Waveguide | Horn | Tip |

FIGURE 1.33 Schematic of an ultrasonic Terfenol-D device.

operates at 0.5 revolutions per minute and generates a torque of 3 Newton-meters has been built using the inchworm approach [23]. Another device of the inchworm type offers a speed of 0.5 revolutions per minute (rpm), but it generates an extremely high torque of 12 Newton-meters and accuracy microsteps of 800 μ rad [26]. In spite of their excellent positional precision and strong holding torques, the inchworm-type rotating motors that are currently on the market are notoriously inefficient. In the resonant rotating motor that Claeyssen and his colleagues [27] suggested, a significant portion of the efficiency constraint has been eliminated. Two linear Terfenol-D actuators are utilized to create elliptic vibrations on a circular ring that functions as a stator. The circular ring then transfers the vibrations to rotating rotors that are pushed up against the ring. It is said that the prototype is capable of reaching a top speed of 17 revolutions per minute and a peak torque of 2 newton meters.

The field of ultrasonic rotational motors is now the focus of a significant amount of research as well as business interest. These motors find use in a broad variety of applications, from the autofocusing lenses of cameras to the manipulators of robotic systems. A rotating actuator that uses Terfenol-D to reach a comparatively high speed of 13.1 revolutions per minute and a maximum torque of 0.29 Newton-meters was created by Akuta [23]. This particular motor, which is seen in Figure 1.34, makes use of two Terfenol-D exciter rods in order to cause revolutions in the shaft.

1.7.1.4   Active Vibration Control

Magnetostrictive materials are also divided into three categories when utilized for vibration control. There are three potential ways to handle vibrations: actively, passively, and partially. In most cases, these directions are used to dampen vibrations of low amplitude and high frequency. A magnetostrictive material's loss factor is

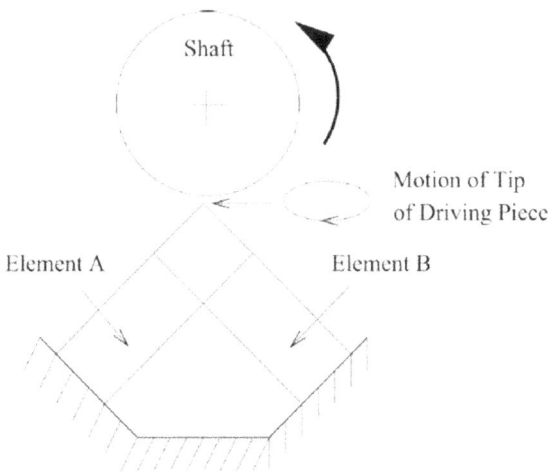

FIGURE 1.34   Rotational ultrasonic motor.

much higher than that of a piezoelectric material. Vibrational energy is dissipated by passive ways primarily through the hysteresis effect and the eddy current of ferromagnetic materials. Magnetostrictive materials are well-suited for structural damping applications that need a high degree of stiffness due to their superior stiffness compared to other materials. These disadvantages may be overcome, however, if composite materials are developed by combining many other intelligent materials. Using magnetostrictive materials for AVC complements more traditional techniques of control, such as applying dynamic forces or deformation to attenuate vibration.

Piezo-actuators have shown to be the most effective option for this application, which involves the use of smart materials in high-frequency settings. Magnetostrictive materials, on the other hand, because to their high strokes and greater energy density, drew the attention for that sort of high frequency actuation [28] (1.4-2.5 J/m3) Terfenol-D was one of the compounds that received a significant amount of research during the early years of the 21st century as part of a number of European Union (EU) projects that were designed to find new approaches to the problem of vibration control in aeronautical applications. In these projects, Terfenol-D played a significant role.

Active suspension systems are a kind of active control system that make use of the kinetic energy of the vehicle in order to mitigate the negative effects that vibrations might have on the driver and passengers.

Vibrations that aren't wanted might result in unwanted noise, poor performance, or even serious damage. While passive damping materials have been used successfully for a very long time, there are numerous situations in which they are not sufficient or effective. This is particularly true for vibrations with a low frequency and a small amplitude. As a result, approaches such as AVC with feedback control have recently started to be used in order to fulfill performance criteria. With the invention of "smart" materials, researchers have been concentrating their efforts on developing matching actuators for AVC. This field of study has been going strong for years. In this context, magnetostrictive devices are regarded to be one of the most attractive opportunities in AVC actuators technology. This is in contrast to other types of actuators, such as piezoelectric, shape-memory, and magnetorheological. Magnificent output forces may be generated by Mas, and they can travel considerable distances (compared to other emerging actuator technologies). Furthermore, they can be driven at very high frequencies.

These features make them ideal for usage in a number of vibration control contexts, such as (1) machine tools; (2) ship structures; (3) helicopter blades; (4) aircraft; (5) cylindrical structures; (6) linear structures; (7) cables; (8) multi-dof systems.

In order to make efficient use of MAs, one has to have an accurate understanding of the dynamics of the device in addition to a mathematical model to put control techniques into action. Because of the interplay of electrical, magnetic, and mechanical phenomena, the effect of magnetostriction, which is the foundation of the operating principle upon which these actuators are based, is exceedingly complicated. The nonlinear behavior of the material, which is often Terfenol-D [7] or Galfenol [29], is noticed at high operating frequencies [30], which adds another layer of complication to the situation. In particular, research has been conducted to examine the behavior of Terfenol-D throughout a wide range of working circumstances [9] and for a variety of thicknesses.

There are now a few different models in the scientific literature that describe magnetostriction. Many studies have been conducted with the goal of modeling the hysteresis of magnetostrictive materials using one of two methodologies, either physics-based or phenomenon-based hysteresis models. Since they are often derived from fundamental physical assumptions, physics-based hysteresis models make it possible to acquire a clear scientific understanding and explanation of the phenomena with which they are concerned. The disadvantage of these models is that they need for a significant amount of physics knowledge and are unique to each systems; as a result, they are not as widely used as phenomenon-based models. One of the most well-known models of hysteresis that is based on physics is called the Jiles-Atherton model of ferromagnetic hysteresis. This model is a quantitative representation that is predicated on a macromagnetic formulation. The model addresses polycrystalline materials that are isotropic, and it identifies domain wall motion as the primary magnetization mechanism. On the other hand, phenomenon-based models do not provide any new insights into the way the material behaves, but they are often used in contexts where the behavior of the material is not the primary emphasis. Both the Presaich model and the Prandtl-Ishlinskii model are able to accurately explain or forecast the hysteretic behavior of a material that is well-controlled and consistent. None of these models is reliant on any particular application. Current research has focused on the creation of intricate numerical methods, such as neural networks and genetic algorithms, in order to determine hysteresis parameters.

In spite of the fact that there are a number of models in published works, the nonlinearities and complexities of these models often prevent their practical implementation in control applications. In point of fact, it is important to keep in mind that the conventional control theory is predicated on linear time invariant systems. So, having an excessively deep understanding of the dynamic behavior of the system is not only unnecessary, but it may even prohibit the design of classical controllers like PID controllers. This is because such information is not needed. Also, the complexity of the model may result in difficult hardware measurements (and hence expensive prices) for real-time applications, but it may not result in any noticeable boost in performance. This is because of the nature of the applications. As a result, the purpose of this study is to create a model of an MA that is as simple as possible so that it may be used in various applications successfully.

1.7.1.5   Modeling Inertial Magnetostrictive Actuators
1.7.1.5.1 Functioning Principle of Magnetostrictive Inertial Actuators

Figure 1.35 depicts a conventional example of an MA. The active component is a massive magnetostrictive rod that is encased in a coil and is made to interact with a magnetic field that is produced by permanent magnets that are attached to the bar's ends. A change in the electric field that occurs when the magnetostrictive material is traversed by the flowing current in the coil causes a change in the magnetic field that acts to oppose this change. This ultimately results in the

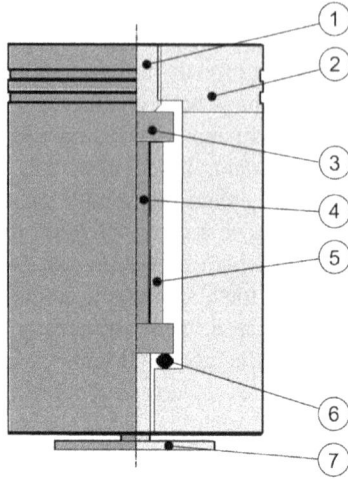

FIGURE 1.35 There are a number of parts that make up an inertial linear MA, including (1) a prestressing screw, (2) an inertial mass, (3) permanent magnets, (4) a magnetostrictive rod, (5) a coil, (6) a spring, and (7) an end effector.

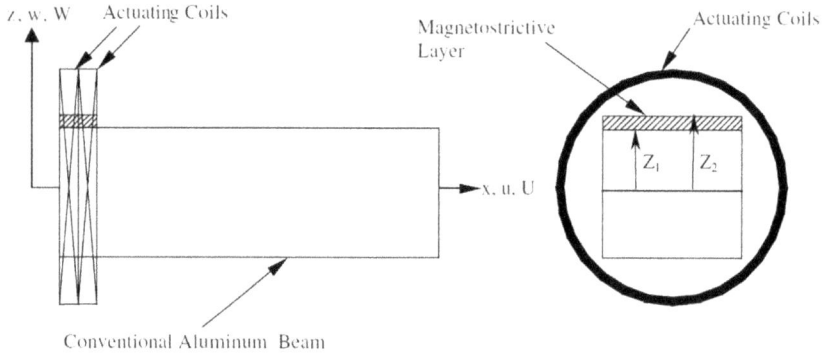

FIGURE 1.36 Common aluminum coated in a magnetostrictive material, with an actuation coil on the outside.

alignment of the magnetic domains inside the material, which, in turn, causes the magnetostrictive rod to either lengthen or shorten and results in the production of a significant amount of force. The magnetostrictive rod is compressed by a prestress mechanism, which is often a screw. This causes the rod to be pushed on an elastic component. During operation, it is necessary for the active material to be mechanically compressed, and this need has two aspects. While the tensile strength of the material is low (less than 28 MPa), the efficiency and coupling factors are both significantly improved when the material is compressed. In most cases, the part that is elastic has a low stiffness. The prestress mechanism does not considerably impact the overall effective stiffness since the magnetostrictive rod and the elastic element are arranged in parallel.

1.7.1.5.2 Equations Describing the Magnetomechanical Connection

The behavior of magnetostrictive materials is nonlinear; yet, it is possible, under certain circumstances, to approximatively model this behavior as linear. The primary hypotheses are as follows: (1) low operating frequencies, (2) reversible magnetostriction processes (no power losses), and (3) stress and strain distribution that is uniform throughout the magnetostrictive rod.

The linear magnetomechanical equations are used to depict the coupling that exists between the mechanical strain of the material and the magnetization of the material when certain assumptions are made.

$$S = S^H T + dH \tag{1.47}$$

$$B = d^* T + \mu^T H \tag{1.48}$$

In this equation, S represents the strain, T represents the stress, S^H represents the mechanical compliance at a constant applied magnetic-field strength H, d and d^* represent the linear piezomagnetic cross-coupling coefficients, μ^T represents the magnetic permeability at a constant stress, and B represents the magnetic flux occurring within the material. $d^* = d$ is the result that is obtained when the magnetostrictive process is considered to be reversible. In most cases, this is going to be the case for driving forces or fields at a low level. Terfenol-D [9] and Galfenol are the two types of magnetostrictive materials that are most often used. Table 1.6 lays up the basic qualities of these two materials. The values included inside the brackets are those that were used in the mathematical model that will be explained later.

A significant amount of interest from a variety of researchers has been directed into the suppression of vibration in buildings by using smart materials. For usage as sensors and/or actuators, smart materials including shape memory alloys, piezoelectric materials, and electro-rheological fluids are often put to use.

TABLE 1.6

Principal Similarities and Differences Between Galfenol and Terfenol-D

Symbol	Terfenol-D $Tb_{0.3} Dy_{0.7} Fe_2$	Galfenol FEGA
S^H /10^{-11} [m²/N]	3.3–4.5 (4.5)	3.1
d/10^{-8} [m/A]	2.0–2.1 (2.1)	3.4–4.2
μ^T / μ^0	12	290

Much research has been done on the potential uses of piezoelectric materials in intelligent structural applications. Recently, magnetostrictive materials have been drawing attention as potentially useful options due to their dual application in the fields of sensor technology and actuator technology. Magnetostrictive materials are becoming more popular for use in smart structure applications as a result of the availability of the magnetostrictive material Terfenol-D in a variety of forms, including powders.

1.7.1.6   Miscellanea

Magnetostrictive materials' application framework is not limited to the aforementioned problems; rather, it offers tailored answers to unique actuation challenges. Needle actuation for fuel injectors is one of the most intriguing MA uses; it has also been used for micro-pumps and acoustic purposes. One of the most intriguing uses of MA is in needle actuation for fuel injectors. MAs may be found in a variety of designs and implementations, but in general, they are composed of two primary components.

1.7.1.7   Actuator Capable of Self-Sensing

The goal of the concept known as a self-sensing actuator is to realize full system integration by integrating the actuator and the sensor into a single unit. Magnetostrictive materials, particularly Terfenol-D, are seeing a rise in the number of actuator applications as a result of its high energy density, high force, wide frequency bandwidth, and quick reaction. Terfenol-D may be expensive now, but if production methods are optimized and costs are reduced, its usefulness will certainly expand. Actuators with the configuration shown in Figure 1.37 have many applications, including sonar, chatter control of boring tools, high-precision micropositioning, borehole seismic sources, geological tomography, hydraulic valves for fuel injection systems, deformable mirrors, hydraulic pumps, bone-conduction hearing aids, exoskeletal telemanipulators, self-sensing actuators, and degassing in manufacturing processes like rubber vulcanization. The four main application subgroups that will be examined further in the chapter in light of current transducer design are sonar transducers, linear motors,

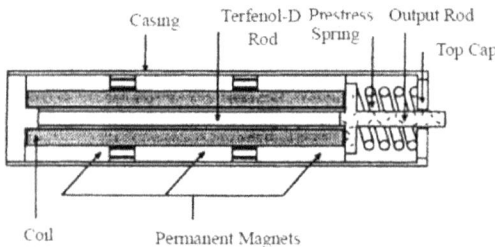

FIGURE 1.37   Picture (a) layout for (b) and typical actuators made from magnetostrictive materials.

rotational motors, and hybrid smart material transducers. More information may be found in [7, 9].

1.7.2 SENSORS

The discovery of Terfenol-D, a magnetostrictive supermaterial, has led to the creation of a new kind of sensitive material that may be used in the construction of magnetostrictive force sensors. Because of the unique properties of this material, such as its high magnetostrictive coefficient at room temperature, which is anywhere from 100 to 1,000 times more than the magnetostrictive properties of typical materials. Terfenal-D material's strong electromechanical coupling coefficient and fast reaction time (approximately a few microseconds) have garnered a lot of interest in academic settings as well as in industrial communities.

1.7.2.1 Sensor-Based Software Applications

Magnetostrictive materials are being used in a broad range of sensor designs, as is indicated by the expanding number of papers and patents pertaining to these materials. In this overview, the term "sensor" is being used in a general sense to refer to the characteristics of magnetostrictive materials that make it possible to generate electrical signals in response to mechanical or magnetic excitations such as force, strain, and torque. Specifically, the term "sensor" is being used to indicate these characteristics because they make it possible to generate electrical signals. Changes in the magnetoelastic state caused by these factors (or a combination of them) may create observable shifts in the magnetization anisotropy because of the magneto-mechanical connection. It is common practice to wrap a pick-up coil around the magnetostrictive material in order to detect changes in the magnetization of the material. This serves the dual purpose of completing the sensing mechanism and providing a method for transitioning from the magnetic to the electrical regimes. The Faraday-Lenz law of electromagnetic induction is the principle that connects the magnetization in the material to the voltage V that is generated across a pick-up coil.

$$V = -NA\frac{dB}{dt} \tag{1.49}$$

The magnetization state is quantified by the following equation, where N and A are the number of turns and the constant cross-sectional area of the coil, respectively, and B is the magnetic induction:

$$B = \mu_0 \left(H + M \right) \tag{1.50}$$

Conversely, interferometry methods may be employed to measure the changes in wave velocity caused by the magnetostrictive material's reaction to excitations from the outside world. For example, the stiffness changes that are associated with the ΔE effect can be detected using these techniques. The following section provides an overview of sensor designs, with the primary focus being given on the primary operating principles that allow the sensors to function properly. The list is not exhaustive; nonetheless, it is important to note that a large number of different designs are conceivable on the basis of the basic operating principles that have been provided here. The cited sources include more reading material for your perusal.

1.7.2.2 Torque Sensors

Magnetostrictive noncontact torquemeters were developed on the basis of the idea that the torque applied to a shaft causes stresses of the opposite sign, $+\tau$ and $-\tau$, which are orientated at ±45 degrees from the axis of the shaft. The magnetic characteristics will alter in the directions of $+\tau$ and $-\tau$ whether a magnetostrictive amorphous ribbon is connected to the shaft or the shaft itself is magnetostrictive. Either in a differential form using a series of perpendicular coils, as illustrated in Figure 1.38a, or using a single Hall effect or equivalent magnetic field intensity sensor, these features are able to be detected. In the automobile sector, for example, systems that use fly-by-wire steering might benefit from the utilization of this sort of sensor. A second category of noncontact torquemeters is based on the changes in permeability that occur in a magnetostrictive material when that material is exposed to torsional stress. In particular, applications that need less sensitivity may benefit from the increased mechanical strength that magnetic steels or alloys can give. This is because magnetic steels or alloys have a higher magnetic susceptibility. One such example is shown in Figure 1.38b, in which the operating torque of a drill bit is measured using two sensing coils that are linked in series. One of the coils is positioned over the flutes, and the other is positioned over the shank (the permeability of the shank is less sensitive to changes in torque than the flutes.)

The AC magnetic field is excited by a coil, and the proportional output of the sensor is the differential voltage created by the sensing coils in response to the changing permeability of the bit caused by the applied torque. The AC magnetic field is excited by an excitation coil.

1.7.2.3 Sensors That Detect Both Deformation and Position

Magnetostrictive ribbons or wires that have been annealed in a transverse field provide for very sensitive strain gauges. A sensor of this sort was fabricated by annealing strips of Metglas 2605SC in a magnetic field of 208 kA/m (2.6 kOe) for ten minutes at a temperature of 390 degrees Celsius and then quickly cooling them in a saturation field. Since magnetomechanical coupling depends on the state of strain in the material, the sensor reacts to changes in the permeability of the ribbon, which, in turn, depends on the state of strain in the material. Using the definition of a dimensionless gauge factor as the fractional change in the measured parameter (permeability in this case) by the change in strain, $F = \dfrac{\left(\frac{\partial\mu}{\partial S}\right)}{\mu}$, this sensor has a F equal to about 250,000, which compares exceptionally well with

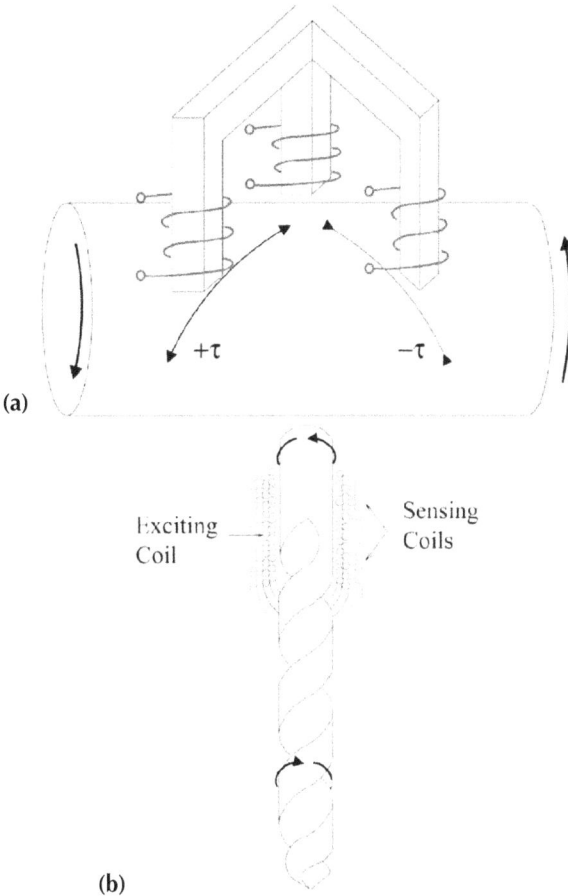

(a)

(b)

FIGURE 1.38 Two types of differential readings are possible: (a) readings in directions at an angle of 45 degrees to the shaft axis and (b) readings of the permeability changes experienced by a drill bit when applied to a torque.

resistive strain gauges, which have a F of 2, and semiconductor gauges, which have a F of 250. Normal thermal expansion has the potential to cause the sensor to become saturated, which is one of the issues that might arise with this device. Despite the fact that this solution restricts operation to AC regimes, it is possible to circumvent this issue by bonding the material with a very viscous liquid. A magnetostrictive material functioning as an auditory waveguide may be used to create a position detector if the material is strong enough. The components of the device shown in Figure 1.39 include a permanent magnet that is attached to the target and moves along the length of the waveguide, an emitter/receiver head that sends and receives an acoustic or current pulse down the waveguide, and a damper that prevents unwanted wave reflections. The functioning of the sensor is

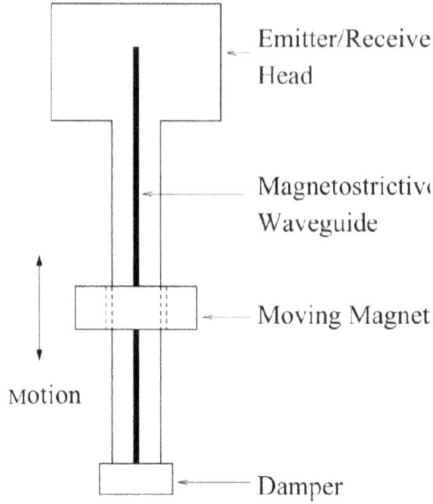

FIGURE 1.39   Magnetostrictive waveguide position sensor.

based on a very straightforward principle: when the magnet comes into contact with the magnetostrictive waveguide, the waveguide's material characteristics are altered in a localized region. These changes in the material properties may be identified in a variety of different methods. In one implementation, an acoustic pulse that was transmitted by the emitter is partly reflected back due to the stiffness discontinuity that was caused by the magnet (the Δ E effect). A second implementation of this concept involves the emitter sending a continuous current pulse down the waveguide. This results in the creation of a circumferential magnetic field, which then interacts with the axial field produced by the magnet. The Wiedemann effect is the twisting of the wire that occurs as a consequence of the ensuing helical field and is sent back to the receiver head. In any case, the length of time it takes for the original pulse and the reflected pulse to pass through the waveguide may be used as a measurement of where the magnet is located along the waveguide. This sensor may be used to detect fluid levels by connecting the magnet to a float, or it can be used for general position sensing of up to 50 m with an accuracy of ⁺₋1 mm (both of these applications need the use of a float).

1.7.2.4 Magnetometers

If the magnetostriction of a material can be understood as a function of the magnetic field, then measuring the magnetic field may be as simple as measuring the length of a magnet. A laser interferometer, an optical fiber, a strain gauge, a capacitor, or a known-length material like a piezoelectric compound may all be used to provide an accurate length reading. Two plates made of piezoelectric and magnetostrictive materials, for instance, connected through a thin electric field, form a fairly basic structure (see Figure 1.40a). The piezoelectric material induces

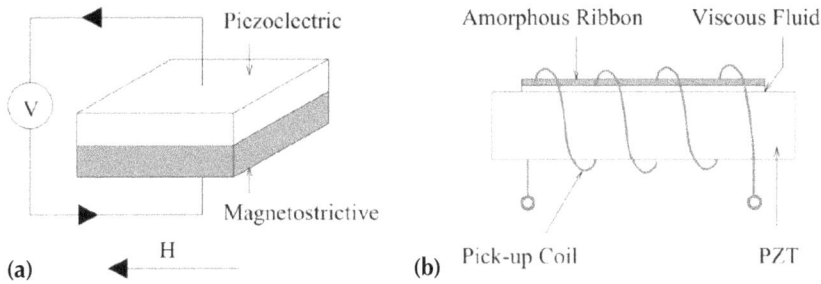

FIGURE 1.40 Magnetostrictive/piezoelectric hybrid sensors. An alternating carrier emf is generated by the piezoelectric plate during its resonance, and this emf may be recovered in one of two ways: (a) as a voltage across the piezoelectric plate or (b) as an emf induced in a pick-up coil located in close proximity to the piezoelectric plate.

a voltage proportional to the strain applied to it when subjected to a magnetic field, which, in turn, causes the magnetostrictive material to stretch. An alternative method for building a magnetometer involves using a viscous fluid to bond a field-annealed metallic glass ribbon to a resonant PZT plate (see Figure 1.40b). Another approach is also doable. By applying an AC voltage to a PZT plate, a longitudinal stress field is created. When properly bonded, the dynamic stress in the metallic ribbon is identical to that in the PZT, while the static stress is filtered out by the viscous fluid. The prevailing dynamic stresses trigger an oscillating electromotive force (emf) in the nearby pick-up coil thanks to the Villari effect. When the coil is exposed to low-frequency magnetic fields, an emf. is generated; this emf is then isolated from the carrier emf using conventional phase-sensitive detection techniques. The detection limit may reach 6.9×10^{-6} A/m at 1 Hz, which is on par with that of fluxgate magnetometers. Another sort of magnetometer uses an optic fiber coated with a magnetostrictive layer. When the sensor is exposed to magnetic fields, the magnetostrictive material becomes magnetostrictive, causing it and the optic fiber to deform. As a consequence, laser beams experience slight changes in their optical path length as they travel through an optic fiber, which may be detected and measured using an interferometer. High-sensitivity metallic glass ribbons have been utilized in such devices, resulting in quasistatic resolutions of 1.6×10^{-3} to 8.0×10^{-3} A/m. At last, length changes in a Terfenol-D rod caused by a magnetic field have been measured using a diode laser interferometer. Even though significant nonlinear dependencies were discovered, a maximum sensitivity of 160×10^{-6} A/m was achieved. To achieve this optimum sensitivity, however, the sensor must be operated within the zone of its ideal mechanical preload.

1.7.2.5 Force Sensors

The Villari effect may be used with crystal or amorphous magnetostrictive materials to provide a simple and durable force sensor. As a matter of fact, this is possible. The magnetostrictive feature is the basis for the sensor's working principle

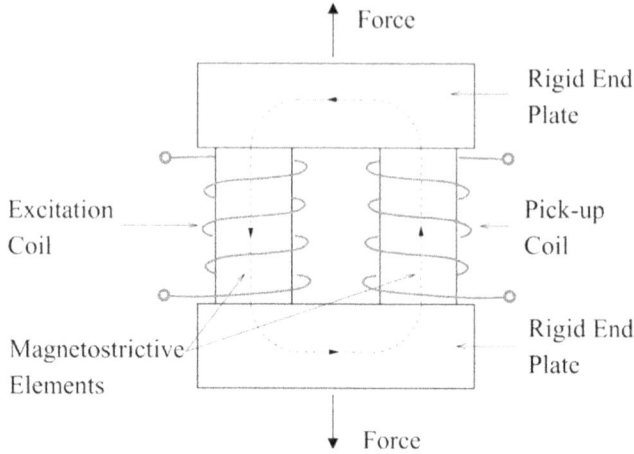

FIGURE 1.41   Based on the Villari phenomenon, a magnetostrictive force sensor.

due to the fact that magnetization is dependent on the material's stress state. The term "operational principle" describes this aspect. The construction shown in Figure 1.41 consists of two magnetostrictive components, one encircled by an excitation coil and the other by a pick-up coil, and two stiff end plates. In one mode of operation, an alternating current voltage is applied to the excitation coil, causing a magnetic flux inside the sensor and a corresponding voltage within the sensing coil. Magnetostriction happens in the elements when a force is applied, which modifies the magnetic flow. An equivalent voltage shift in the pick-up coil may be seen whenever the magnetic flux changes. In constant flux, the second mode of operation, the excitation voltage might fluctuate but the pick-up coil output voltage stays the same. Force changes are then linked to the unique excitation voltage. This new kind of force sensor has many advantages over the standard strain gauge-based force sensors now on the market. As with the Villari effect sensor, suggests amorphous ribbons as a future study platform. Numerous other concepts, such as percussion sensors, pressure sensors, and force sensors based on magnetoelastic strain gauges, have been explored or patented in recent years.

1.7.3   AEROSPACE

There was a period of rapid growth in the aerospace industry over the course of the past few decades. More and more strict standards are being applied to the material characteristics utilized in aviation, whether for civilian, military, or space-based applications. When it comes to the continuous attempts to enhance the aviation sector, smart materials and structures are a crucial concern and research direction. All of this is a part of a bigger plan to boost the aviation sector. So-called smart materials are substances that can, to some extent, adapt their physical properties in response to external stimuli such as electrical currents. Nowadays, magnetostrictive

(a)

(b)

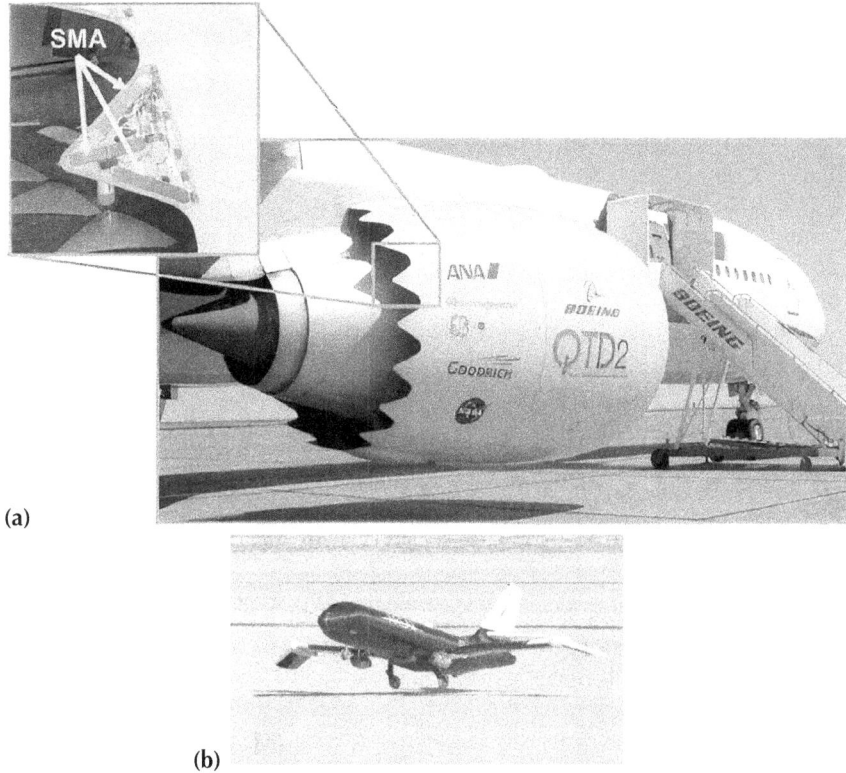

FIGURE 1.42 (a) SMA-based shape control in the engines of Boeing 777-300ER aircraft; (b) SMA torque tubes positioned in the chord-wise direction along the hinge-line to regulate the articulation of the outboard wing portion.

materials are the most researched and applied smart materials. Moreover, a few other forms of intelligence-based materials need consideration. In order to detect a wide range of signals, smart materials must be able to sense their environment. Among these signals are things like heat, pressure, magnetism, electricity, luminosity, radioactivity, and other forms of electromagnetic radiation. Many forms of smart materials have the potential to modify a broad range of signals and attributes, such as shape, stress and strain, current and voltage, and damping coefficient. There are several approaches of bringing about these modifications.

The smallest amount of vibration might eventually disable the structure, which would cause considerably bigger issues. Aerospace engineers have known for a long time that structural vibration poses serious risks to aircraft. As an example, it is challenging to arrange a conventional vibration control structure for the spindle structure, which is commonly used in aerospace engineering and is susceptible to vibration caused by external excitation as it performs the fundamental tasks of aircraft positioning and stabilization. This is due to the fact that dampening vibrations in the spindle construction is a challenging task. Yet, the newly designed wireless piezoelectric stack actuator may be able to effectively dampen vibrations while keeping the plane intact. There are three basic methods for dampening vibrations; they are known as passive vibration control, semiactive vibration control, and active vibration control. To use the material properties of the building itself to convert vibrational energy into internal energy or other forms of energy like electrical energy, and then to use that energy, is a realization notion of passive vibration control. Spreading sensors to monitor the target structure's vibration and deploying actuators to fine-tune dampening or create dynamic force are both essential components of AVC systems. Vibration may also be actively controlled by doing things like actively changing dampening or actively providing dynamic force. Semi-active regulation lies somewhere in the center.

1.7.4 Bio

Biomedical science is an interdisciplinary area that applies the principles and techniques of contemporary physical science and engineering to the study of how biological things are put together and how they work, such as the human body and other forms of life. In most contexts, terms like "biomedical measurement and monitoring," "biomedical information processing," "biomedical treatment," and so on, relate to the same set of activities. It's clear that the biomaterial properties outlined before play a significant role in the many biological applications considered so far. Histocompatibility, hemocompatibility, and other measures of biocompatibility are important, but these materials should also be non-toxic and not cause immune rejections or inflammatory responses. For the purposes of biomedical measuring and monitoring and biomedical information processing, biomaterials should also be responsive to electrical, magnetic, or mechanical signals. Biomedical therapies need materials with specific therapeutic effects, mechanical properties, and resistance to biological aging. Because of their unusual properties, magnetostrictive materials have garnered a lot of interest for use in healthcare. They can detect physical quantities like as temperature, pressure, liquid viscosity, and flow velocity, as well as chemical quantities such as pH and carbon dioxide. They're also put to use as flow sensors. This has led to a plethora of research on magnetostrictive alloys, particularly their applications in medicine. Such research has looked at phenomena including cellular behavior and the mechanics behind magnetostriction. Researchers have also looked at the cutting-edge uses of magnetostrictive alloys, wireless implanted devices, magnetic field sensors, and remote cell drives.

1.7.4.1 Behavior of Individual Cells
1.7.4.1.1 Cell Viability
Cell viability is defined as the percentage of original cells that are still alive after being co-cultured with biomaterials. Cells may be affected by their surrounding environment in a number of ways, including but not limited to the parameters in cell culture, stimulation by drugs and growth factors, and reactivity to a wide range of disorders. If one wants to know how healthy cells are, one must first

assess their viability. Vargas-Esteve et al. [24] performed research on the impact of Fe-Ga alloy films and microparticles on cell survival using macrophages, osteoblasts, and osteosarcoma cells. After introducing Fe-Ga alloy microparticles to the macrophages, the findings revealed that the Fe-Ga alloy films had no negative effect on the survival of the three cell types. Researchers also observed no significant variations in cellular activity between the two groups (Figure 1.43a,b). Holmes et al. [25] reported using the magnetostrictive alloys $Fe_{88} Ga_{12}$, $Fe_{71} Ga_{29}$, and $Fe_{40} Ni_{38} Mo_4 B_{18}$ in biodegradable implants. Cytotoxicity results revealed that $Fe_{40} Ni_{38} Mo4 B_{18}$ was not biocompatible, whereas $Fe_{88} Ga_{12}$ and $Fe_{71} Ga_{29}$ degradation products did not affect the viability of L929 fibroblast cells. Moreover, it was shown that a magnetic field could be applied remotely to regulate the deterioration rates of magnetostrictive alloys by inducing the alloys to produce low-magnitude vibrations that accelerated their degradation rates. Because of this, magnetostrictive alloys' deterioration rates might be adjusted from a distance. Magnetostrictive alloys, which can be controlled magnetically and naturally deteriorate, were shown to have significant promise as active implants. Both the Fe-Pd alloy and the Tb-Dy-Fe alloy were shown to be non-toxic in further studies, independent of whether or not an external magnetic field was present. Nevertheless, studies showed that 24 hours after exposure to Fe-Co alloy, cell viability dropped significantly.

1.7.4.1.2 Cell Proliferation

Cell division is critical to the maintenance and development of all living things. Cells proliferate and spread via a process called cell division. In unicellular species, reproduction occurs by cell division, whereas in multicellular organisms, cell division is used to replace worn-out or dead cells. Wang et al. [31] looked at the in vitro cytocompatibility of three different types of Fe-Ga alloys as potential biodegradable metallic materials, namely $Fe_{81} Ga_{19}$, $(Fe_{81} Ga_{19})_{98}B_2$, and $(Fe_{81}Ga_{19})_{99.5}(TaC)_{0.5}$ alloys. The possibility of using these alloys as biodegradable metals was investigated. Furthermore, MC3T3-E1 cells showed great adherence and proliferation on the Fe-Ga alloys surfaces after only four and 24 hours of culture. Studies on ferrite showed that, similar to Fe-Ga alloys, this ferromagnetic material had little effects on cell proliferation behavior, and cells proliferated normally on the material surface. Cell growth was likewise not affected by Terfenol-D/poly(vinylidene fluoride-co-trifluoroethylene) composite films, according to a separate investigation. Most recent studies, however, have shown that an external magnetic field greatly enhances cells' capacity to proliferate (Figure 1.43c,d).

1.7.4.1.3 Cell Morphology

One of the most crucial factors in assessing the biocompatibility of various materials is their ability to interact with various cell shapes. Several cell morphologies, such as spherical, polyhedral, spindle, and cylindrical, are associated with certain cellular tasks. As an added bonus, a cell's shape may often tell us something about its physiological state. Scanning electron microscopy and fluorescence

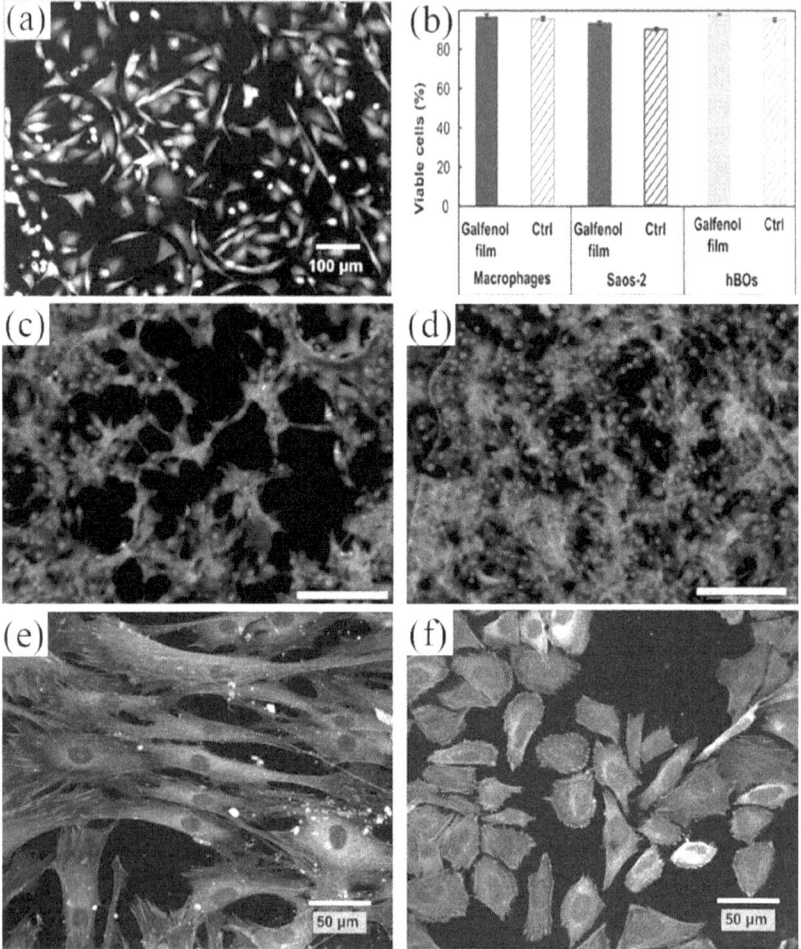

FIGURE 1.43 (a) Green (alive) and brown (dead) (red) macrophages, Saos-2 cells, and hOBs that were cultured on top of Fe-Ga alloy films or glass coverslips and (b) the proportion of those cells that survived. Images showing pre-osteoblasts cultivated for 72 hours under static (c) and dynamic (d) conditions on Terfenol-D/poly(vinylidene fluoride-co-trifluoroethylene) composite films (nucleus stained with DAPI-blue and cytoskeleton stained with TRITC-red). Vinculin immunodetection (green), stress fiber distribution (red), and nuclei staining (blue) in confocal laser scanning microscopy (CLSM) images of (e) hOBs and (f) Saos-2 cells cultured on Fe-Ga alloy sheets.

microscopy are often employed to try to identify cell morphology. After five days of co-cultivation with Fe-Pd alloy extracts, MG-63 cells showed typical growth characteristics, including a fusiform or polygonal shape and filopodia with spikes, in a research on the alloys. The presence or absence of filopodia served as a proxy for cell shape. The effect of ferrite on the form of cultured cells was studied using experiments quite similar to those described earlier. Primitive human fetal osteoblasts, designated fHOb (406-05f), had the typical polygonal or spindle form. Human osteoblast cells (hOBs) and a human osteoblast-like cell line (Saos-2) were able to form focal contact with Fe-Ga alloy sheets and had a distinct stress fiber profile, indicative of healthy cell shape (Figure 1.43e,f). Scientists made the discovery after seeing that hOBs and a human osteoblast-like cell line (Saos-2).

1.7.4.2 The Impact That Magnetostriction Has on the Cell's Effect Mechanisms

1.7.4.2.1 Magnetostriction-Induced Mechanical Stimulation as a Stimulus

According to the Joule effect and the Villari effect, magnetostrictive materials may transform magnetic energy into mechanical energy. Magnetostrictive materials respond to an alternating current (AC) or direct current (DC) field by inducing static or dynamic strain, respectively, which may be translated into mechanical stimulation. Remote control of the corresponding strain is possible through a magnetic field due to the fact that magnetostrictive strain is formed as a function of the initial length of the material and the intensity of the applied magnetic field. This is so because the amount of magnetostrictive strain developed is proportional to the intensity of the magnetic field and the initial length of the material. In addition, magnetostrictive strain is typically below 1 ppm, therefore the associated mechanical stimulation is unlikely to significantly degrade the cells' physical condition. This has led many to believe that magnetostrictive alloys would play a pivotal role in a wide range of medical and health-related fields.

1.7.4.2.2 The Effects That the Processes of Mechanical Stimulation Have on Cells

Molecular studies have shown that connections between mechanical and chemical signals at the cellular level are essential to the reciprocal mechanochemical transformation pathways (Figure 1.44). Mechanical signals may be generated by both intracellular (F_i) and extracellular (F_e) forces, and these may influence cellular activities (Figure 1.44a). F_i may be transmitted to neighboring cells via intercellular connections or by pulling on integrin-coupled extracellular matrix ECM adhesion molecules. Intercellular junctions and the use of mechanical pressures to the extracellular matrix are two ways in which F_i in different cells may be brought together. Normal methods of applying iron to cellular structures include shearing, stretching, or compression. Cells are able to sense shifts in iron content by mechanical gating of ion channels, changes in cytoskeleton and primary cilium deformations, and alterations in receptor-ligand interactions. As shown in Figure 1.44b, the cytoskeleton is involved in the production and transmission of forces from membrane proteins to intracellular structures, as well as the control of cellular activity via a cascade of mechanochemical signal

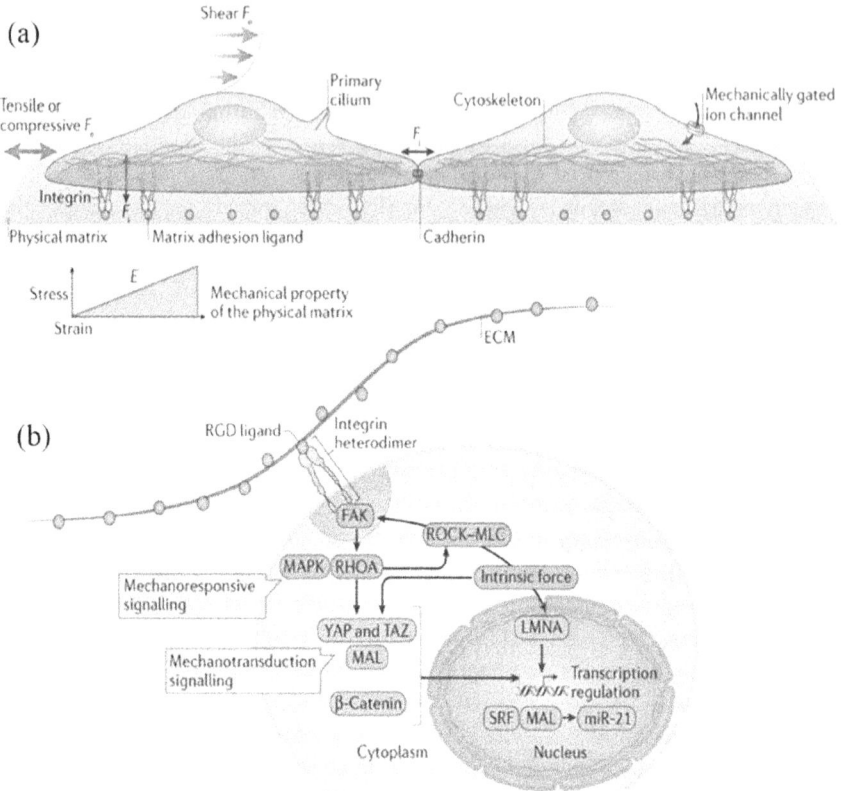

FIGURE 1.44 This includes (a) the method of action and transmission of Fi and Fe, and (b) the transduction and modification of mechanochemical signals.

transduction and transformation. Several studies have shown that the kind of mechanical stress given to cells may alter their growth and function. Human cortical bone remodeling was demonstrated to not occur at strain concentrations below 200 ppm. Human physiology is represented by strains between 200 ppm and 2,000 ppm, and this range is what was used to create the bone strain map. And at stress levels over 2,000 ppm, bone formation outpaced bone resorption, resulting in bone regeneration. In a study with turkeys, researchers found that a strain signal at 30 Hz and 100 ppm was adequate to preserve bone density, whereas in a study with rats, researchers found that a signal at 2 Hz and 930 ppm was sufficient to stimulate bone formation. Two investigations showed that this range of values for a strain signal was adequate to maintain bone mass stability. These results revealed that several characteristics, including the loading types, amplitude, frequency, and duration of the mechanical stress, interact to determine the cellular response to the stressor. In addition, if the mechanical stress is applied at a high enough frequency, even a little amount of force may elicit a beneficial response from the cells.

1.7.4.3   Remote Cell Actuation

Magnetostrictive materials have gained considerable interest in recent years as a possible strategy for long-distance stimulation of cells in biomedicine. The effects of an external magnetic field on interactions between cells and particles made of Fe-Ga alloy were studied by Vargas-Estevez et al. The macrophages and the alloy particles were co-cultured, and subsequently the particles were injected into the cells. Static magnetic fields were used to create particle chains on the cell membrane and inside the cells. The magnetic field vector allowed for the remote regulation of these chains without impacting cell shape (Figure 1.45a,b). Ribeiro et al. [32] grew MC3T3-E1 pre-osteoblast cells on sheets of Terfenol-D/ poly(vinylidene fluoride-co-trifluoroethylene). For the course of the incubation process, we provided a magnetic field (up to 110 ppm) to stimulate the cells mechanically (Figure 1.45c,d). The study found that mechanical stimulation increased the number of osteoblasts by around 20%, revealing the amazing potential of magnetostriction in distant cell actuation. Hart et al. also applied an external magnetic field at 30 Hz and 170 kA/m to a pig tibia after it had been directly bonded with Terfenol-D composites. When a strain of more than 900 ppm was detected on the surface of the bone (Figure 1.45e,f), it was shown that Terfenol-D composites are effective in promoting bone tissue formation through distant magnetostrictive actuation.

1.7.4.4   Sensors That Detect Magnetic Fields

There is promise for the development of magnetostrictive-based sensors to detect magnetic fields in living organisms. Noninvasive techniques such as magnetoencephalography (MEG) and magnetocardiography (MCG) have allowed researchers to learn more about higher-level brain functions in humans. Measuring the biomagnetic field is a powerful tool for clinical diagnosis. The sensor must have a low detection limit in the low frequency band in order to monitor the biological magnetic field. The amorphous soft magnetic alloys FeGaB and FeCoSiB have been developed as magnetic field sensors for the detection of biological magnetic fields because of their strong magnetostriction and high permeability. Magnetic fields in living systems may be measured using these sensors. Yu et al. developed a microelectromechanical system (MEMS) resonant magnetic field sensor based on an AlN/FeGaB bilayer nanoplate resonator (Figure 1.46a). With a detection limit of ~ 1 nT, this sensor is sensitive to frequencies as low as ~ 1 Hz. By evaluating a thin-film magnetoelectric (ME) sensor, Reermann and coworkers were able to quantify R-waves in human hearts (Figure 1.46b). An exchange bias stack composed of 20 times 5 nm Ta/3 nm Cu/8 nm MnIr/200 nm FeCoSiB was sprayed onto the underside of the sensor to increase its sensitivity. Using FeCoSiB's magnetostrictive properties was the motivation for this. Moreover, Ren et al. used the special qualities that enable the measurement of the degradation rate of artificial bone in vitro (Figure 1.46c). Results suggest that magnetostrictive materials have

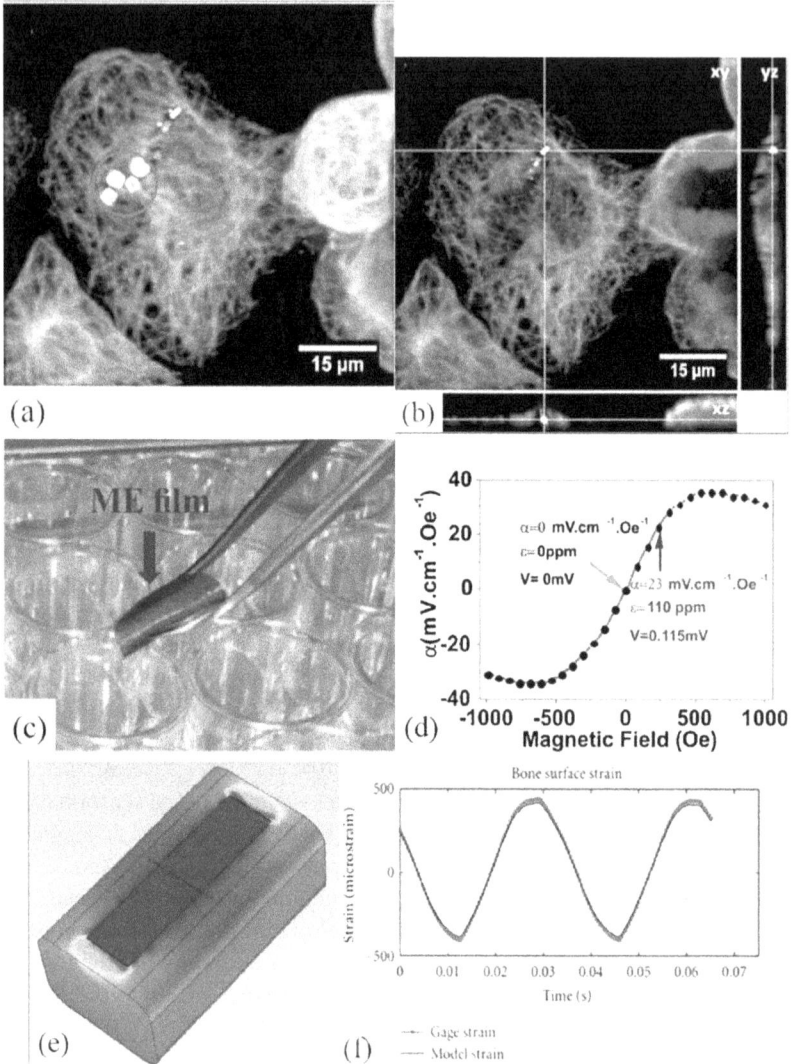

FIGURE 1.45 Magnetically exposed macrophage cytoskeleton interacting with Galfenol-based microparticles. Orthogonal picture (a) of the same macrophages after 1 minute in a magnetic field and 24 hours in culture. Six Fe-Ga alloy microparticles were found in one macrophage, with four located inside the red circle and two indicated by red arrows. Two of these microparticles, as shown in the *yz* picture on the right and the *zx* image at the bottom, were within the cell, as determined by orthogonal projection. The green-stained microtubules were evenly spread out. Blue represented nuclei stained with Hoechst, whereas white represented microparticles visible by reflection visualization (b) and 3D reconstruction. Terfenol-D/poly(vinylidene fluoride-co-trifluoroethylene) film (c) and a magnetoelastic characterization (d) with values for the magnetoelastic coefficient, strain, and voltage (V) created by the dynamic cell culture. When a Terfenol-D composite is bonded to a bone and then driven at 30 Hz and 170 kA/m, we can observe (e) the strain distribution along the longitudinal direction of the bone model as determined by finite element analysis, and (f) the predicted strain from the model versus gage strain on the bone surface.

considerable promise for enhancing the detection capabilities of magnetic sensors used to assess biomagnetic fields.

1.7.4.5 Wireless Implantable Devices

One of the limitations of typical electromagnetically based wireless implants is the size of the antennas required [180,181/143,144]. In contrast, the performance of the ME antenna, which is made of piezoelectric material and surrounded by a magnetostrictive layer, is relatively unaffected by its reduction in size. In order to create a ME heterostructural antenna for usage in wireless connection applications, Rangriz et al. simulated a combination of a piezoelectric AlN film and a magnetostrictive FeGaB film (Figure 1.46d). The results showed that the ME antenna exhibited low loss, high matching, and good efficiency when utilized as an implanted device. In addition, Zaeimbashi et al. [33] developed NanoNeuroRFID, a novel, wireless, ultra-compact device that may be implanted in the brain. The ME

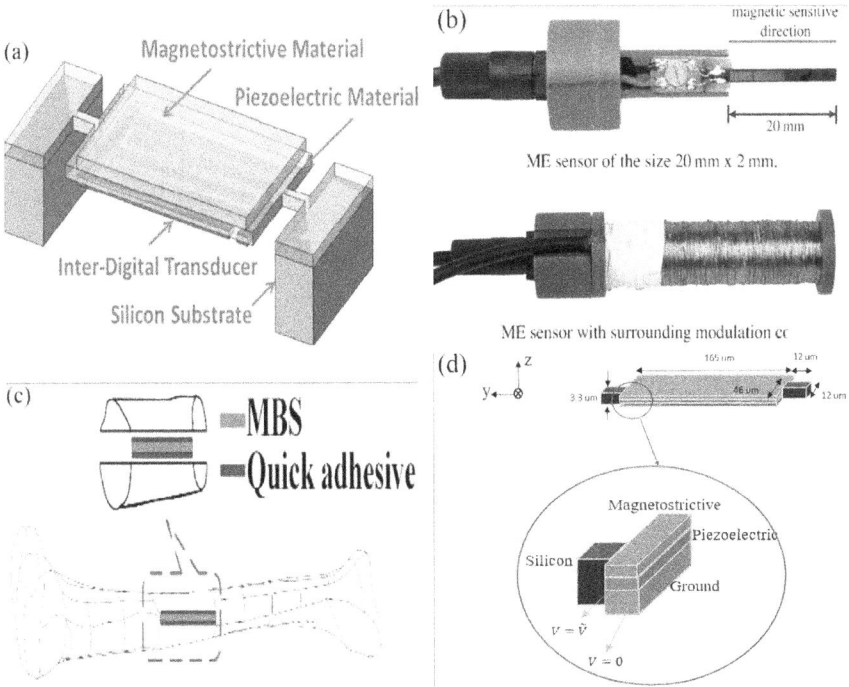

FIGURE 1.46 Biomagnetic field measurement and wireless implanted devices are two fields where magnetostrictive materials have found a number of promising uses. (a) This is a 3D diagram of the MEMS resonant magnetic field sensor (b) The magnetic resonance electrocardiogram (ME) sensor (c) Model for monitoring and evaluating the degradation rate of artificial bone. (d) A magnetoelectric (ME) heterostructure made up of a piezoelectric AlN film and a magnetostrictive FeGaB film.

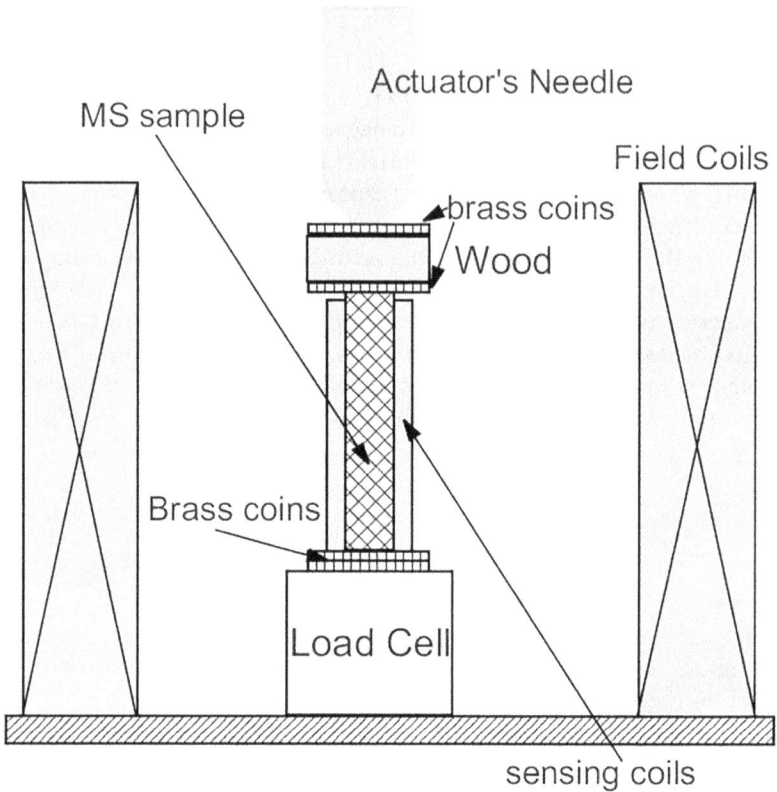

FIGURE 1.47   A diagram of the lab bench used to study magnetostrictive harvesting.

antenna used in this system was structurally identical to that developed by Rangriz et al. [34]. Contactless induction of brain magnetic fields, communication with external transceivers, and self-power supply are only a few of the many tasks that this equipment is capable of doing. Magnetostrictive materials have shown strong promise as a viable means of enhancing the functionality of wireless implanted devices. One way to do this is by making the gadget function with less resistance.

1.7.5 ENERGY HARVESTING

The potential for harvesting energy from vibrations has been offered in a variety of contexts in recent years. The primary mode of operation of many smart materials, materials like magnetostrictive, may convert mechanical energy from vibrations in the surroundings into electrical energy. To do this, researchers take use of the fact that certain materials exhibit coupling between mechanical variables and electric or magnetic fields. Intelligent materials are used to achieve this goal. Whereas there is a considerable body of experimental testing available in the scientific literature relating to piezoelectric-based systems, magnetostrictive-based harvesting devices have received very less attention. Consider the active material to have linear properties, which allowed us to implement some of the key phenomena involved in energy harvesting systems.

Energy harvesting's potential uses in the aviation industry are only just being investigated. Energy harvesting technology was developed to address the problem of powering sensors and actuators that cannot function without external power. The second goal is to establish a self-power supply of distributed sensor actuators that is more cost-effective than current systems (which need frequent battery replacement). The sensor will be placed in a variety of locations on the target structure throughout the design phase of the development study for SHM technology within the aviation sector. This will allow for a more thorough and consistent method of monitoring the intended structure. More sensors mean less overall system reliability in modern aviation systems. This is because the dependability of sensors is crucial to today's aviation systems. Thus, it is sometimes necessary to build up multiple times the number of redundancies to address the issue of sensor failure and lessen the effect of sensor failure on the functioning of the system. Recent years have seen extensive study and use of SHM technology, but energy harvesting technology has also attracted considerable attention and research due to the need of using it to address the energy supply issue of a large number of widely scattered sensors. Energy harvesting systems are based on the principle of converting and storing different types of ephemeral or external energy once they have been turned into electricity using the appropriate sensors. An energy harvesting system consists of the load, the energy converter, and the intended use of the collected energy. Large-scale targets may be powered by a variety of energy sources, including as the sun, the wind, the tides, geothermal heat, and so on. In contrast, the term "small-scale energy" refers to a wide variety of energies that are not applicable to large-scale applications, such as thermal energy, mechanical energy, electromagnetic radiation, and others.

When they are exposed to mechanical strain, the magnetostrictive materials that have lately found usage in industry create a magnetic field. Alterations in the voltage that is applied result in corresponding shifts in the flux of the magnetic field, and vice versa. The Villari effect is the name given to the strain that occurs in certain materials as a result of the application of an external magnetic field to the materials in question. Altering the magnetic field may result in the production of an electric current if a coil is wrapped around the object in question. Magnetic pickup devices are built on a foundation that is quite comparable to that of electrical pickup devices. with the key distinction being that magnetostrictive layers are responsible for generating the changing magnetic field. The schematic of these harvesters is depicted in Figure 1.48.

1.7.6 CIVIL

In response to a change in their magnetic state, magnetostrictive materials suffer a change in their elasticity. They might replace current intelligent materials used in civil structures and become the standard.

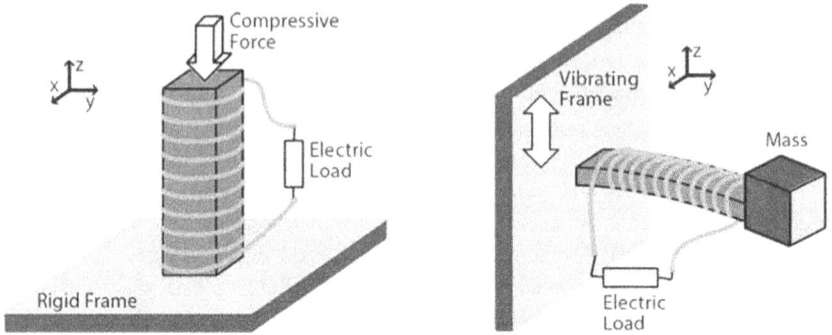

FIGURE 1.48   Magnetostrictive energy converters schematic.

Due to the growth of large-scale civil engineering structures and infrastructures, traditional ways of enhancing the safety of structures by increasing their bearing capacity and resistance have encountered a bottleneck (such as large-scale bridges, super high-rise buildings, long-span steel framing, large dams, oil production platform, etc.). Due to the congestion, these approaches are no longer viable. Formerly used in aeronautical, mechanical, and biomedical engineering, intelligent materials are currently being used to the field of civil engineering. Incorporating structural health monitoring and active, semi-active, or passive vibration management based on smart materials is essential for reviving inert structures and making them functional. It has the ability to self-diagnose, self-treat, self-adapt, and self-calibrate, among other things. Two kinds of ultrasonic device based on magnetostrictive materials were developed for non-destructive inspection of concrete constructions. Ultrasonic transducers were used to investigate the link between the concrete's internal qualities and the resulting ultrasonic signals (see Figure 1.49). In addition, an ultrasonic tomography system was created to properly monitor the quality and assess the safety of concrete bridges.

1.7.7 INJECTORS

Because of their great power and consistent performance, diesel engines find widespread use in a variety of areas, including agriculture, industry, and others. Yet, as a result of the ongoing depletion of fossil fuels and the worsening of issues related to pollution of the environment, nations all over the world have gradually but steadily implemented highly stringent limitations on vehicles. This had a significant bearing on the progress that was made with diesel engines. Researchers from both the United States and other countries have put a lot of effort into finding ways to increase diesel engines' fuel efficiency and cut down on the pollution they produce in order to make them more environmentally friendly. The fuel injector, which is the terminal of the diesel injection system and the core component of

FIGURE 1.49 Methods for determining the composition and internal characteristics of concrete without breaking them.

the entire fuel injection system, has become a research hotspot for many scholars. This is because the fuel injector is the component that is at the center of the entire fuel injection system. Non-direct-drive injectors and direct-drive injectors are the two groups of injectors that may be differentiated from one another on the basis of whether or not the injector needle is driven directly by the actuator. It has always been difficult to effectively solve the inherent problems of non-direct-drive structures, such as long hydraulic circuits and needle displacement lag, despite the relative maturity of the research on non-direct-drive injectors at the present time and the very fruitful results achieved by relevant scholars. This is because non-direct-drive structures have more moving parts than direct-drive injectors do. The actuator of the direct-drive injector directly controls the needle, and the stroke of the needle valve may be precisely adjusted by the user. As this is happening, the length of the oil tube is substantially shortened, and the structure of the injector is simplified; this demonstrates a significant increase in the potential for application. For the most part, the energy-converting material that allows the injector to open and close in a direct-drive system is piezoelectric ceramics. This material is employed in the injector. The injector that was developed with piezoelectric materials demonstrates excellent characteristics in response speed. On the other hand, the output force of piezoelectric materials is relatively low, and there is a problem with depolarization. These factors prevent the piezoelectric injector from achieving further improvements in its injection performance.

GMMs are a relatively new form of intelligent magnetic material that have found widespread use in a variety of sectors, including precision actuation, active control, and others. In contrast to piezoelectric materials, it has a number of glaring benefits, including the following:

1. Since the magnetostriction coefficient is high, which is three to five times higher than that of piezoelectric materials, once it is identified, it will become a focus of research activity.

2. The efficiency of energy conversion for GMM ranges from 49% to 56%, while the efficiency of energy conversion for piezoelectric materials ranges from 23% to 52%. Due to the improved energy conversion efficiency, GMM has the potential to be used in a variety of different ways within the electromechanical industry.

3. The Curie temperature, which is 380 degrees Celsius, is much greater than that of piezoelectric materials, which range from 40 to 180 degrees Celsius, which is advantageous for improving the device's capacity to adapt to its surroundings.

4. Because of the high energy density, which is anywhere from 12 to 38 times higher than that of piezoelectric materials, the creation of high-power devices is made much simpler.

5. The GMM has a quick reaction speed that can go down to the microsecond range, and the electromechanical conversion is virtually finished at the same time.

Giant magnetostrictive actuator (GMA) construction developed for direct-drive fuel injector using GMM as energy conversion material is shown in Figure 1.50. This design was accomplished by basing the GMA on the structure of the direct-drive fuel injector. The GMM is intended to have a cylindrical construction, and in conjunction with the plunger having the form of a "T," the GMA is able to fulfill the criteria for the operating mode of the fuel injector even when the injector is switched off and closed normally.

Current is used as the independent variable in a mathematical model, with GMA displacement serving as the dependent variable. Models of the magnetic circuit, magnetization, and displacement output are constructed in this sequence. Structure-wise, this model borrows from GMA, whereas GMM's material characteristics form the basis of this one. Because of factors like the power amplifier power limitations and the GMA coil impedance limitation, it is challenging to obtain high values for the experimental current and operating frequency, and it is also challenging to measure certain intermediate variables. In order to better understand how the GMA functions, the Advanced Modeling Environment for Conducting Simulation of engineering Systems (AMEsim) is being evaluated as a simulation platform. In this study, we use the GMA mathematical model as a foundation and merge its functional properties with those of AMEsim to create the AMEsim numerical model, which is a simplification of the GMA model. Prototypes of GMAs were built, and trials confirmed the accuracy of the previously constructed AMEsim numerical model. We simulated the GMA's magnetic potential distribution, step response, and frequency characteristics in AMEsim.

Using GMM as the energy-transducing material, the creation of massive magnetostrictive actuators for injectors has been one of the most active fields of research in recent years. As machining and manufacturing technologies progress, some studies have considered the feasibility of producing giant magnetostrictive actuators (GMAs) for direct-drive injectors. Fuel injectors maintain their non-operating state when the power is cut. The purpose of this is to ensure the gadget

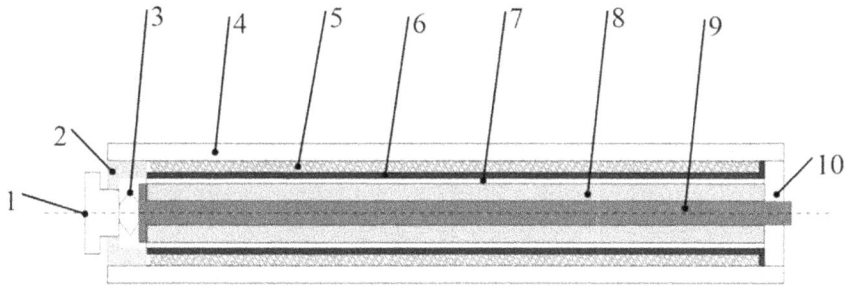

FIGURE 1.50   The direct-drive injector's GMA structure seen in schematic form. First, a nut that's tightened, then a top cap, then a disk spring, then a shell, then an excitation coil, and finally a skeleton of coils. Air gap number 7, GMM cylinder number 8, "T" plunger number 9, and bottom cover number 10.

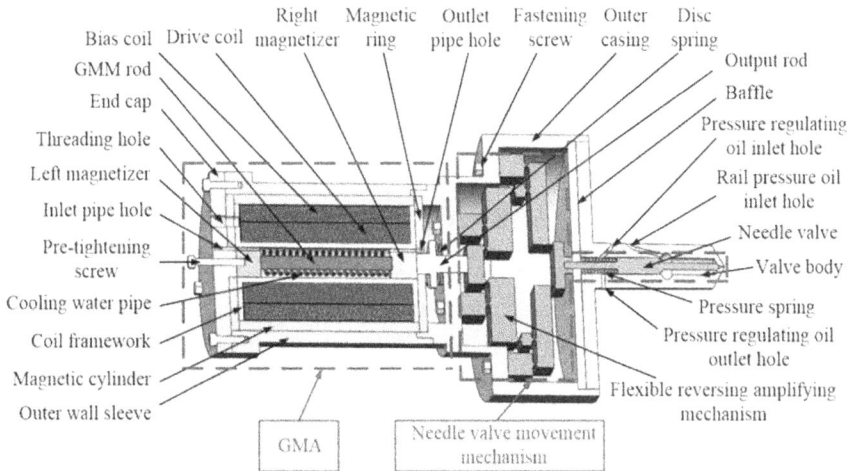

FIGURE 1.51   Structure of the GMI.

doesn't shake. One kind of fuel injector is the direct-drive variety. To modify the actuator's directional displacement output, it is customary to install a displacement conversion mechanism on the actuator's internal circuitry. The axial dimension of the actuator is little impacted by the displacement conversion mechanism; even so, the mechanism's nesting with the GMM cylinder affects the distribution of the magnetic field inside the material, and hence the displacement output of the GMM cylinder. The GMA outer shell prevents damage from the air medium, but its material properties also affect the distribution of magnetic flux inside the GMM cylinder. Analysis of the influence of the mechanism and the GMA shell material on the magnetic field distribution of the GMM cylinder, and optimization

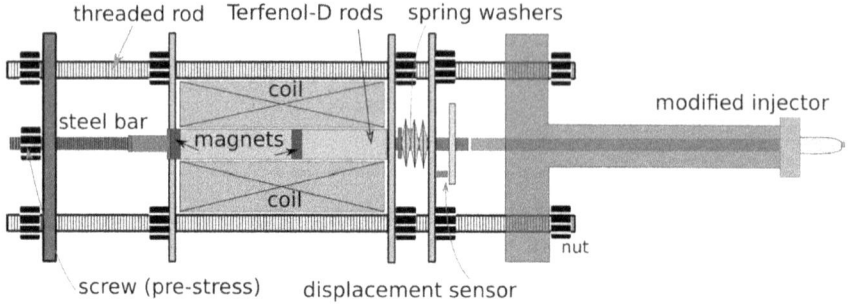

FIGURE 1.52 Modified fuel injector layout showing the addition of a bespoke magnetostrictive actuator.

of the conversion mechanism and the shell material in the design, are prerequisites for moving forward with actuator design and practical implementation. This will result in a large and uniform magnetic density on the GMM cylinder, a large output displacement, and no sudden change in stress.

High-pressure common rail injection systems are built on a crucial component known as the electronic control fuel injector. It does this by managing the driver in accordance with the signal that is sent from the electronic control unit, which, in turn, determines whether or not the needle valve is open or closed. At this time, the injector is responsible for delivering the high-pressure oil that is contained in the common rail pipe to the combustion chamber. In this particular method of operation, the performance of the fuel injection system is directly impacted not only by the kind of actuator used but also by the driving mode of the needle valve and the amount of needle valve lift. The electronic control fuel injector may be classified into two different types, the electromagnetic type and the piezoelectric type, depending on the kind of actuator it uses. These two categories of gasoline injectors have been around for a while and now have almost the entire market share for injectors. The electromagnetic fuel injector has a number of flaws, the most notable of which are a poor reaction speed and an unpredictable needle valve lift. The piezoelectric fuel injector has a short needle valve lift, a high driving voltage, and a low Curie temperature, which are all negatives. Nevertheless, the piezoelectric fuel injector has a rapid response speed. As a result, there is a barrier to progression when it comes to enhancing the fuel injection performance of these two distinct kinds of fuel injectors. The GMM is a new category of intelligent material that has exceptional qualities such as a big strain, a quick reaction speed, a large magneto-mechanical coupling coefficient, and a high energy density. It finds widespread use in the domains of ultra-precision machining, intelligent sensor technology, and fluid element driving. Because of this, the use of the GMM in the electronic fuel injector may result in improved driving effects. At the moment, academics from both the United States and other countries have conducted study on the huge magnetostrictive injector. Xue et al. [35] investigated

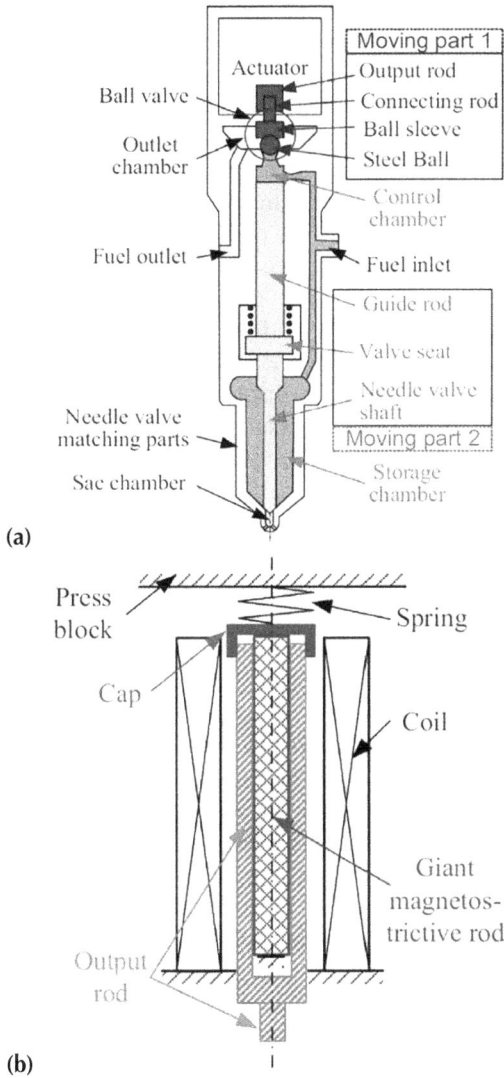

(a)

(b)

FIGURE 1.53 The GMI structural profile, including (a) a drawing of the whole injector and (b) a drawing of the actuator.

the design, modeling, and driving technique of the giant magnetostrictive injector (GMI); Tanaka et al. [36] finished the development of the GMI and investigated its fuel injection characteristics; Yan et al. [37] conceived of the GMI concept machine and were responsible for its design. Also, they modeled and assessed the magnetic field, valve core stroke, and fuel injection rate inside the structure. In contrast, the study described earlier found that the driving mode of the GMI needle valve was an indirect drive. This means that in order to drive the needle valve, the actuator had to work in conjunction with the hydraulic servo control circuit. As a direct consequence of this, the fuel injector has a slower reaction speed, and the steadiness is not as good as it might be. In addition, only the opening and shutting of the needle valve can be regulated, but the precise location of the needle valve as well as its speed cannot be reliably controlled. As a result, exact quantitative fuel injection cannot be achieved. Thus, in this work, a structural scheme of the direct-drive type GMI is provided in order to fulfill the driving requirements of the needle valve of the fuel injector. This will allow the requirements to be met. The needle valve is directly driven by the cooperation of the two mechanisms. The driving element is a GMA, and the flexible reversing amplification mechanism is a micro-displacement transmission mechanism. Together, these two mechanisms form the micro-displacement transmission mechanism. It was decided to construct the GMA magnetic field model, the hysteresis nonlinear model, and the magnetostrictive model, in that order. In this study, we simulate and investigate the magnetic field model, the hysteresis nonlinear model, and the magnetostrictive model. The magnetostrictive model is also solved numerically and validated experimentally. In the last step of the process, an electro-magnetic-mechanical coupling simulation study is performed on the whole output model of the GMI using the program COMSOL Multiphysics. The validity of the model is validated by theoretical calculation and experimental testing, and the practicability of the design of the GMI structure is validated through simulation in order to give a reference for GMI research.

From the beginning of the 21st century, one of the most important research topics has been the creation of novel actuators that are able to give continuous control of the fuel injection rate throughout the fuel combustion process. This control should be targeted at achieving a NOx-soot trade-off. A job of this kind demands an appropriate structuring of the injection-rate profile throughout the operation of the engine. This may be accomplished by maintaining a constant and regulated needle lift, which controls the fuel flow through the injector nozzle. In theory, this paradigm might be put into practice with the assistance of multi-functional materials such as piezo-ceramics, magnetostrictives, or other alloys that are able to immediately convert an electric signal into a displacement. Using an appropriate actuation mechanism that is able to give a fully regulated continuous motion of the needle while withstanding the high pressures produced by the fuel is one way to bring about the desired effect.

Control of the injection rate of fuel has come a long way in fuel injectors during the course of their history. The very first diesel injectors had no moving parts and were completely passive. In point of fact, during the delivery phase of this type of atomizer, an increase in the fuel pressure inside the atomizer causes a needle to be pushed upward, and then after that, it stays in its neutral position thanks to the calibrated spring's response force. The spring tension, the component masses, and the internal fuel pressures all play a role in this atomizer's straightforward mechanical and hydraulic functioning. The injector's opening and closing times, as well as the injection time, are thus defined by the aforementioned parameters, which cannot be altered after assembly. This is because it is evident that neither the injection time nor the opening and closing can be controlled accurately. The following is a description of how the newer electromagnetic fuel injectors used in diesel engines work on a basic level. With common rail systems, the fuel pump creates and regulates the high pressure that is applied to the electromagnetic injector. Within the injector, the fuel flux is divided in two: one branch supplies fuel to the injector itself, while the other regulates the opening and closing of the needle through a rod. The last fork works directly onto a "control volume," which keeps the rod and needle at rest through a pressure balance at a deactivated electromagnetic actuator. As soon as the actuator coil is activated, the control volume becomes imbalanced, and the spring that is working on the rod opens the needle that is located in the nozzle. This allows the fuel to flow through the nozzle. After the electromagnetic actuator has been turned off, the injection procedure comes to an end. This is a somewhat complicated system that gives the user choice over the timing of the fuel injection, even if the opening and closing profiles are to some extent predetermined. Terfenol-D, an alloy of terbium, iron, and dysprosium, is the most well-known material for this use. It displays gigantic magnetostriction up to 1,500 ppm, which is equivalent to around 150 m with a 100 mm rod length. In the grand scheme of things, this makes it possible to have a relatively small MS-based injector system. When a magnetic field is applied to large magnetostrictive materials, the resulting deformation is a function of both the magnetic field and the mechanical stress that the sample is subjected to. In most cases, the maximum stroke behavior has a maximum obtained at a compressive stress in the region of 5–20 MPa, and then it declines after that. This is because the maximum stroke behavior is proportional to the stress. Therefore, in order to make full use of the magnetostriction for linear actuation, it is necessary to have a pre-stress mechanical system, which will be discussed in more detail in the following paragraphs.

In order to combat global issues such as the depletion of fossil fuel resources and the pollution of the air, the automotive industry has to develop more efficient cars that use less fuel and produce fewer harmful emissions. In recent years, more stringent emission rules for cars have been created, and the fast growth of electrically propelled vehicles has led to greater criteria being imposed for fuel ignition vehicles. Fuel ignition engines can only remain competitive with vehicles that operate on alternative fuels if they have been able to significantly reduce their fuel consumption and pollutant output. Finding alternative fuels and refining the fuel injection system are two of the most important things that can be done to create a cleaner burning gasoline ignition engine. Using GMM as its central component, GMI may be able to solve the issues that arise in electromagnetic injectors (such as slow response and low controllability) and piezoelectric injectors (such as the need for a hydraulic amplifier and the ease with which their effectiveness degrades) and thus be a good candidate for electronic controlled injectors.

In addition to its high coupling coefficient, rapid response speed, large output force, and high Curie temperature, GMMs have a variety of other advantages as an energy material. AVC, driving hydraulic devices, ultra-precise machining, and other niche uses have all benefited from this material's versatility. Comparing the giant magnetostrictive kind of injector to the electromagnetic injector, one can see how the latter may be seen as an enhanced version of the former due to the latter's continuously changeable displacement in the driving part. The enormous magneto-strictive kind of injector maintains the advantages of high working stability, strong anti-interference capabilities, and a large injection rate, while also achieving a more variable injection rate and having the potential of a direct acting structure.

Because of their great thermal efficiency and versatility, diesel engines find widespread usage in a variety of fields, including industry, agriculture, transporta-tion, and the development of national defense facilities. Yet, beginning in the 1970s and continuing to the present day, nations all over the world have gradually intro-duced more stringent emission laws as a response to the depletion of fossil fuels and the worsening of environmental degradation. China has released the fifth and sixth standards of vehicle emission in order to address the issue of exhaust emissions from vehicles. These standards were developed in accordance with China's national requirements. The number of heavy-duty diesel vehicles in China has surpassed 11 million as of the first day of July 2021. Although this represents only 4.4% of the total number of vehicles in China, the emissions of nitrogen oxides and particulate matter produced by these vehicles account for 85% and 65%, respectively, of the total emissions produced by vehicles. The criteria of the National Standard Six for heavy diesel cars are more stringent when compared to the national standard five, and the limitations of nitrogen oxides and particulate matter are decreased by 77% and 67%, respectively. New and significant obstacles are being thrown into the path of diesel engine development. The question of how to further enhance the diesel engine's fuel economy and build a "green diesel engine" that satisfies the criteria of high power, low fuel consumption, and low pollution has emerged as a prominent research priority among industry professionals and academics. The fuel system is often referred to as the "heart" of a diesel engine, and the fuel injector, which serves as the terminal of the fuel system, is considered to be the "core component" of the entire fuel system. The fuel injector is what ultimately determines how fuel is injected into the cylinders, and its performance in this regard has a significant impact on the diesel engine's overall efficiency.

1.7.8   HYBRID MAGNETOSTRICTIVE/PIEZOELECTRIC DEVICES

The great degree of technological interest associated with hybrid smart material actuators suggests that they need their own category, apart from those devoted to sonar, linear, or rotary applications. Due to their complementary electrical properties, magnetostrictive materials and piezoelectric components may gen-erate a resonant electric circuit when used together in a device. A combination of inductive magnetostrictive materials and capacitive piezoelectric components may achieve this. A device of this kind, when driven at resonance, behaves as a perfect resistive load, requiring no further energy input beyond that which is effectively translated to mechanical motion or lost to internal losses. This greatly simplifies the amplifier's design and aids in achieving high efficiencies. To get over the difficulties of a Tonpilz piston-type sonar transducer moving at just one end, a hybrid device consisting of a stack of piezoelectric Navy type I ceramic rings linked to a quarter wavelength Terfenol-D composite tube has been developed (see Figure 1.54a). The challenges of achieving motion at merely It is possible to accomplish addition at one end and cancellation at the other due to the inherent 90-degree phase shift between the magnetostrictive and piezoelectric transduction processes and the quarter-wavelength design of the components. Device's mechanical functioning is unidirectional; array-baffling is required for unidirectionality in terms of sound. When employed singularly, the device yields a front-to-rear pressure ratio of 5 dB; when the array is loaded, this value jumps to 15 dB. Both rotational motors and linear inchworm motors have benefited from

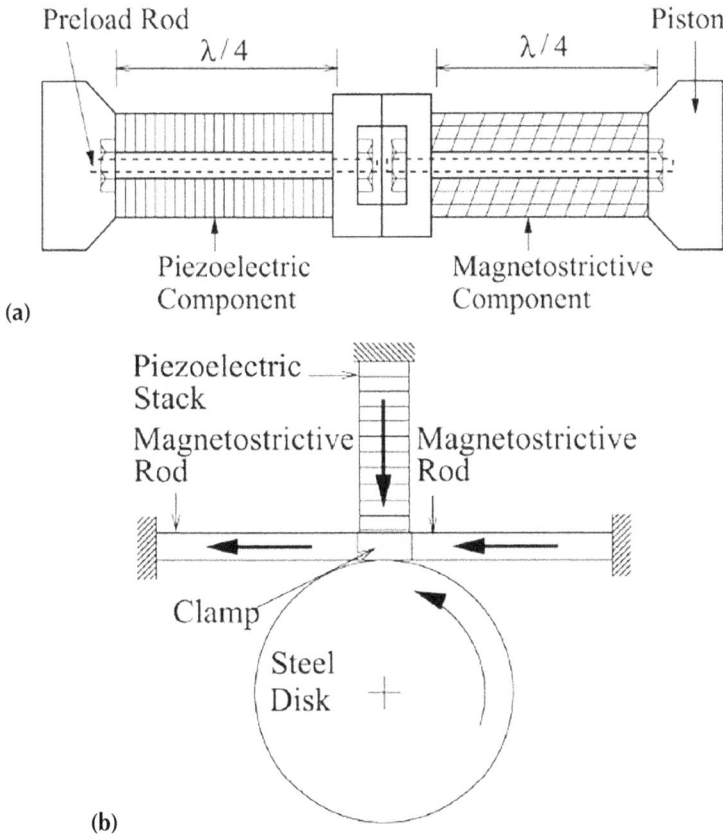

FIGURE 1.54 Transducers that combine magnetostrictive and piezoelectric technologies are called hybrids: (a) Rotating motor and (b) sonar projector.

the hybrid piezoelectric/magnetostrictive transduction technique. Research by Venkataraman and colleagues in 1998, for instance, the prototype features a setup identical to that seen in Figure 1.54b, except that piezoelectric stacks are used for clamping and Terfenol-D rods are used for translation. The magnetic bias on the Terfenol-D elements allows the direction of motion to be easily altered, while the natural driving time for the inchworm is given by the intrinsic 90-degree phase lag that occurs between the two types of elements. The zero-load speed of this motor is 25.4 mm/s, and its stall load is 115 N.

As a follow-up to the proof-of-concept transducer that was shown in Figure 1.54a,b presents an illustration of a hybrid magnetostrictive/piezoelectric rotating motor. The rotational motion is created by a piezoelectric stack that clamps a piece of friction material onto a spinning disk. At the same time, two magnetostrictive rods move the clamp tangentially to the disk, which results in the rotational motion. As was said before, the sequence of the motion is decided by the natural time of the piezoelectric and magnetostrictive response. This timing is what determines the order of the motion. With excitation voltages ranging from 30 to 40 V and frequencies ranging from 650 to 750 Hz, the device generates a speed of four revolutions per minute (rpm).

1.7.9　Monitoring the Condition of Structures

To monitor the desired state of structural health, a structural health monitoring technology must be used. The goal of this technology is to lessen the amount of times that structural faults need to be repaired and the number of times that performance maintenance needs to be performed. The maintenance techniques with and without SHM are shown in Figure 1.55. If this technique is used in conjunction with other types of repairs, the health of the target structure may be maintained at a high level, which results in increased safety and a longer life cycle.

As a whole, SHM technology keeps tabs on large structures like airplane skins, thus the best detector for it should be spread out to cover as much ground as possible. Additionally, aerospace applications have stringent requirements for sensor robustness, so piezoelectric materials have been widely used and studied in structural health monitoring technology due to their ability to operate solely off of their inherent polarization rather than requiring external voltage sources or magnetic fields. This is due to the fact that piezoelectric materials may function independently of the polarization state of the material itself.

The past several decades have seen a significant increase in the amount of work that has been put into developing more effective and timely methods for monitoring and controlling the aging of civil constructions, infrastructures, historical buildings, structures, and sites. In the end, the purpose of the monitoring is to anticipate any problems that may arise later in life or in the near future. The majority of the time, this is accomplished with the assistance of a visual examination that is carried out by trained professionals. Yet, it is abundantly clear that this strategy is not only expensive but also prone to unreliability and, eventually, impossible to implement for the extensive cultural legacy of nations such as Italy.

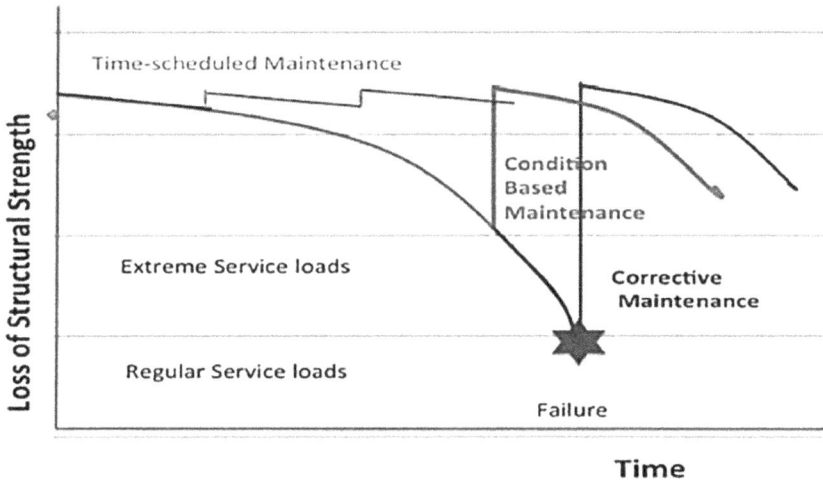

FIGURE 1.55 Methods of upkeep with and without predictive SHM.

FIGURE 1.56 Rendering of a walkway in an archeological site with an embedded energy harvester.

1.7.10 Transducers

Magnetostrictive transducers are the first kind of transducers. When compared to other types of transducers, magnetostrictive transducers have the benefit of being able to be driven by conventional low impedance amplifiers. This is especially useful at frequencies well below resonance, when the low impedance of a magnetostrictive transducer enables driving voltages to be kept low. This has the potential to be helpful in medical applications and may, more generally, considerably simplify the design of amplifiers. The measured complex electrical impedance frequency response function $Zee = V/I$ is shown in Figure 1.57 for a Terfenol-D

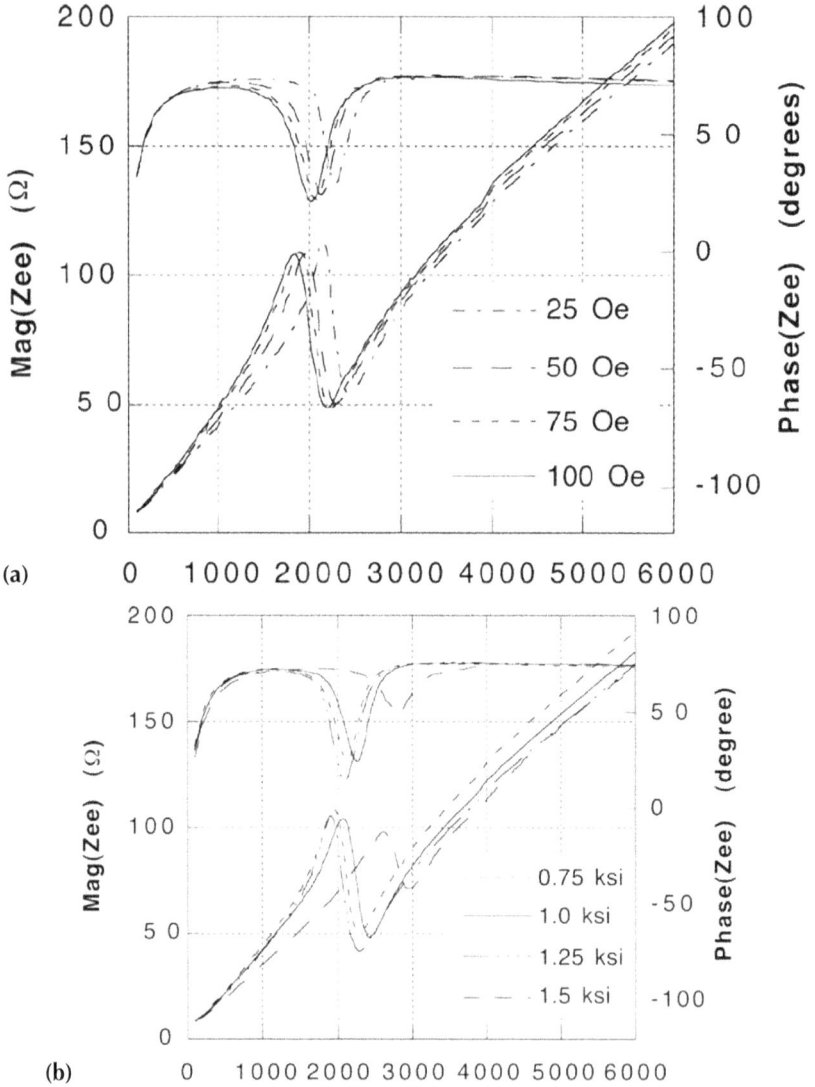

FIGURE 1.57 The magnitude and phase of the total electrical impedance, Zee, as a function of frequency. (a) a constant AC drive level of 8 kA/m (100 Oe) and varying bias circumstances and (b) a bias condition of 5.2 MPa, 24 kA/m (0.75 ksi, 300 Oe).

Prestress Bolt   Permanent Magnet   Solenoid   Compression Spring

Magnetic Coupler   Magnetostrictive Rod   Displacement Plunger

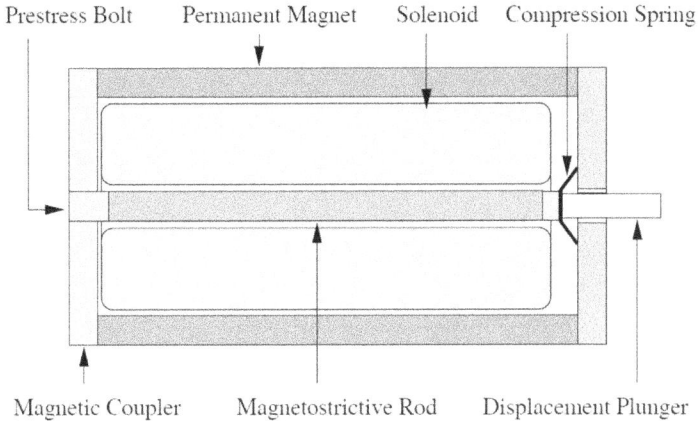

FIGURE 1.58 Magnetic transducer cross section.

transducer that was developed in accordance with the general configuration that is seen in Figure 1.58. This transducer consists of a magnetostrictive rod of cylinder shape, a copper-wire solenoid enclosing it, a preload mechanism comprising bolt and spring washer, magnetic couplers, and a permanent magnet of barrel shape providing the bias magnetization. While the individual design specifics will vary depending on the application of the smart structure, this configuration illustrates the fundamental components that are necessary to get the highest possible level of performance from the magnetostrictive material. Since magnetostriction is caused by the rotation of magnetic moments, a magnetostrictive transducer that is driven by an alternating current magnetic field would vibrate at a frequency that is twice as fast as the drive frequency, but the motion will only go in one direction. This is seen in Figure 1.59, where the solid lines indicate the unbiased input and the strain output that corresponds to it. The performance increases that were gained by introducing a magnetic bias to the material are shown by the dashed lines in the graph. The frequency of the input has been maintained, the output is now bidirectional, and the ratio of output to input has been increased. It is common practice to make use of a permanent magnet in conjunction with a static field created by feeding a DC through the solenoid. This configuration is used in an effort to precisely center functioning around the desired bias point. It should be emphasized that while sole use of permanent magnets for biasing provides the benefit of significant reductions in power consumption, it also has the drawback of increased bulk and weight. On the other hand, DC result in significant power losses due to ohmic heating, but they enable significant reductions in both size and weight. The magnetic biasing may also be given by magnets that are piled on top of one another and placed in series with the rod or rods; this design is referred to as a stacked-magnet setup. As compared to the barrel-magnet arrangement, it has been discovered that the stacked-magnet configuration may give improvements in the magneto-mechanical coupling of up to 5% for large rods (L > 20 cm,

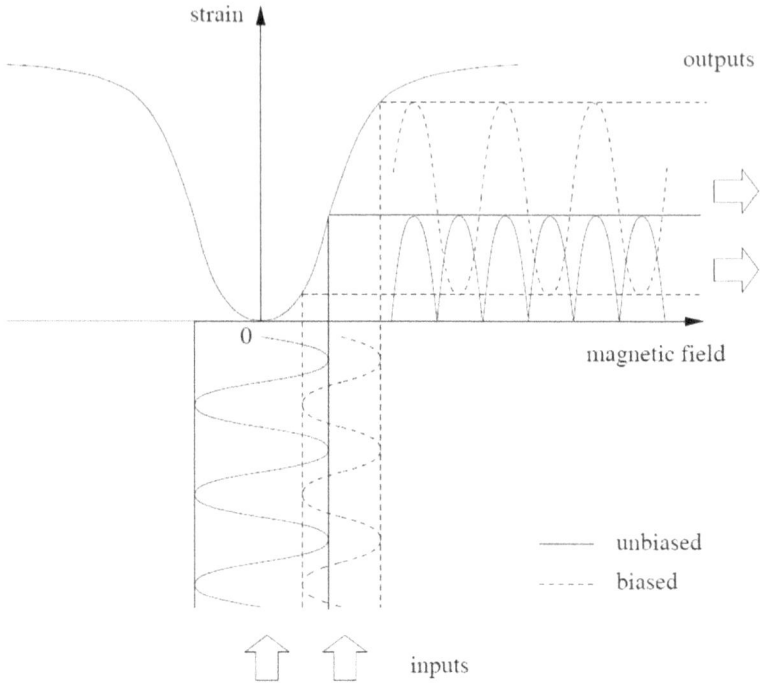

FIGURE 1.59   The influence of magnetic bias on the strain generated by a magnetostrictive transducer.

D > 2.5 cm). With stacked-magnet systems, however, it is not uncommon for there to be collateral issues such as saturation effects and resonance frequency changes. It is essential for there to be effective magnetic flux closure inside the circuit produced by the rod itself, the couplers, and the permanent magnets, and this can only be achieved with transducers that have been carefully constructed. To conclude, even if current magnetostrictive materials like Terfenol-D are produced with the magnetic moments approximately perpendicular to the rod axis, it is still necessary to provide a static stress, also known as mechanical preload, in order to achieve complete alignment of all the moments. Since the moments that were originally aligned with the rod axis do not contribute to the magnetostriction, a mechanically free rod will only generate around half of its maximal magneto-striction. This is due to the fact that the moments that were positioned randomly when the rod was first created.

Also, the stress anisotropy that is produced by the static compression (or ten-sion in the case of materials with negative magnetostriction) would improve the overall magnetoelastic state of the material in the way that has been explained. During dynamic transducer operation, it is important to keep in mind that the stress in the magnetostrictive rod might experience substantial variation in com-parison to the nominal preload in systems that make use of linear washers for preloading. This point is stressed. Because of the magnetomechanical connection, this has the potential to have a significant effect on the performance of the magnetostrictive transducer and the driving electronics by having an effect on the magnetic state and, as a result of that, the electrical regime (see Figure 1.57). In this study, we looked at how the performance of a Terfenol-D transducer was affected by two factors: mechanical preload and magnetic bias. To prevent operating the rod in tension is a second reason for using a mechanical preload. This is especially important when driving brittle materials such as Terfenol-D ($\sigma_t = 28$ MPa, $\sigma_c = 700$ MPa) at or near mechanical resonance.

1.7.10.1 Sonar Transducers

Efficient sonar transducers need the generation of high mechanical power at low frequencies. This is sometimes accompanied by the specification that the frequency bandwidth, or its quantitative analogue, the quality factor Q, be maintained to a minimum. Due to its low magnetomechanical coupling coefficient $k = 0.30$, nickel required a high Q to obtain acceptable efficiency when utilized in sonar applications during World War II. Despite this, nickel saw extensive usage in sonar devices of the era. Modern huge magnetostrictive materials, on the other hand, have a massively increased coupling coefficient of over $k = 0.70$, allowing the transducer to operate at low Q while still producing tremendous power outputs. A Terfenol-D Tonpilz transducer with specs like those shown in Figure 1.60a may produce 200 Hz of bandwidth at 2 kHz of resonance ($Q = 10$) and 200 dB of source level relative to 1 μ Pa at 1 meter. Another Terfenol-D transducer has been reported to be capable of generating a maximum output of 206 dB referenced to 1 μ Pa at 1 m and functioning throughout a broad usable bandwidth of 50 Hz to 5 kHz. Figure 1.60b,c shows alternative configurations in which the linear motion of cylindrical magnetostrictive rods is used to flex an adjacent shell or to generate radial vibrations in a tube or ring (c).

1.8   LITERATURE REVIEW

Magnetostrictive materials have gained a lot of interest in recent years due to their adaptable characteristics. By varying the control gain and strength of the applied magnetic field, we may influence the vibration behavior of a variety of geometries using these intelligent materials. In the early 2000s, it was shown that the vibrations of a multilayered system could be regulated more effectively if thin magnetostrictive plies were situated at a greater distance from the sandwich's core [38]. The generalized differential quadrature method (GDQM) was used to track the transient response of multi-layered magnetostrictive plates [39]. DQM was also used to the first-order shear deformation theory (FSDT) to study the dynamic reactions of magnetostrictive tapered plates [40]. In addition, [41] created a nonlinear 2D model to investigate the mechanical behaviors of laminates with magnetostrictive layers. In 2016 [42], the magnetostriction-assisted vibration management of the smart sandwich plates under a follower load was investigated using a higher-order shear deformation theory (HSDT). The same group of researchers attempted to demonstrate the impact of orthotropic springs in a stiff medium on the frequency characteristics

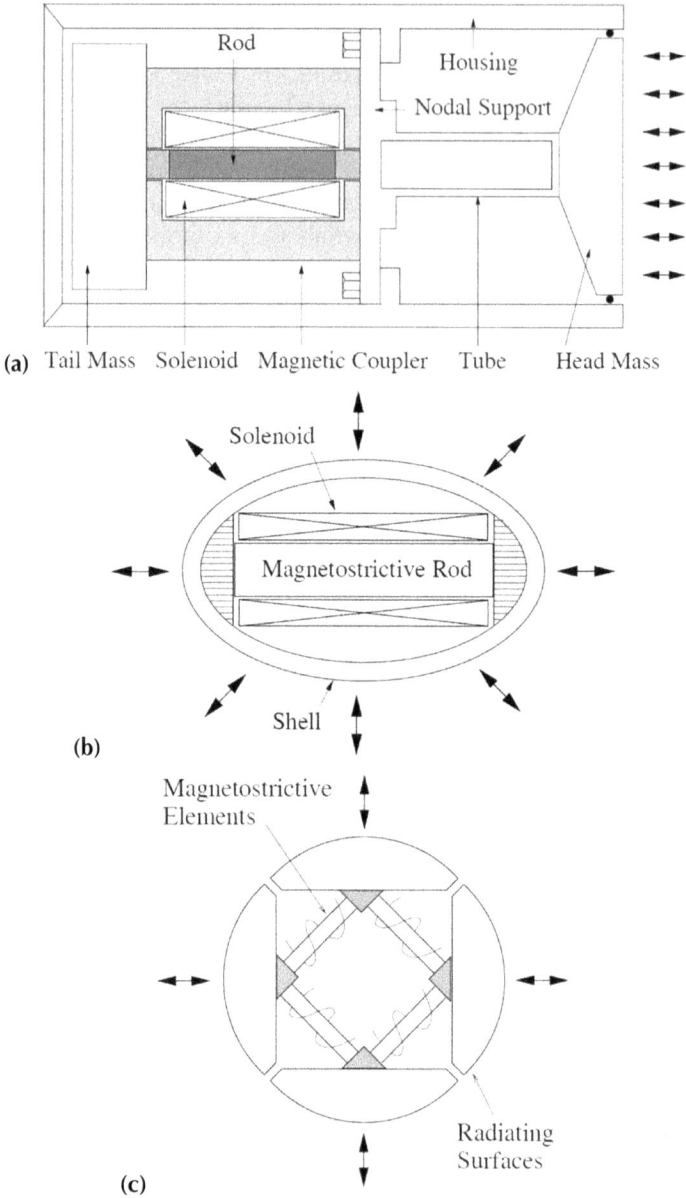

(a) Tail Mass   Solenoid   Magnetic Coupler   Tube   Head Mass

(b)

(c)

FIGURE 1.60   Three types of magnetostrictive sonar transducers are described in detail (a) the Tonpilz, (b) the flextensional, and (c) the square ring.

under control of smart magnetostrictive plates in a different investigation [43]. With the help of the DQM, Hong [44] analyzed the thermo-mechanical resonance of laminated magnetostrictive plates. Under the influence of thermo-magnetic fields, the wave propagation characteristics of magnetostrictive sandwich nanoplates (MSNPs) were investigated [45]. In this chapter, we use an analytical solution to the governing equations to get the frequency and phase velocity of the waves. It is worth noting that the Kirchhoff plate theory provided the basis for the governing equation in the aforementioned investigation. The same group of researchers used the nonlocal strain gradient theory (NSGT) to investigate the wave movement in an MSNP [46]. In 2019 [47], the dynamic responses of a microplate with a magnetostrictive layer were investigated using FSDT employing the modified couple stress theory and DQM. In 2020 [48], the authors used HSDT to tackle the dynamic stability issue of nanocomposite shells with a magnetostrictive layer. In 2020 [49], they presented a system for controlling the oscillatory behaviors of a laminated composite plate using intelligent magnetostrictive layers. Smart magnetostrictive facesheets were used in another study to fine-tune the FGM microshells' dynamic stability [50]. In 2021 [51], it was stated that a quasi-3D theory was used to investigate the controlled vibrations of a magnetostrictive laminated composite plate with a viscoelastic core. Recent references [52] Zenkour and El-Shahrany and Ebrahimi, Dabbagh et al. [53] detail the effects of humidity and temperature on the free vibrations and wave propagation behaviors of smart magnetostrictive plates.

REFERENCES

1. Bozorth, R.M., *Ferromagnetism*. 1993: IEEE Press.
2. Grunwald, A. and A.-G. Olabi, *Design of a magnetostrictive (MS) actuator*. Sensors and Actuators A: Physical, 2008. **144**(1): p. 161–175.
3. Calkins, F.T., *Design, analysis, and modeling of giant magnetostrictive transducers*. 1997: Iowa State University.
4. Yanan, M. and F. Li. Application and development research on giant magnetostrictive apparatus. In *2010 2nd International conference on mechanical and electronics engineering*. 2010: IEEE.
5. Jiles, D., *Theory of the magnetomechanical effect*. Journal of Physics D: Applied Physics, 1995. **28**(8): p. 1537.
6. Dapino, M.J., *On magnetostrictive materials and their use in adaptive structures*. Structural Engineering and Mechanics, 2004. **17**(3–4): p. 303–330.
7. Apicella, V., et al. Review of modeling and control of magnetostrictive actuators. In *Actuators*. 2019: MDPI.
8. Restorff, J., M. Wun-Fogle, and A. Clark, *Temperature and stress dependences of the magnetostriction in ternary and quaternary Terfenol alloys*. Journal of Applied Physics, 2000. **87**(9): p. 5786–5788.
9. Goldman, A., *Handbook of modern ferromagnetic materials*. Vol. 505. 2012: Springer Science & Business Media.
10. Lee, E.W., *Magnetostriction and magnetomechanical effects*. Reports on Progress in Physics, 1955. **18**(1): p. 184.
11. Atulasimha, J. and A.B. Flatau, *A review of magnetostrictive iron-gallium alloys*. Smart Materials and Structures, 2011. **20**(4): p. 043001.
12. Mohamed, A., et al., *The Fe-Ga phase diagram: Revisited*. Journal of Alloys and Compounds, 2020. **846**: p. 156486.
13. Huang, M. and T.A. Lograsso, *Short range ordering in Fe-Ge and Fe-Ga single crystals*. Applied Physics Letters, 2009. **95**(17): p. 171907.
14. Clark, A., et al., *Magnetostriction of ternary fe-ga-x (x = c, v, cr, mn, co, rh) alloys*. Journal of Applied Physics, 2007. **101**(9): p. 09C507.

15. Restorff, J., et al., *Magnetostriction of ternary Fe-Ga-X alloys (X = Ni, Mo, Sn, Al).* Journal of Applied Physics, 2002. **91**(10): p. 8225–8227.

16. Kubota, T., et al., *Large magnetostriction in rapid-solidified ferromagnetic shape memory Fe-Pd alloy.* Journal of Magnetism and Magnetic Materials, 2002. **239**(1–3): p. 551–553.

17. Clark, A.E., *Magnetostrictive rare earth-Fe2 compounds.* Handbook of Ferromagnetic Materials, 1980. **1**: p. 531–589.

18. Clark, A., *Magnetostrictive RFe2 intermetallic compounds.* Handbook on the Physics and Chemistry of Rare Earths, 1979. **2**: p. 231–258.

19. Olabi, A.-G. and A. Grunwald, *Design and application of magnetostrictive materials.* Materials & Design, 2008. **29**(2): p. 469–483.

20. Turtelli, R.S., et al., Co-ferrite-A material with interesting magnetic properties. In *IOP conference series: Materials science and engineering.* 2014: IOP Publishing.

21. Zhang, Y., et al., *Design of compensation coils for EMI suppression in magnetostrictive linear position sensors.* Sensors, 2012. **12**(5): p. 6395–6403.

22. Engdahl, G., *Design procedures for optimal use of giant magnetostrictive materials in magnetostrictive actuator applications.* Parameters, 2002. **1**: p. 2.

23. Akuta, T., Rotational type actuators with Terfenol-D rods. In *Proceedings 3rd international conference on advances in sensors, actuators.* 1992.

24. Vargas-Estevez, C., et al., *Study of Galfenol direct cytotoxicity and remote microactuation in cells.* Biomaterials, 2017. **139**: p. 67–74.

25. Holmes, H.R., et al., *Biodegradation and biocompatibility of mechanically active magnetoelastic materials.* Smart Materials and Structures, 2014. **23**(9): p. 095036.

26. Vranish, J., et al., *Magnetostrictive direct drive rotary motor development.* IEEE Transactions on Magnetics, 1991. **27**(6): p. 5355–5357.

27. Claeyssen, F., et al., *Design and construction of a resonant magnetostrictive motor.* IEEE Transactions on Magnetics, 1996. **32**(5): p. 4749–4751.

28. Clark, A., et al., *Temperature dependence of the magnetic anisotropy and magnetostriction of Fe 100− x Ga x (x = 8.6, 16.6, 28.5).* Journal of Applied Physics, 2005. **97**(10): p. 10M316.

29. Engdahl, G. and I.D. Mayergoyz, *Handbook of giant magnetostrictive materials.* Vol. 386. 2000: Elsevier.

30. Flatau, A.B., M.J. Dapino, and F.T. Calkins, *High bandwidth tunability in a smart vibration absorber.* Journal of Intelligent Material Systems and Structures, 2000. **11**(12): p. 923–929.

31. Wang, G., et al., *A co-dispersion nanosystem of graphene oxide@ silicon-doped hydroxyapatite to improve scaffold properties.* Materials & Design, 2021. **199**: p. 109399.

32. Ribeiro, C., et al., *Proving the suitability of magnetoelectric stimuli for tissue engineering applications.* Colloids and Surfaces B: Biointerfaces, 2016. **140**: p. 430–436.

33. Zaeimbashi, M., et al., *NanoNeuroRFID: A wireless implantable device based on magnetoelectric antennas.* IEEE Journal of Electromagnetics, RF and Microwaves in Medicine and Biology, 2019. **3**(3): p. 206–215.

34. Rangriz, F., A. Khaleghi, and I. Balasingham. Wireless link for micro-scale biomedical implants using magnetoelectric antennas. In *2020 14th European conference on antennas and propagation (EuCAP).* 2020: IEEE.

35. Xue, G., et al., *Design method and driving voltage waveform of giant magnetostrictive actuator used on electronic controlled injector.* Transactions of the Chinese Society for Agricultural Machinery, 2017. **48**(6): p. 365–372.

36. Tanaka, H., Y. Sato, and T. Urai, *Development of a common-rail proportional injector controlled by a tandem arrayed giant-magnetostrictive-actuator.* SAE Transactions, 2001: p. 2010–2014.

37. Yan, R.G., Z.J. Wang, and L.H. Zhu. Research of the giant magnetostrictive fule injector. In *Advanced materials research*. 2014: Trans Tech Publ.

38. Pradhan, S., et al., *Control of laminated composite plates using magnetostrictive layers*. Smart Materials and Structures, 2001. **10**(4): p. 657.

39. Hong, C., *Transient responses of magnetostrictive plates by using the GDQ method*. European Journal of Mechanics-A/Solids, 2010. **29**(6): p. 1015–1021.

40. Ghorbanpour Arani, A., H. Khani Arani, and Z. Khoddami Maraghi, *Vibration analysis of rectangular magnetostrictive plate considering thickness variation in two directions*. International Journal of Applied Mechanics, 2015. **7**(04): p. 1550059.

41. Santapuri, S., J. Scheidler, and M. Dapino, *Two-dimensional dynamic model for composite laminates with embedded magnetostrictive materials*. Composite Structures, 2015. **132**: p. 737–745.

42. Arani, A.G. and Z.K. Maraghi, *A feedback control system for vibration of magnetostrictive plate subjected to follower force using sinusoidal shear deformation theory*. Ain Shams Engineering Journal, 2016. **7**(1): p. 361–369.

43. Arani, A.G., Z.K. Maraghi, and H.K. Arani, *Orthotropic patterns of Pasternak foundation in smart vibration analysis of magnetostrictive nanoplate*. Proceedings of the Institution of Mechanical Engineers, Part C: Journal of Mechanical Engineering Science, 2016. **230**(4): p. 559–572.

44. Hong, C.-C., *Thermal vibration of laminated magnetostrictive plates without shear effects*. International Journal of Electrical Components and Energy Conversion, 2017. **3**(3): p. 63.

45. Ebrahimi, F. and A. Dabbagh, *Thermo-magnetic field effects on the wave propagation behavior of smart magnetostrictive sandwich nanoplates*. The European Physical Journal Plus, 2018. **133**: p. 1–12.

46. Ebrahimi, F. and A. Dabbagh, *Wave propagation analysis of magnetostrictive sandwich composite nanoplates via nonlocal strain gradient theory*. Proceedings of the Institution of Mechanical Engineers, Part C: Journal of Mechanical Engineering Science, 2018. **232**(22): p. 4180–4192.

47. Ghorbanpour Arani, A., H. Khani Arani, and Z. Khoddami Maraghi, *Size-dependent in vibration analysis of magnetostrictive sandwich composite micro-plate in magnetic field using modified couple stress theory*. Journal of Sandwich Structures & Materials, 2019. **21**(2): p. 580–603.

48. Ghorbani, K., et al., *Investigation of surface effects on the natural frequency of a functionally graded cylindrical nanoshell based on nonlocal strain gradient theory*. The European Physical Journal Plus, 2020. **135**(9): p. 1–23.

49. Zenkour, A.M. and H.D. El-Shahrany, *Control of a laminated composite plate resting on Pasternak's foundations using magnetostrictive layers*. Archive of Applied Mechanics, 2020. **90**(9): p. 1943–1959.

50. Yuan, Y., et al., *Dynamic stability of nonlocal strain gradient FGM truncated conical microshells integrated with magnetostrictive facesheets resting on a nonlinear viscoelastic foundation*. Thin-Walled Structures, 2021. **159**: p. 107249.

51. Zenkour, A.M. and H.D. El-Shahrany, *Quasi-3D theory for the vibration of a magnetostrictive laminated plate on elastic medium with viscoelastic core and faces*. Composite Structures, 2021. **257**: p. 113091.

52. Zenkour, A.M. and H.D. El-Shahrany, *Hygrothermal vibration of adaptive composite magnetostrictive laminates supported by elastic substrate medium*. European Journal of Mechanics-A/Solids, 2021. **85**: p. 104140.

53. Ebrahimi, F., A. Dabbagh, and T. Rabczuk, *On wave dispersion characteristics of magnetostrictive sandwich nanoplates in thermal environments*. European Journal of Mechanics-A/Solids, 2021. **85**: p. 104130.

2 Static Analysis of Magnetostrictive Materials and Structures

Background

The phenomenon known as buckling was first brought to the attention of researchers working in the area of mechanics close to a century ago. At that time, a significant amount of investigation has been carried out in this area, the initial focus being on columns, followed by shells, and finally plates. When compressive in-plane pressures are applied to the plate, the plate first maintains its flat position and then simply contracts. On the other hand, the plate will buckle when the forces acting upon it exceed a certain threshold level. Understanding the buckling behavior of plates requires taking a number of important preparatory steps, one of which is to do research on the buckling behavior of a plate around the margins of the support. If a structure is subjected to load, then under light loads, the structure will only very slightly deform. If the load that is being applied continues to grow, eventually the structure will reach what is known as its critical value, at which time it will no longer be able to support the force that is being exerted on it, and from that point on, the structure will experience significant deformation. In many instances, the buckling load of the structure will be less than the yield stress value. To put it another way, when a plate is subjected to direct pressure, bending, shear, or a combination of these stresses in its plane, the buckling load of the plate will be less than the yield stress value. This occurs when the thickness of the structure is not very great. It is possible for the plate to bend locally before the whole member becomes unstable or before the material's yield stress is achieved.

Buckling is the sudden change in shape (deformation) of a structure under load, such as the bending of a column under pressure or the folding of a plate under shear. Buckling is a term that comes from the field of structural engineering. It is possible for a member of a structure that is being subjected to an ever-increasing load to suddenly deform when the load reaches a critical level. In this scenario, the structure is said to have buckled. Buckling can happen even when the stresses created in the structure are much lower than those required to cause failure in the material. Failure in the material is required in order for there to be buckling.

DOI: 10.1201/9781003355427-2

Additional loading could cause significant deformations, some of which could be unpredictable. On the other hand, if the post-buckling deformations do not bring about the complete failure of the member, then the member will still be able to support the buckling load.

Static Analysis of Magnetostrictive Materials and Structures

The buckling behavior of magnetostrictive materials and structures is analyzed using a mathematical framework presented in this chapter. Ultimately, the impacts of various parameters and external influences on the buckling behavior of magnetostrictive structures are supplied statistically within the context of a series of drawings. This allows for a more thorough understanding of the phenomenon.

2.1   BUCKLING ANALYSIS OF MAGNETOSTRICTIVE PLATES

In the current research, the buckling behavior of magnetostrictive materials that have been merged with functionally graded facesheets is the primary area of investigation. The power-law model is used to determine the functionally graded layer's effective material qualities, which are obtained by the layer. It has been determined via the use of Eringen's nonlocal theory how many small-scale parameters there are. On the other hand, to consider the elastic medium, the suggested approach is based on the foundation laid by Winkler and Pasternak. Higher-order sinusoidal shear deformation theory was applied in order to get at the governing equation, and the governing equation that was arrived at was solved analytically using the Galerkin solution for various boundary conditions. It is envisaged that the current study would assist engineers and designers in understanding and predicting buckling reactions and that it will be helpful in the design of nanoscale systems like sensors and actuators, which are now the most in-demand technology.

2.1.1   HOMOGENIZATION

In the field of materials science, functionally graded materials, also known as FGMs, may be distinguished from other types of materials by the progressive volume changes in composition and structure that result in matching changes in material characteristics. The mechanical properties of functionally graded materials, such as Young's modulus of elasticity, Poisson's ratio, shear modulus of elasticity, and material density, change slowly in different planes along the thickness of the plane in functionally graded materials, which are heterogeneous materials. For making functionally graded materials, a blend of ceramic and metal, or a combination of many distinct kinds of materials, is often employed. It is highly vital to investigate and understand the static and dynamic features of smart functional structures such as beams, plates, and shells because of the large diversity of materials and the growing uses of these materials in a number of scientific and industrial sectors. When determining the mechanical characteristics of graded materials, the volume percentage of the component materials is one of the most important factors to consider. It is presumable, in accordance with Witte's law, that Young's modulus and the density of the material will vary depending on the plane in which they are measured over the whole volume fraction range. In addition to this, it is assumed that Poisson's ratio stays the same all the way through the plate's thickness.

2.1.1.1 The Process of Creating Functionally Graded Materials

It has been found that many various types of materials have distinct spatial microstructures. These microstructures cause spatial changes in the characteristics of the material's elasticity, which necessitates the use of a non-homogeneous model. Rock and soil, for instance, generally exhibit depth-dependent characteristics that are the consequence of the overburden of material above a certain site in geomechanical investigations. These qualities are caused by the accumulation of material above the place in question. Microstructural grading is also widely observed in biological cellular materials such as wood and bone. As a result of biological adaptation, the microstructures that are the most resistant to stress have been distributed to the areas of the material that are subjected to the most stress. The creation of graded materials that include spatial characteristics that have been deliberately designed to increase mechanical performance has garnered a lot of attention as of late, which has sparked great interest in the field. Throughout the 1990s, the primary concern about graded materials was the management of thermal stresses in buildings that were subjected to high temperatures as well as surface degradation. As a result of this research, a new category of engineered materials known as functionally graded materials has been produced. These materials have been created with varying spatial features that have been suited to certain uses. Advanced manufacturing processes, including as powder metallurgy, chemical vapor deposition, centrifugal casting, solid mold building, and other designs, are often used to regulate the graded composition of such materials. These techniques are also referred to as designs. Several different hypotheses have been proposed as potential solutions to the problem of stress concentration in plates constructed from functionally graded materials. For instance, the Mori-Tanaka theory, the Reddy theory, and the Halpin-Tsai theory, each of which will be discussed in a little more detail further in the text.

2.1.1.1.1 Mori-Tanaka Theory

Estimating the effective characteristics of portions of the microstructure of functionally graded materials that have a continuous matrix and a discontinuous particle phase may be done with the help of the Mori-Tanaka model if the material in question has a continuous matrix. This takes into consideration the way in which elastic fields interact with one another in components that are next to one another. In the equations that follow, the letters m and c denote the characteristics of metals and ceramics, respectively, which are represented by the subscripts. Effective local bulk modulus (K_f) and shear modulus of functionally graded materials may be stated as follows, in accordance with the Mori-Tanaka model [1–3]:

$$V_c = \left(\frac{z}{h} + \frac{1}{2} \right)^p \tag{2.1}$$

$$\frac{K_f(z) - K_m}{K_c - K_m} = \frac{V_c(z)}{1 + \left[1 - V_c(z)\right] \dfrac{3(K_c - K_m)}{3K_m + 4G_m}} \tag{2.2}$$

$$\frac{G_f(z) - G_m}{G_c - G_m} = \frac{V_c(z)}{1 + \left[1 - V_c(z)\right] \dfrac{3(G_c - G_m)}{3G_m + f_1}} \tag{2.3}$$

$$f_1 = \frac{G_c \left[9K_c + 8G_c\right]}{6\left[K_c + 2G_c\right]} \tag{2.4}$$

In Eq. (2.1), p denotes a non-negative variable parameter known as a gradient index. This index is responsible for determining the qualities of the material as a function of thickness.

2.1.1.1.2 Reddy Theory

The power law model is one of the most desirable models for plates made of functionally graded materials. In this model, it is assumed that the effective material properties of the plates are continuous in the thickness direction (z-axis direction) according to the power function of different volume fractions. This model is one of the reasons why the power law model is one of the most desirable models for plates made of functionally graded materials. The power law model may be used to analyze the constituent components of effective materials, which can be broken down into the following categories:

$$P = P_u V_u + P_l V_l \tag{2.5}$$

where P_u and P_l represent the characteristics of the material at the lower and higher levels, respectively, and where V_u and V_l represent the corresponding volume fractions, which are defined as follows:

$$V_u = \left(\frac{z}{h} + \frac{1}{2} \right)^p,$$
$$V_l = 1 - V_u \tag{2.6}$$

2.1.1.1.3 The Halpin-Tsai Theory

There have been multiple theoretical frameworks developed in order to predict the properties of functionally graded materials based on the properties of the pure components and the morphology of the composite material. These frameworks have been developed in order to predict the properties of functionally graded materials. Each component of the composite is assumed to behave independently of the others, which is a fundamental presumption underlying all of these theories. These theories provide a simple route to evaluate the individual contributions of component properties such as matrix and modulus of the reinforcing phase, volume fraction of the reinforcing phase, aspect ratio, Enables amplifier phase direction, and more. While the general goal of such theories is to predict composite material performance for a set of components, these theories provide a simple route to evaluate the individual contributions of component properties. Halpin and Tsai are responsible for the development of a well-known composite theory that predicts the stiffness of unidirectional composites based on the aspect ratio. The early micromechanical studies of Hermans and Hill serve as the foundation for this hypothesis. Hermans expanded the form of Hill's self-consistent theory by contemplating a single fiber contained inside a cylindrical shell of matrix and then placing this structure in an endless medium with the composite material's average attributes. Hermann's findings have been simplified into an analytical form that can be used to a wide range of amplifier geometries by Halpin and Tissa. This form is applicable to discontinuous amplifiers as well as continuous amplifiers. On the basis of this concept, the general composite material moduli (E_c) may be calculated as follows [4]:

$$\frac{E_c}{E_m} = \frac{1+\xi\eta\phi_f}{1-\xi\eta\phi_f} \tag{2.7}$$

The following is the definition of η according to the equation that was just given:

$$\eta = \frac{E_f/E_m - 1}{E_f/E_m + \xi} \tag{2.8}$$

Nevertheless, this strategy is predicated on a number of assumptions, including the following: (i) the reinforcement and matrix are linearly elastic, isotropic, and tightly bonded; (ii) the filler is perfectly aligned; (iii) the filler is asymmetric; and (iv) particle-particle interactions are not explicitly considered. Naturally, in the case of all theories concerning composites, it is believed that the characteristics of the matrix and the reinforcement are the same as those of the pure components. It is necessary to emphasize that when comparing the theoretical properties of the composite material with the experimental data of the composite material, many complications arise. This is especially true in the case of layered silicate nanocomposites. As a result, it is necessary to emphasize this point. The choice of composite theory also has a role in determining how well the anticipated and observed qualities line up with one another, in addition to the physical gaps that exist between theory and experiment. Because of its accuracy in forecasting the hardness of composites reinforced with glass fibers and its compatibility with various geometries of the reinforcing phase, particularly discs, the model developed by Halpin-Tsai has garnered a lot of attention recently.

2.1.2 POROSITY

Throughout the manufacturing process, there are a number of faults that cannot be avoided. These errors may have a negative impact on the effectiveness, performance, and quality of systems and structures. One of these flaws that have a substantial impact on the mechanical qualities of the structure is called porosity. Porosity is a flaw that may have an impact on the system's ability to execute mechanical functions. A continuous system made up of porous (see Figure 2.1) materials would typically have a reduced level of stiffness as a consequence of the presence of porosity. As of right now, the frequency of the system, as well as its stability, will progressively decline with the growth in the porosity coefficient. Researchers have lately been interested in this phenomenon since it is unavoidable for smart materials to have porosity. The existence of porosity in smart materials is what drew their attention [5, 6].

(a) Even distribution
(b) Uneven distribution

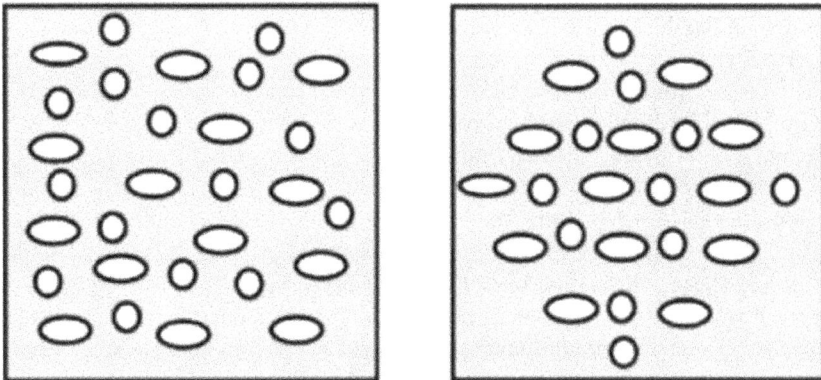

a) Even distribution   b) Uneven distribution

FIGURE 2.1 Schematic of (a) even and (b) uneven porosity distribution.

2.1.3 FORMULATION

It is unavoidable to research original building designs as a direct consequence of the recent structural revolution. Buckling analysis of these types of structures is required for use in engineering-related domains due to this fact.

The power-law equation states that the characteristics of the top and bottom layers may be calculated by employing the volume fraction of the constitutive materials. This is possible since volume fractions are proportional to power laws. When k equals zero, it represents a nano-surface entirely made of aluminum (Al). When k is less than zero, it indicates a material entirely made of aluminum oxide (Al_2O_3). In homogeneous Young's modulus and mass density of the facesheets are calculated using the power-law technique and are as follows [7–9]:

For z $\epsilon[h_0,h_1]$

$$\Gamma(z) = (\Gamma_t - \Gamma_b)\left(\frac{z - h_0}{h_1 - h_0}\right)^k + \Gamma_b \tag{2.9}$$

For z $\epsilon[h_2,h_3]$

$$\Gamma(z) = (\Gamma_t - \Gamma_b)\left(\frac{z - h_3}{h_2 - h_3}\right)^k + \Gamma_b \tag{2.10}$$

The symbol $\Gamma(z)$ for Young's modulus and mass density, parameter z, varies with thickness. Moreover, it is assumed that Poisson's ratio remains constant throughout the thickness of the material.

The Cartesian coordinate system is employed in this investigation, and the origin of the coordinate system is located at one of the plate's corners (as indicated in Figure 2.2). The foundation effect has been taken into account using the medium developed by Winkler and Pasternak.

The higher-order sinusoidal shear deformation plate theory, initially presented by Touratier [11–13], has been used in this chapter. Normal to the midline (neutral surface) stays normal before and after deformation, according to the higher-order sinusoidal shear deformation plate theory. Reddy's theory is widely regarded as the most well-known higher-order plate theory, and it is the one most often used while doing research on multilayer plates constructed from composite materials. A novel theory of shear deformation of plates was provided by Shi [14] in the year 2007. This theory was comparable to the theory that was presented by Reddy in the sense that the transverse shear strain varied in a parabola. Both the Reddy and Shi theories are Third-Order Plane Theories, which implies that they both believe that displacement can be characterized by a function of the third order [15]. Higher-order deformation theories enable displacements to vary not just linearly

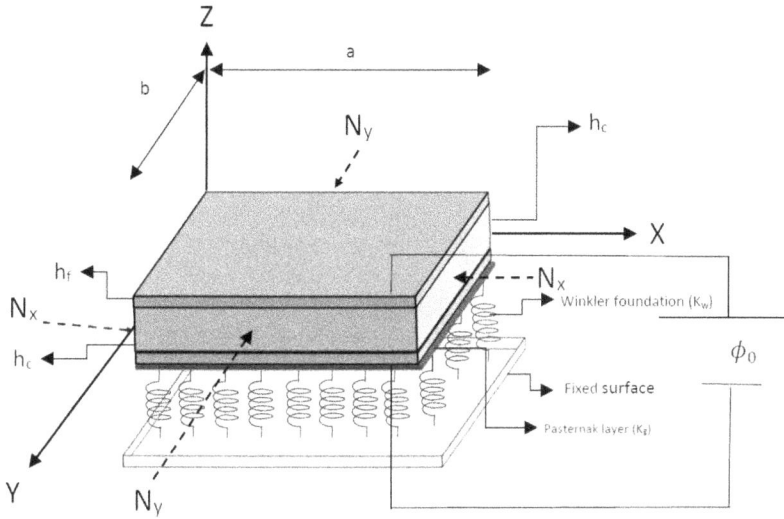

FIGURE 2.2 Sandwich composite arrangement [10].

but also as cubic functions, making it possible for these types of changes to occur. It is important to note that the in-plane cubic strains, denoted by the symbol "z^3," as well as shear stresses are of the second order in the third order theory of shear.

The presumptions that underlie the high-order shear plane theory:

1. The shear strain, and hence the shear stress, does not remain constant all the way through the plate's thickness.
2. There is no need to include a shear correction component in the strain calculations.
3. The displacement field is characterized by a shift in transverse shear that is quadratic along the plate thickness.

Theory of high-order sinusoidal shear

Touratier was the first person to put up the idea of sinusoidal shear deformation [11–13], and further research, such as that conducted by Vidal and Polit [16], helped to expand the concept further. According to the principle of sinusoidal shear deformation, the transverse shear stress is removed from the plate's top and lower surfaces.

The displacement field may be depicted as follows [17] based on the presumptions of the higher-order sinusoidal shear deformation plate theory:

$$U_x(x,y,z,t) = u_0(x,y,t) - z\frac{\partial w_0(x,y,t)}{\partial x} + f(z)\phi_x \qquad (2.11)$$

$$V_y\left(x,y,z,t\right) = v_0\left(x,y,t\right) - z\frac{\partial w_0\left(x,y,t\right)}{\partial y} + f\left(z\right)\phi_y\left(x,y,t\right) \qquad (2.12)$$

$$W_z\left(x,y,z,t\right) = w_0\left(x,y,t\right) \qquad (2.13)$$

The shear correction factor is not necessary for higher-order shear deformation theory, as opposed to first-order shear deformation plate theory. Moreover, the function f(z), which has the following definition, characterizes the displacement field along the thickness direction: $f\left(z\right) = \frac{H}{\pi}\sin\left(\frac{\pi z}{H}\right) = g\left(z\right) = \frac{df\left(z\right)}{dz} = \cos\left(\frac{\pi z}{H}\right)$

2.1.3.1 Theories That Are Not Classical

Due to the many applications that nanostructures have in nanoproblems, such as nano mechanical resonators, nano scale mass sensors, electro mechanical nano actuators and nano generators, understanding the mechanical behavior of nanostructures is of the utmost importance. Continuous modeling of nanostructures has garnered a lot of interest recently. This might be attributed to the difficulties associated with performing precise experimental measurements at nanoscales as well as the high processing costs associated with molecular dynamics simulation. At the nanoscale level, the size effects of the problem geometry play an essential role in the mechanics of structures. To take these effects into account in traditional continuum-based theories, research has been conducted. Estimating the mechanical behavior of nanostructures has been accomplished with the assistance of a number of different theories that are based on the modified continuum. These theories include the theory of nonlocal elasticity, the nonlocal theory of strain gradient elasticity, and the theory of coupled stress elasticity.

1- Eringen's nonlocal theory
2- Nonlocal strain gradient theory
3- Theory of elasticity of coupled stress

2.1.3.1.1 The Nonlocal Elasticity Hypothesis Proposed by Eringen

Eringen [18, 19] presented the notion of nonlocal elasticity over 20 years before the creation of carbon nanotubes (CNTs). Yet, prior to the development of nanostructures such as carbon nanotubes and graphene sheets (GS), this insightful idea did not get a great deal of attention. The idea that this theory might be used to the analysis of the size-dependent mechanical response of nanostructures was initially offered by Peddieson et al. [20]. The stress at a place is solely dependent on the strain at that point, according to the classical theory of elasticity, which is incapable of predicting size effects. On the other hand, according to the nonlocal elasticity theory, strains at all places influence the stress at an arbitrary position, as illustrated in Figure (2.3). Because of these preliminary assumptions, the theory is able to take into consideration intermolecular interactions. When volume pressures are not taken into account, the constitutive relation of the nonlocal integral may be stated using the following relation.

$$\sigma_{ij}^{nl} = \iiint_v \varphi\left(\left|x - x'\right|, \zeta\right)\sigma_{ij}^{n} dV \tag{2.14}$$

In the equation that was just presented, the variables denoted by σ_{ij}^{nl} ، σ_{ij}^{n} ، φ and ζ, respectively, stand for nonlocal stress, local stress, kernel function, and small-scale coefficient. Likewise, $\left|x - x'\right|$ is the volume symbol and indicates the distance between points x and x'. It also represents the distance between points x and x'. It is important to keep in mind that the nonlocal coefficient is determined by the following relation:

$$\zeta = \frac{e_0 a}{L} \tag{2.15}$$

In the equation that was just presented, the internal and exterior longitudinal properties of the calibration coefficient are denoted by the variables e_0, a, and L, respectively. Each nanostructure has a length that is distinctive on both the inside and the outside. With carbon nanotubes, for instance, the length of the c-c bond is often selected to be used as a criterion for determining the internal length (see Figure 2.3). The calibration factor may be derived either by experimental observations or through the molecular dynamics (MD) of the system.

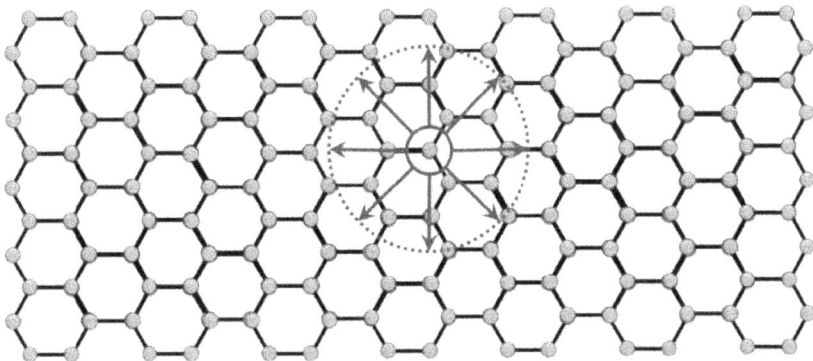

FIGURE 2.3 According to the nonlocal tension hypothesis, the strain at one place in a nanostructure is reliant on the stress at all other sites in the structure.

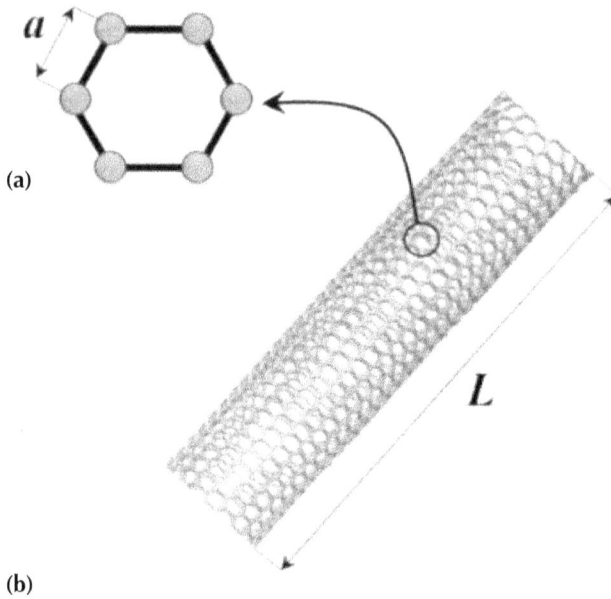

FIGURE 2.4 Both the length of the internal features (a) and the length of the external features (b) are taken into consideration (L).

2.1.3.1.2 The Hypothesis of Nonlocal Strain Gradient Theory

The nonlocal elasticity hypothesis that was covered in the preceding part has two weaknesses that need to be addressed. To begin, the impacts that are not local vanish once a given distance has been traveled. For instance, scale effects on the axial vibration of uniform nanorods, as anticipated by nonlocal elasticity, vanish when the rod length is more than 20 nm. Second, nonlocal elasticity can only be used to anticipate the decrease in stiffness that occurs in structures on a small scale. On the other hand, there is evidence that the stiffness of certain small-scale structures increases with increasing length, particularly when the structures are longer. It is possible to quantify this stiffness hardening by integration with surface effects or strain gradients. The NSGT was introduced by Lim et al. [21] as a means of overcoming the deficiencies of the nonlocal elasticity theory. The theory makes use of two kernel functions. The new theory is able to capture the weakening and strengthening of a material's stiffness at scales that are rather modest. As compared to the nonlocal effects, the scale effect, which is anticipated by the NSGT, arises in a greater range of lengths. The NSGT is used to characterize the connection between stress and strain according to the following relationship:

$$\left(1-\left(e_1 a\right)^2 \nabla^2\right)\left(1-\left(e_0 a\right)^2 \nabla^2\right)\sigma_{ij} = \left(1-\left(e_1 a\right)^2 \nabla^2\right)\left[C_{ijkl}\varepsilon_{kl}\right] \qquad (2.16)$$

2.1.3.1.3   The Idea of Coupled Stress Elasticity

Experiments have shown that the micron-scale deformation behavior of metals and polymers depends on the size of the problem geometry. This behavior cannot be described by standard theories of mechanics, hence the reliance of geometry size on deformation behavior was explained by the theory of linked stress. In the next paragraphs, the distinction between coupled stress theories of mechanics and conventional theories of mechanics will be addressed. In the field of particle mechanics, the forces that are exerted on the individual material particles cannot be directly seen. Observations on the motion of individual matter particles are the only way to establish the existence of these forces. The variation in the kinetic energy of material particles is another method that may be used to ascertain these values. A particle of matter in a deformable continuum may only experience linear displacement according to the principles of Newtonian physics. The only thing that can alter the velocity of the particles of matter is the force that is acting on them, and according to Newton's second law, this force is connected to the acceleration that the particles of matter are experiencing. In the theory of Newtonian mechanics, the motion of a particle of matter is caused by an external force. In the classical couple stress theories for linear elastic materials, the loads that are applied to the material particles include not only the forces to move the material particles for displacement but also a couple to apply rotation to them. This is because in these theories, linear elastic materials are assumed to have a couple stress relationship between them. has been applied. Under the confines of this traditional conception, the only equilibrium relations of forces and momentum(s) that are relevant are the common ones, and the couplet is rendered irrelevant in the lack of higher order equilibrium prerequisites. Using the coupled stress elasticity, the connection between stress and strain may be described according to the following equation:

$$U = \frac{1}{2}\int_{\Re}\left(\sigma_{ij}\varepsilon_{ij} + m_{ij}\mathcal{X}_{ij}\right)dV \qquad (i,j=1,2,3) \qquad (2.17)$$

The Cauchy stress tensor, the classical strain tensor, the deviatoric part of the Couple stress tensor, and the symmetric bending tensor are respectively represented by the symbols, σ, ε, m and $\mathcal{X}$. Moreover, the definitions of the strain and bending tensors are derived from the following relation:

$$\varepsilon_{ij} = \frac{1}{2}\left(u_{i,j} + u_{j,i}\right)$$

$$\mathcal{X}_{ij} = \frac{1}{2}\left(\theta_{i,j} + \theta_{j,i}\right) \qquad (2.18)$$

Also, $\theta_i = \dfrac{1}{2} e_{ijk} u_{k.j}$, which e_{ijk} is the symbol of Permutation. Continuous equations can be written according to Eq. (2.19):

$$\sigma_{ij} = \lambda \epsilon_{kk} \delta_{ij} + 2\gamma \epsilon_{ij} \tag{2.19}$$

$$m_{ij} = 2\gamma l^2 X_{ij}$$

It is important to note that the symbol for Kronecker's delta is δ_{ij}, and the length scale parameter of the material, l, is what represents the influence of the Couple stress.

The stress at a reference point (x) in an elastic continuum is reliant not only on the strain of point x but also on the strain of all the bystander points, as stated by Eringen's nonlocal theory [18], which is typically applicable to circumstances on the micro and nanoscale:

$$\left(1 - \mu^2 \nabla^2\right)\sigma^{nl} = \sigma^{cl} \tag{2.20}$$

where σ^{nl} represents the nonlocal stress tensor and σ^{cl} represents the classical stress tensor. Apart from that, the Laplacian operator is represented by the ∇^2. It turns out that $\mu = e0\,a$ also holds true for the nonlocal parameter.

Due to its distinctive qualities, such as the ability to transform one kind of energy (magnetic, electric, or mechanical) into another, magnetostrictive material has attracted much attention in engineering applications. The core layer's stress, strain, and magnetic properties may be expressed as follows [10, 22]:

$$
\left(1 - \mu^2 \nabla^2\right)
\begin{Bmatrix}
\sigma_{xx}^{c,nl} \\
\sigma_{yy}^{c,nl} \\
\tau_{yz}^{c,nl} \\
\tau_{xz}^{c,nl} \\
\tau_{xy}^{c,nl}
\end{Bmatrix}
=
\begin{bmatrix}
C_{11}^c & C_{12}^c & 0 & 0 & 0 \\
C_{21}^c & C_{22}^c & 0 & 0 & 0 \\
0 & 0 & C_{44}^c & 0 & 0 \\
0 & 0 & 0 & C_{55}^c & 0 \\
0 & 0 & 0 & 0 & C_{66}^c
\end{bmatrix}
\begin{Bmatrix}
\varepsilon_{xx}^c \\
\varepsilon_{yy}^c \\
\gamma_{yz}^c \\
\gamma_{xz}^c \\
\gamma_{xy}^c
\end{Bmatrix}
- H_{zz}
\begin{Bmatrix}
\bar{e}_{31} \\
\bar{e}_{32} \\
0 \\
0 \\
0
\end{Bmatrix}
\tag{2.21}
$$

where $\sigma_{ij}, \varepsilon_{ij}\,(i, j = x, y, z)$ are the stress and strain constants, respectively. In addition, H_{zz} represents a magnetic field. In addition to this, it is important to understand that the abbreviation C_{ij} stands for the elastic stiffness coefficient. If it is assumed that there is no thickness stretching ($\varepsilon_{zz} = 0$), then the plane stress elastic constants may be calculated as [20]:

$$C_{11}^c = \frac{E_c}{1-\vartheta_c^2} \qquad C_{12}^c = \frac{\vartheta_c E_c}{1-\vartheta_c^2} \qquad C_{66}^c = \frac{E_c}{2(1+\vartheta_c)} \qquad C_{55}^c = C_{44}^c = C_{66}^c \qquad (2.22)$$

where E_c and v_c stand for the core layer's modulus of elasticity and Poisson's ratio, respectively, and $\overline{e}_{ij}$ are the transformed magnetostrictive moduli that may be achieved as a result of the previous transformations [22–24]:

$$\overline{e}_{31} = e_{31} \cos^2 \theta + e_{32} \sin^2 \theta$$
$$\overline{e}_{32} = e_{32} \cos^2 \theta + e_{31} \sin^2 \theta \qquad (2.23)$$
$$\overline{e}_{34} = (e_{31} - e_{32}) \sin \theta \cos \theta$$

The magnetic field, denoted by the symbol H_{zz}, may be calculated using the following relation [22–25]:

$$H_{zz} = k_c I(x,y,t)$$

$$I(x,y,t) = C(t) \frac{\partial w_0(x,y,t)}{\partial t} \qquad (2.24)$$

The width, radius, and number of turns of a coil all contribute to the constant of the coil, indicated by the symbol k_c. Moreover, the current through the coil $I(x,y,t)$ and the gain ($C(t)$) are shown. It's important to realize that we often treat the control gain as a continuous value. Moreover, we may express the gain from velocity feedback as $K_c C(t)$. It is possible to make the following representation of the connection between the coil constant and the coil's properties:

$$K_c = \frac{n_c}{\sqrt{b_c^2 + 4r_c^2}} \qquad (2.25)$$

In the equation just presented, the terms n_c, b_c, and r_c refer, respectively, to the number of turns of the coil, the width of the coil, and the radius of the coil.

It is possible to characterize the stresses of the magnetostrictive layer as [26]:

$$
\begin{Bmatrix} \varepsilon_{xx}^f \\ \varepsilon_{yy}^f \\ \varepsilon_{yz}^f \\ \varepsilon_{xz}^f \\ \varepsilon_{xy}^f \end{Bmatrix} = \begin{Bmatrix} \epsilon_{xx}^c \\ \epsilon_{yy}^c \\ \epsilon_{yz}^c \\ \epsilon_{xz}^c \\ \epsilon_{xy}^c \end{Bmatrix} + z \begin{Bmatrix} K_{xx}^c \\ K_{yy}^c \\ K_{yz}^c \\ K_{xz}^c \\ K_{xy}^c \end{Bmatrix} + f(z) \begin{Bmatrix} \chi_{xx}^c \\ \chi_{yy}^c \\ \chi_{yz}^c \\ \chi_{xz}^c \\ \chi_{xy}^c \end{Bmatrix} + f(z)' \begin{Bmatrix} X_{xx}^c \\ X_{yy}^c \\ X_{yz}^c \\ X_{xz}^c \\ X_{xy}^c \end{Bmatrix}
\tag{2.26}
$$

where

$$
\begin{Bmatrix} \epsilon_{xx}^c \\ \epsilon_{yy}^c \\ \epsilon_{yz}^c \\ \epsilon_{xz}^c \\ \epsilon_{xy}^c \end{Bmatrix} = \begin{Bmatrix} \dfrac{\partial u_0}{\partial x} \\[2mm] \dfrac{\partial v_0}{\partial y} \\[2mm] 0 \\ 0 \\ \left(\dfrac{\partial u_0}{\partial y} + \dfrac{\partial v_0}{\partial x} \right) \end{Bmatrix}, \quad \begin{Bmatrix} K_{xx}^c \\ K_{yy}^c \\ K_{yz}^c \\ K_{xz}^c \\ K_{xy}^c \end{Bmatrix} = \begin{Bmatrix} -\dfrac{\partial^2 w_0}{\partial x^2} \\[2mm] -\dfrac{\partial^2 w_0}{\partial y^2} \\[2mm] 0 \\ 0 \\ -\dfrac{\partial^2 w_0}{\partial x^2} + -\dfrac{\partial^2 w_0}{\partial y^2} \end{Bmatrix}, \quad \begin{Bmatrix} \chi_{xx}^c \\ \chi_{yy}^c \\ \chi_{yz}^c \\ \chi_{xz}^c \\ \chi_{xy}^c \end{Bmatrix} = \begin{Bmatrix} \dfrac{\partial \phi_x}{\partial x} \\[2mm] \dfrac{\partial \phi_y}{\partial y} \\[2mm] 0 \\ 0 \\ \left(\dfrac{\partial \phi_x}{\partial y} + \dfrac{\partial \phi_y}{\partial x} \right) \end{Bmatrix}
$$

$$
\tag{2.27}
$$

It is possible to describe the relationship between stress and strain for the face-sheets using the formula [27]:

$$
(1 - \mu^2 \nabla^2) \begin{Bmatrix} \sigma_{xx}^{f,nl} \\ \sigma_{yy}^{f,nl} \\ \tau_{yz}^{f,nl} \\ \tau_{xz}^{f,nl} \\ \tau_{xy}^{f,nl} \end{Bmatrix} = \begin{bmatrix} C_{11}^f & C_{12}^f & 0 & 0 & 0 \\ C_{21}^f & C_{22}^f & 0 & 0 & 0 \\ 0 & 0 & C_{44}^f & 0 & 0 \\ 0 & 0 & 0 & C_{55}^f & 0 \\ 0 & 0 & 0 & 0 & C_{66}^f \end{bmatrix} \begin{Bmatrix} \varepsilon_{xx}^f \\ \varepsilon_{yy}^f \\ \gamma_{yz}^f \\ \gamma_{xz}^f \\ \gamma_{xy}^f \end{Bmatrix}
\tag{2.28}
$$

TABLE 2.1
Magnetostrictive Core Layer and FG Facesheet Material Qualities

Symbol	Aluminum	Aluminum oxide	Magnetostrictive
E (GPa)	70	380	30
ϑ	0.3	0.3	0.25
$p\left(kg\ m^{-3}\right)$	2,702	3,800	9,250
$e_{31}N\left(mA\right)^{-1}$	-	-	442.55
$e_{32}N\left(mA\right)^{-1}$	-	-	442.55
$e_{33}N\left(mA\right)^{-1}$	-	-	-212

The following are some examples of how elastic functions may be defined:

$$C_{11}^f = \frac{E(z)}{1-\vartheta_f^2} \qquad C_{12}^f = \frac{\vartheta_f E(z)}{1-\vartheta_f^2} \qquad C_{44}^f = \frac{E(z)}{2\left(1+\vartheta_f\right)} \qquad C_{55}^f = C_{44}^f = C_{66}^f \qquad (2.29)$$

The elastic modulus of the facesheets, E_f, and v_f, the poison's constant of the facesheets, are specified as the variables. The material parameters of the layers that follow one another are detailed in Table 2.1.

It is possible to describe the nonzero stresses of both the top and lower layers as [28]:

$$\begin{Bmatrix} \mathcal{E}_{xx}^f \\ \mathcal{E}_{yy}^f \\ \mathcal{E}_{yz}^f \\ \mathcal{E}_{xz}^f \\ \mathcal{E}_{xy}^f \end{Bmatrix} = \begin{Bmatrix} \epsilon_{xx}^c \\ \epsilon_{yy}^c \\ \epsilon_{yz}^c \\ \epsilon_{xz}^c \\ \epsilon_{xy}^c \end{Bmatrix} + z \begin{Bmatrix} \kappa_{xx}^f \\ \kappa_{yy}^f \\ \kappa_{yz}^f \\ \kappa_{xz}^f \\ \kappa_{xy}^f \end{Bmatrix} + f(z) \begin{Bmatrix} \chi_{xx}^c \\ \chi_{yy}^c \\ \chi_{yz}^c \\ \chi_{xz}^c \\ \chi_{xy}^c \end{Bmatrix} + f(z)' \begin{Bmatrix} X_{xx}^c \\ X_{yy}^c \\ X_{yz}^c \\ X_{xz}^c \\ X_{xy}^c \end{Bmatrix} \qquad (2.30)$$

in which

$$\left\{\begin{matrix}\varepsilon_{xx}^f\\\varepsilon_{yy}^f\\\varepsilon_{yz}^f\\\varepsilon_{xz}^f\\\varepsilon_{xy}^f\end{matrix}\right\}=\left\{\begin{matrix}\dfrac{\partial u_0}{\partial x}\\\dfrac{\partial v_0}{\partial y}\\0\\0\\\left(\dfrac{\partial u_0}{\partial y}+\dfrac{\partial v_0}{\partial y}\right)\end{matrix}\right\},\left\{\begin{matrix}\kappa_{xx}^f\\\kappa_{yy}^f\\\kappa_{yz}^f\\\kappa_{xz}^f\\\kappa_{xy}^f\end{matrix}\right\}=\left\{\begin{matrix}-\dfrac{\partial^2 w_0}{\partial x^2}\\-\dfrac{\partial^2 w_0}{\partial y^2}\\0\\0\\-\left(\dfrac{\partial^2 w_0}{\partial x^2}+-\dfrac{\partial^2 w_0}{\partial y^2}\right)\end{matrix}\right\},\left\{\begin{matrix}\chi_{xx}^f\\\chi_{yy}^f\\\chi_{yz}^f\\\chi_{xz}^f\\\chi_{xy}^f\end{matrix}\right\}=\left\{\begin{matrix}\dfrac{\partial\phi_x}{\partial x}\\\dfrac{\partial\phi_y}{\partial y}\\0\\0\\\left(\dfrac{\partial\phi_x}{\partial y}+-\dfrac{\partial\phi_y}{\partial x}\right)\end{matrix}\right\},\left\{\begin{matrix}\chi_{xx}^f\\\chi_{yy}^f\\\chi_{yz}^f\\\chi_{xz}^f\\\chi_{xy}^f\end{matrix}\right\}=\left\{\begin{matrix}0\\0\\\phi_x\\\phi_y\\0\end{matrix}\right\}$$

$$(2.31)$$

2.1.3.2 Governing Equation

The first variation of Hamilton's principle can be expressed as follows [29]:

$$\int_0^t \delta(U-T+V)dt=0 \qquad (2.32)$$

where U,T and V represent the strain energy, the kinetic energy, and the external work, respectively. The first strain energy variation may be written as [30]:

$$\delta U^c=\int_0^a\int_0^b\left(N_{xx}^c\left(\delta\varepsilon_{xx}^c\right)+M_{xx}^c\left(\delta\kappa_{xx}^c\right)+N_{yy}^c\left(\delta\varepsilon_{yy}^c\right)+M_{yy}^c\left(\delta\kappa_{yy}^c\right)+N_{xz}^c\left(\delta\varepsilon_{xz}^c\right)\right.$$
$$\left.+M_{xz}^c\left(\delta\kappa_{xz}^c\right)+N_{yz}^c\left(\delta\varepsilon_{yz}^c\right)+M_{yz}^c\left(\delta\kappa_{yz}^c\right)+N_{xy}^c\left(\delta\varepsilon_{xy}^c\right)+M_{xy}^c\left(\delta\kappa_{xy}^c\right)\right)dA \qquad (2.33)$$

$$\delta U^f=\int_0^a\int_0^b\left(N_{xx}^f\left(\delta\varepsilon_{xx}^f\right)+M_{xx}^f\left(\delta\kappa_{xx}^f\right)+N_{yy}^f\left(\delta\varepsilon_{yy}^f\right)+M_{yy}^f\left(\delta\kappa_{yy}^f\right)+N_{xz}^f\left(\delta\varepsilon_{xz}^f\right)\right.$$
$$\left.+M_{xz}^f\left(\delta\kappa_{xz}^f\right)+N_{yz}^f\left(\delta\varepsilon_{yz}^f\right)+M_{yz}^f\left(\delta\kappa_{yz}^f\right)+N_{xy}^f\left(\delta\varepsilon_{xy}^f\right)+M_{xy}^f\left(\delta\kappa_{xy}^f\right)\right)dA \qquad (2.34)$$

where

$$\left(N_{ij}^c,M_{ij}^c,F_{ij}^c,G_{ij}^c\right)=\int_{-\frac{h_c}{2}}^{\frac{h_c}{2}}\sigma_{ij}^c\left(1,z,f(z),f(z)'\right)dz,\quad (i,j=x,y,z)$$

$$\left(N_{ij}^f,M_{ij}^f,F_{ij}^c,G_{ij}^c\right)=\int_{\frac{h_c}{2}}^{\frac{h_c}{2}+h_f}\sigma_{ij}^f\left(1,z,f(z),f(z)'\right)dz$$

$$+\int_{-\left(\frac{h_c}{2}+h_f\right)}^{-\frac{h_c}{2}}\sigma_{ij}^f\left(1,z,f(z),f(z)'\right)dz,\quad (i,j=x,y,z)$$

$$(2.35)$$

The terms "core layer" and "facesheets" are denoted by the superscripts "c" and "f," respectively, in the equations shown above. The following is the nonlocal relations of forces and moments with strains:

$$\left(1-\mu^2\nabla^2\right)\begin{Bmatrix}\{N\}\\\{M\}\\\{F\}\end{Bmatrix}=\begin{bmatrix}[A]&[B]&[S]\\[B]&[D]&[M]\\[S]&[M]&[O]\end{bmatrix}\begin{Bmatrix}\{\epsilon\}\\\{\kappa\}\\\{\chi\}\end{Bmatrix}$$

$$\left(1-\mu^2\nabla^2\right)\begin{Bmatrix}G_{yz}\\G_{xz}\end{Bmatrix}=\begin{bmatrix}0&L_{44}\\L_{55}&0\end{bmatrix}\begin{Bmatrix}\S_{yz}\\\S_{xz}\end{Bmatrix} \tag{2.36}$$

where

$$\{N\}=\{N_{xx}\quad N_{yy}\quad N_{xy}\}^T\quad \{M\}=\{M_{xx}\quad M_{yy}\quad M_{xy}\}^T\quad \{F\}=\{F_{xx}\quad F_{yy}\quad F_{xy}\}^T$$

$$\{\epsilon\}=\{\epsilon_{xx}\quad \epsilon_{yy}\quad \epsilon_{xy}\}^T\quad \{\kappa\}=\{\kappa_{xx}\quad \kappa_{yy}\quad \kappa_{xy}\}^T\quad \{\chi\}=\{\chi_{xx}\quad \chi_{yy}\quad \chi_{xy}\}^T$$

$$\tag{2.37}$$

where

$$\begin{bmatrix}A_{11}&B_{11}&D_{11}&S_{11}&M_{11}&O_{11}&L_{11}\\A_{22}&B_{22}&D_{22}&S_{22}&M_{22}&O_{22}&L_{22}\\A_{12}&B_{12}&D_{12}&S_{12}&M_{12}&O_{12}&L_{12}\\A_{44}&B_{44}&D_{44}&S_{44}&M_{44}&O_{44}&L_{44}\\A_{55}&B_{55}&D_{55}&S_{55}&M_{55}&O_{55}&L_{55}\\A_{66}&B_{66}&D_{66}&S_{66}&M_{66}&O_{66}&L_{66}\end{bmatrix}$$

$$=\int_{-\frac{hc}{2}}^{\frac{hc}{2}}\begin{bmatrix}1\\z\\z^2\\f(z)\\f(z)^2\\f(z)^{\cdot2}\end{bmatrix}^{TT}\begin{bmatrix}C_{11}^c\\C_{22}^c\\C_{12}^c\\C_{44}^c\\C_{55}^c\\C_{66}^c\end{bmatrix}dz+\int_{\frac{h_c}{2}}^{-\frac{h_c}{2}+hf}\begin{bmatrix}1\\z\\z^2\\f(z)\\f(z)^2\\f(z)^{\cdot2}\end{bmatrix}^T\begin{bmatrix}C_{11}^f\\C_{22}^f\\C_{12}^f\\C_{44}^f\\C_{55}^f\\C_{66}^f\end{bmatrix}dz+\int_{-\left(\frac{h_c}{2}+h_f\right)}^{-\frac{h_c}{2}}\begin{bmatrix}1\\z\\z^2\\f(z)\\f(z)^2\\f(z)^{\cdot2}\end{bmatrix}^T\begin{bmatrix}C_{11}^f\\C_{22}^f\\C_{12}^f\\C_{44}^f\\C_{55}^f\\C_{66}^f\end{bmatrix}dz$$

$$\tag{2.38}$$

In this investigation, external forces come from two sources:

- Winkler and Pasternak foundation
- Uniformly distributed in-plane loads
- Hygro-thermal loads

A force acting on a nanoplate may either transfer energy into it or remove it in the form of work. Work is defined as the energy delivered by external forces along the axis of the nanoplate that produces displacement and deformation in the nanoplate in the examined system, which is a sandwich nanoplate made of magnetostrictive and functionally graded material. The first external work variant may be written as [5, 31]:

$$\delta V = \int_0^a \int_0^b \left[k_w . \delta w_0 - kg \left(\frac{\partial^2 \delta w_0}{\partial x^2} + \frac{\partial^2 \delta w_0}{\partial y^2} \right) - N_{xx} \left(\frac{\partial^2 \delta w_0}{\partial x^2} \right) - N_{yy} \left(\frac{\partial^2 \delta w_0}{\partial y^2} \right) \right] \quad (2.39)$$

where k_w and k_p are the Winkler and shear layer stiffnesses, respectively. Moreover, the dimensionless version of the Winkler and Pasternak foundation is described as follows:

$$K_w = \frac{k_w a^4}{D_{11}} \qquad K_g = \frac{k_g a^2}{D_{11}} \qquad (2.40)$$

The pre-buckling resultant forces are assumed to be as follows, as illustrated in Figure 2.5:

$$N_{xx} = \gamma_1 P \qquad N_{yy} = \gamma_2 P \qquad (2.41)$$

The first kinetic energy variation may be written as:

$$\delta T = \frac{1}{2} \int \int \left[-I_0 \left(\frac{\partial^2 u_0}{\partial t^2} \right) \delta u_0 - I_1 \left(\frac{\partial^3 u_0}{\partial x \partial t^2} \right) \delta w_0 - I_3 \left(\frac{\partial^2 u_0}{\partial t^2} \right) \delta \phi_x + I_1 \left(\frac{\partial^3 w_0}{\partial x \partial t^2} \right) \delta u_0 \right.$$

$$+ I_2 \left(\frac{\partial^4 w_0}{\partial x^2 \partial t^2} \right) \delta w_0 + I_4 \left(\frac{\partial^3 u_0}{\partial x \partial t^2} \right) \delta \phi_x - I_3 \left(\frac{\partial^2 \phi_x}{\partial t^2} \right) \delta u_0 - I_4 \left(\frac{\partial^3 \phi_x}{\partial x \partial t^2} \right) \delta w_0 - I_5 \left(\frac{\partial^2 \phi_x}{\partial t^2} \right) \delta \phi_x$$

$$- I_0 \left(\frac{\partial^2 v_0}{\partial t^2} \right) \delta v_0 - I_1 \left(\frac{\partial^3 v_0}{\partial y \partial t^2} \right) \delta w_0 - I_3 \left(\frac{\partial^2 v_0}{\partial t^2} \right) \delta \phi_y + I_1 \left(\frac{\partial^3 w_0}{\partial y \partial t^2} \right) \delta v_0 + I_2 \left(\frac{\partial^4 w_0}{\partial y^2 \partial t^2} \right) \delta w_0$$

$$\left. + I_4 \left(\frac{\partial^3 w_0}{\partial y \partial t^2} \right) \delta \phi_y - I_3 \left(\frac{\partial^2 \phi_y}{\partial t^2} \right) \delta v_0 - I_4 \left(\frac{\partial^3 \phi_y}{\partial y \partial t^2} \right) \delta w_0 - I_5 \left(\frac{\partial^2 \phi_y}{\partial t^2} \right) \delta \phi_y - I_0 \left(\frac{\partial^2 w_0}{\partial t^2} \right) \delta w_0 \right] dA$$

$$(2.42)$$

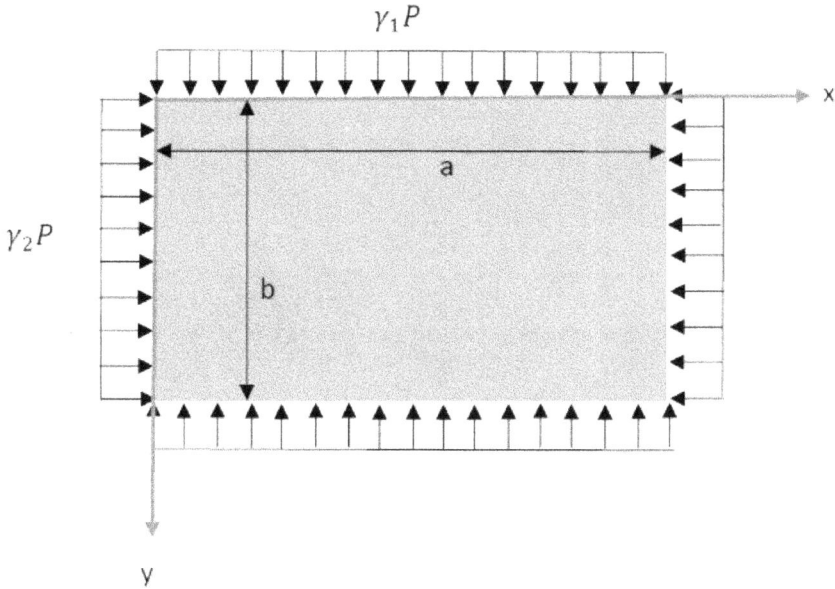

FIGURE 2.5 The rectangular sandwich composite nanoplate that was loaded.

It defines the mass moment of inertia as:

$$I_i = \left(\int_{-h_c/2}^{h_c/2} Y_i \rho_c dz \right) + \left(\int_{-\frac{(h_c+h_f)}{2}}^{-h_c} Y_i \rho_f dz + \int_{\frac{h_c}{2}}^{\frac{(h_c+h_f)}{2}} Y_i \rho_f dz \right)$$

$$Y_i = \left[1, z, z^2, f(z), zf(z), f(z)^2 \right] \dots\dots i = (1,2,3,4,5)$$

(2.43)

2.1.3.2 Equation of Motion

To complete the set of five equilibrium equations, we now plug values for the initial variations of strain energy, work done by applied forces, and kinetic energy into the Euler-Lagrange equation for sinusoidal shear deformation. This is done by inserting Eqs. (2.33) through (2.43) into Eq. (2.32), which yields the following result.

$$\delta u_0 : \frac{\partial N_{xx}}{\partial x} + \frac{\partial N_{xy}}{\partial y} - e_{31} h_c K_c C(t) \frac{\partial^2 w_0}{\partial x \partial t} = I_0 \left(\frac{\partial^2 u_0}{\partial t^2} \right) + I_3 \left(\frac{\partial^2 \phi_x}{\partial t^2} \right) - I_1 \left(\frac{\partial^3 w_0}{\partial x \partial t^2} \right)$$ (2.44)

$$\delta v_0 : \frac{\partial N_{xx}}{\partial y} + \frac{\partial N_{xy}}{\partial x} - e_{32} h_c K_c C(t) \frac{\partial^2 w_0}{\partial y \partial t} = I_0 \left(\frac{\partial^2 v_0}{\partial t^2} \right) + I_3 \left(\frac{\partial^2 \phi_y}{\partial t^2} \right) - I_1 \left(\frac{\partial^3 w_0}{\partial y \partial t^2} \right)$$ (2.45)

$$\delta w_0 : \frac{\partial^2 M_{xx}}{\partial x^2} + \frac{\partial^2 M_{yy}}{\partial y^2} + 2\frac{\partial^2 M_{xy}}{\partial x \partial y} - K_w \cdot w_0 + K_g \left(\frac{\partial^2 w_0}{\partial x^2} + \frac{\partial^2 w_0}{\partial y^2} \right) - N_{xx} \left(\frac{\partial^2 w_0}{\partial y^2} \right) - N_{yy} \left(\frac{\partial^2 w_0}{\partial y^2} \right)$$

$$-e_{31} h_c K_c C(t) \frac{\partial^2 U_x}{\partial x \partial t} - e_{32} h_c K_c C(t) \frac{\partial^2 V_y}{\partial y \partial t} = I_0 \left(\frac{\partial^2 w_0}{\partial t^2} \right) + I_1 \left(\frac{\partial^3 u_0}{\partial x \partial t^2} \right) - I_2 \left(\frac{\partial^4 w_0}{\partial y^2 \partial t^2} \right) - I_2 \left(\frac{\partial^4 w_0}{\partial x^2 \partial t^2} \right)$$

$$I_2 \left(\frac{\partial^4 w_0}{\partial y^2 \partial t^2} \right) + I_4 \left(\frac{\partial^3 \phi_x}{\partial y \partial t^2} \right) + I_4 \left(\frac{\partial^3 \phi_y}{\partial y \partial t^2} \right) \tag{2.46}$$

$$\delta \phi_x : \frac{\partial F_{xx}}{\partial x} + \frac{\partial F_{xy}}{\partial y} - G_{xz} = I_3 \left(\frac{\partial^2 u_0}{\partial t^2} \right) + I_5 \left(\frac{\partial^2 \phi_x}{\partial t^2} \right) - I_4 \left(\frac{\partial^3 w_0}{\partial x \partial t^2} \right) \tag{2.47}$$

$$\delta \phi_y : \frac{\partial F_{yy}}{\partial y} + \frac{\partial F_{xy}}{\partial x} - G_{yz} = I_3 \left(\frac{\partial^2 v_0}{\partial t^2} \right) + I_5 \left(\frac{\partial^2 \phi_y}{\partial t^2} \right) - I_4 \left(\frac{\partial^3 w_0}{\partial y \partial t^2} \right) \tag{2.48}$$

2.1.4 SOLUTION

With Galerkin's technique, the suggested solutions must satisfy the problem's boundary conditions. To proceed, the critical buckling loads will be calculated by replacing the governing equation with the required response (Eq. (2.49)). Galerkin's methodology is a definitive method that can solve partial differential equations with excellent accuracy, convergence, and performance. It has been effectively adjusted to identify various difficulties in buckling analysis.

$$\begin{Bmatrix} u_0(x,y,t) \\ v_0(x,y,t) \\ w_0(x,y,t) \\ \phi_0(x,y,t) \\ \phi_0(x,y,t) \end{Bmatrix} = \begin{Bmatrix} \sum_{m=1}^{\infty}\sum_{n=1}^{\infty} U_{mn} \frac{\partial X_m}{\partial x} Y_n e^{iwt} \\ \sum_{m=1}^{\infty}\sum_{n=1}^{\infty} V_{mn} X_m \frac{\partial Y_n}{\partial y} e^{iwt} \\ \sum_{m=1}^{\infty}\sum_{n=1}^{\infty} W_{mn} X_m Y_n e^{iwt} \\ \sum_{m=1}^{\infty}\sum_{n=1}^{\infty} \phi_{xmn} \frac{\partial X_m}{\partial x} Y_n e^{iwt} \\ \sum_{m=1}^{\infty}\sum_{n=1}^{\infty} \phi_{ymn} X_m \frac{\partial Y_n}{\partial y} e^{iwt} \end{Bmatrix} \tag{2.49}$$

in which U_{mn}, V_{mn}, W_{mn}, ϕ_{xmn}, and ϕ_{ymn} denote the amplitude of the vibration. Moreover, $X_m(x)$ and $Y_n(y)$ are admissible functions that certain boundary requirements, as seen in Table 2.2:

The natural frequencies may be calculated by solving Eq. (2.49), which is obtained by inserting the corresponding terms from Eqs. (2.44) through (2.48), and then isolating the variations in $u_0, v_0, w_0, \phi_x, \phi_y$. The determinant

TABLE 2.2
The Admissible Function $X_m(x)$ and $Y_n(y)$

	Boundary conditions		The function X_m and Y_n	
	At $x = 0, a$	At $y = 0, b$	$X_m(x)$	$Y_n(y)$
SSSS	$X_m(0) = X''_m(0) = 0$	$Y_n(0) = Y''_n(0) = 0$	$\sin(\alpha x)$	$\sin(\beta y)$
	$X_m(a) = X''_m(a) = 0$	$Y_n(b) = Y''_n(b) = 0$		
CCSS	$X_m(0) = X'_m(0) = 0$	$Y_n(0) = Y''_n(0) = 0$	$\sin^2(\alpha x)$	$\sin(\beta y)$
	$X_m(a) = X'_m(a) = 0$	$Y_n(b) = Y''_n(b) = 0$		
CCCC	$X_m(0) = X'_m(0) = 0$	$Y_n(0) = Y'_n(0) = 0$	$\sin^2(\alpha x)$	$\sin^2(\beta y)$
	$X_m(a) = X'_m(0) = 0$	$Y_n(b) = Y'_n(b) = 0$		
CCFF	$X''_m(0) = X'''_m(0) = 0$	$Y_n(0) = Y'_n(0) = 0$	$\cos^2(\alpha x)\left[\sin^2(\alpha x) + 1\right]$	$\sin^2(\beta y)$
	$X''_m(a) = X'''_m(a) = 0$	$Y_n(b) = Y'_n(b) = 0$		

of the bellow equation will be equal to zero in order to capture the natural frequencies.

$$\left\|[K] - \omega^2[M] + i\omega[C]\right\| = 0 \qquad (2.50)$$

Appendix 2A illustrates the relationship between the stiffness, mass, and damping matrices K, M, and C. The natural frequency (ω) must be brought down to zero to achieve the critical buckling stress. ($\omega = 0$).

In this section, we assess the accuracy and usefulness of the recommended system by contrasting the results obtained by the indicated system with the findings obtained by the articles in the relevant body of literature (Table 2.3). The results are first compared to Refs. [32–34]. The hygrothermal state and the magnetostrictive core are disregarded to make the findings more comparable.

Material properties of Al/SiC [32]:

$$E_c = 420\,GPa, \quad E_m = 70\,GPa, \quad \vartheta = 0.3$$

As can be seen, the results converge well, and the amount of error is negligible. It should be mentioned that in the preceding table, critical buckling load is attained according to the following:

$$\bar{P}_{cr} = \frac{12 P_{cr} a^2 \left(1 - \vartheta^2\right)}{E_c h_c^3} \qquad (2.51)$$

TABLE 2.3

Nondimensional Critical Buckling Load Comparison ($\bar{P}_{cr}$) of Al/SiC Under Various Boundary Conditions ($a = b = 10\, h_c$)

Gradient index		$k = 0$		$k = 1$		$k = 2$	
(γ_1, γ_2)	References	CCSS	SSSS	CCSS	SSSS	CCSS	SSSS
(−1,0)	[32] FSDT	63.0039	37.3708	64.8195	37.7132	64.7963	37.7089
	[33] HSDT	63.1628	37.3714	64.9643	37.7172	64.3779	37.5765
	[34]	64.8257	37.3721	66.4778	37.7143	65.9443	37.6042
	Present	63.0668	37.3729	63.6358	37.7101	63.5664	37.669
(−1, −1)	[32] FSDT	33.3206	18.6854	33.9966	18.8566	33.9881	18.8545
	[33] HSDT	33.3392	18.6860	34.0121	18.8571	33.7942	18.8020
	[34]	34.1195	18.6861	34.6939	18.8572	34.5084	18.8021
	Present	36.0382	18.6865	36.3633	18.8551	36.3237	18.8345

2.1.5 RESULTS

Table 2.4 provides further validation to investigate the precision and effectiveness of the proposed method. The findings of the suggested system are contrasted with those of Sobhy [35], Barati et al. [36], and Karimi et al. [37]. A functionally graded nanoplate (Al/Al2O$_3$) is taken into consideration in their study. The suggested approach is effective, as seen in Table 2.4.

In Table 2.5, the non-dimensional buckling load is determined for various values of aspect ratio (a/b) and side-to-thickness ratio (a/h_c) ratios for various values of porosity volume. As can be seen in Table 2.5, an increase in the aspect ratio (a/b) results in a rise in the non-dimensional buckling stress. Increasing the side-to-thickness ratio (a/h_c) also results in an increase in the non-dimensional buckling load. As the value of h_c is increased, the amount of buckling load experienced by the structure is reduced as a result of the influence of velocity feedback gain.

The estimation of the non-dimensional buckling load for simply-clamped and simply-simply boundary conditions is detailed in Table 2.6. According to the data in this table, the buckling load in the SSCC boundary conditions is much higher than that in the SSSS boundary conditions. This is because increased support in boundary conditions tends to make the nanoplate stiffer due to the enhancement of the nanoplate structure. Also, the values of dimensionless buckling load are presented in this table for various values of a nonlocal parameter called Eringen. As previously mentioned, the dimensionless buckling load reduces when the nonlocal parameter is increased. This phenomenon occurs mainly as a consequence of the fact that the nanoplate gets softer when the nonlocal parameter is increased. It is possible that raising the value of the small-scale parameter causes

TABLE 2.4

Comparison of Non-dimensional Buckling Load ($\bar{P}_{cr}$) of Based Al/Al$_2$O$_3$ FG Nanoplate ($a/h = 10$)

k	$\mu = 0\,nm^2$				$\mu = 2\,nm^2$			
0	Sobhy [35]	Barati et al. [36]	Karimi et al. [37]	Present	Sobhy [35]	Barati et al. [36]	Karimi et al. [37]	Present
0.5	18.6876	18.7054	18.68606	18.6865	10.4425	10.4525	10.44166	10.4419
2.5	10.0638	10.0719	10.06316	10.0319	5.6235	5.6281	5.6323	5.63782
5.5	6.2593	6.26108	6.26057	6.26031	3.4976	3.49866	3.49837	3.49822
	5.52	5.52361	5.52039	5.52556	3.0845	3.08656	3.08476	3.08476

TABLE 2.5

Nondimensional Buckling Load for Various Aspect Ratios, SSCC Boundary Conditions, and Even Porosity Distribution ($\mu^2 = 0\,nm^2$, $K_w = K_g = 0$, $k = 0$, $a = 40$ nm, $h_f = 0.2$ nm, $\gamma_1 = -1, \gamma_2 = 0$)

a / b	Porosity	a / h_c			
		5	10	20	30
0.5	0.1	49.8101	85.9312	168.643	266.355
1		36.9724	63.784	125.178	197.707
0.5	0.2	47.7433	81.591	159.097	250.659
1		35.4383	60.5624	118.093	186.056
0.5	0.3	45.6764	77.2508	149.551	234.964
1		33.9041	57.3408	111.007	174.406
0.5	0.4	43.6096	72.9106	140.005	219.268
1		32.37	54.1192	103.921	162.756
0.5	0.5	41.5427	68.5704	130.459	203.573
1		30.8359	50.8977	96.8359	151.105
0.5	0.6	39.4759	64.2303	120.914	187.877
1		29.3017	47.6761	89.75.3	139.455

the system's energy to decrease, which, in turn, causes the system to become less robust.

Because of this, it is plausible to assume that the local system has larger dimensionless buckling loads than the nonlocal system. In other words, the

TABLE 2.6

Nonlocal Buckling Load for SSSS and SSCC Boundary Conditions and Various Values of the Nonlocal Parameter

Nonlocal parameter	(γ_1, γ_2)	SSCC	SSSS
$\mu^2 = 0\ nm^2$	$(-1, 0)$	74.1982	43.9693
$\mu^2 = 1\ nm^2$		73.2939	43.4334
$\mu^2 = 2\ nm^2$		72.4115	42.9105
$\mu^2 = 3\ nm^2$		71.5500	42.4
$\mu^2 = 4\ nm^2$		70.7088	41.9015
$\mu^2 = 0\ nm^2$	$(-1, -1)$	37.0991	21.9846
$\mu^2 = 1\ nm^2$		36.647	21.7167
$\mu^2 = 2\ nm^2$		36.2057	21.4533
$\mu^2 = 3\ nm^2$		35.775	21.02
$\mu^2 = 4\ nm^2$		35.3544	20.9508

overall stiffness is greater than its classical equivalent because the stiffness caused by the pair stress effect is added to the classical stiffness. Conversely, the dimensionless buckling load for uniaxial and biaxial compressive stress is examined in Table 2.6.

Table 2.7 has been created to compare the non-dimensional buckling load for various external loads and uniform and uneven porosity distribution. As can be observed in Table 2.7, the non-dimensional buckling loads are much higher when the distribution is uneven. This is primarily due to the fact that, as a result of the unequal distribution, there are fewer cavities in the nanoplate, which results in the nanoplate having a greater degree of stiffness. Moreover, the Winkler and Pasternak foundation value in Table 2.7 is expected to remain constant for $K_w = 0$ and $Kg = 0$.

On the other hand, to examine the influence of the gradient index on the dimensionless buckling of the nanoplate, Tables 2.8 and 2.9 have been generated for equal and uneven porosity distributions, respectively. These tables are examined for the uniaxial compressive stress, even porosity distribution, and SSCC boundary conditions. Moreover, the velocity feedback gain value is taken into consideration for two distinct values. By analyzing these numbers, we may conclude that a decrease in the dimensionless buckling load occurs as the gradient index increases. The higher Young's modulus of aluminum oxide compared to aluminum may provide a physical explanation for this trend. In addition, as

TABLE 2.7

Nondimensional Buckling Load Comparison Between SSCC and SSSS Boundary Conditions ($\mu^2 = 0\,nm^2$, $\xi = 0.3$, $K_w = K_g = 0$, k = 0, hc = 3 nm, $h_f = 0.2$ nm, $K_c C(t) = 1 \times 10^5$)

Porosity	(γ_1, γ_2)	SSSS	SSCC
Even	(−1, 0)	43.9693	**74.1982**
	(−1, −1)	21.9846	**37.0991**
Uneven	(−1, 0)	58.8503	**99.3098**
	(−1, −1)	29.4251	**49.6549**

TABLE 2.8

Dimensional Buckling Load for Square Nanoplate for Different Gradient Index Values and Even Porosity Distribution Under Simply Boundary Conditions ($\xi = 0.1$, $\mu^2 = 0\,nm^2$, $K_w = K_g = 0$, hc = 3 nm, $h_f = 0.2$ nm, $\gamma_1 = -1, \gamma_2 = 0$)

Non-dimensional buckling load	Velocity feedback gain	k = 0	k = 0.5	k = 1	k = 1.5	k = ∞
$\bar{P}_{cr}$	$k_c C(t) = 1 \times 10^5$	83.0688	66.9483	58.5790	53.4569	32.5308
$\bar{P}_{cr}$	$k_c C(t) = 5 \times 10^5$	19.6879	15.4474	13.3291	12.0601	7.0853

TABLE 2.9

Dimensional Buckling Load for Square Nanoplate for Different Gradient Index Values and Uneven Porosity Distribution Under Simple Boundary Conditions ($\xi = 0.1$, $\mu^2 = 0\,nm^2$, $K_w = K_g = 0$, hc = 3 nm, $h_f = 0.2$ nm, $\gamma_1 = -1, \gamma_2 = 0$)

Non-dimensional buckling load	Velocity feedback gain	k = 0	k = 0.5	k = 1	k = 1.5	k = ∞
$\bar{P}_{cr}$	$k_c C(t) = 1 \times 10^5$	54.1863	42.5932	36.5621	32.8658	17.7099
$\bar{P}_{cr}$	$k_c C(t) = 5 \times 10^5$	12.6798	9.6906	8.1962	7.3004	3.7826

shown in Table 2.8, raising the velocity feedback gain reduces the non-dimensional buckling stress.

Figure 2.6a,b has been modeled in order to investigate how the magnitude of the non-dimensional buckling load is affected by the thickness of the magnetostrictive layer. The Winkler and Pasternak foundation values in these figures are both meant to be zero ($K_w = K_g = 0$). In addition to that, the value of the gradient index must be zero. For six distinct values of the nonlocal parameter, the non-dimensional buckling load is presented for an even and uneven distribution in these figures. These figures demonstrate that, as a result of the combined effects of mass and inertia, increasing the thickness of the magnetostrictive layer results in a reduction in the non-dimensional buckling load. The non-dimensional buckling stress may also be reduced by raising the nonlocal parameter.

To keep an eye on how the thickness of the facesheets affects the non-dimensional buckling loads, the scheme in Figure 2.7a,b uses a constant value of the nonlocal parameter ($\mu^2 = 0$ nm^2). This was done so that the authors could follow this impact. Figure 2.7a,b shows that increasing the thickness of the facesheets results in an increase in the dimensionless buckling load. It should be noted that Figure 2.7a,b is presented for the uniaxial load and simply clamped boundary conditions. The thickness of the nanoplate grows as the value of h_f rises, which, in turn, raises the nanoplate's stiffness, resulting in a larger buckling load.

For the SSSS and SSCC boundary conditions, Figure 2.8a,b was created to differentiate between the impact of the porosity volume on the non-dimensional buckling load. It is clear from Figure 2.8a,b that increasing the porosity volume results in a reduction in the dimensionless buckling load.

Figure 2.9a,b is the final diagram that has been produced to differentiate the impact of the Pasternak foundation on the non-dimensional buckling load. It should be mentioned that Figure 2.9a,b is presented for a constant value of the gradient index ($k = 0$) and a nano-square plane ($a = b = 40$ nm). Both figures that make up Figure 2.9a,b take into account six distinct values for the nonlocal parameters. The dimensionless buckling load is seen to grow when the Winkler foundation parameter is increased in the following figures. This phenomenon arises as a consequence of the fact that an increased quantity of Winkler foundation causes the nanoplate to become more rigid. As a consequence of this, a greater force is necessary for the nanoplate to buckle.

2.1.6 CONCLUDING REMARKS

Employing the higher-order sinusoidal shear deformation plate theory to examine the buckling behavior of a composite magnetostrictive nanoplate integrated with functionally graded facesheets. In addition, Hamilton's principle is used to derive the governing equations, and Galerkin's analytical solution is employed to solve

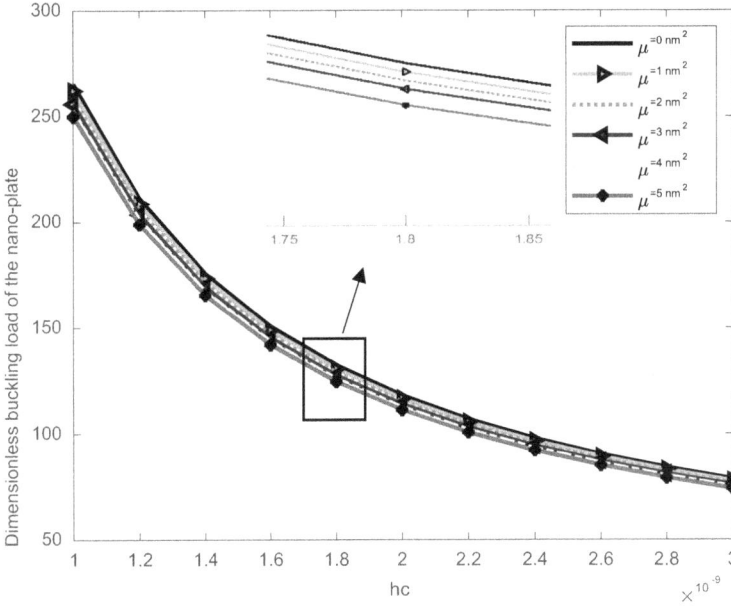

(a) Even distribution

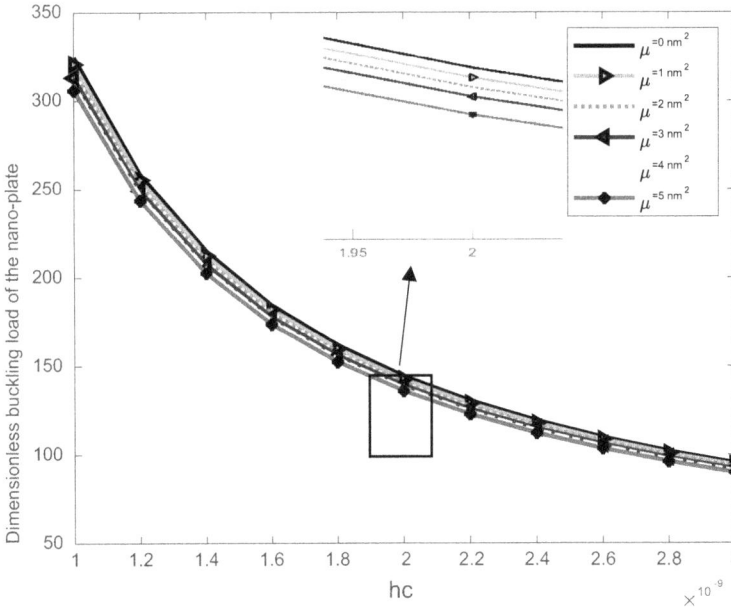

(b) Uneven distribution

FIGURE 2.6 Non-dimensional buckling loads for magnetostrictive materials and their relationship to magnetostrictive layer thickness: (a) Even porosity and (b) uneven porosity.

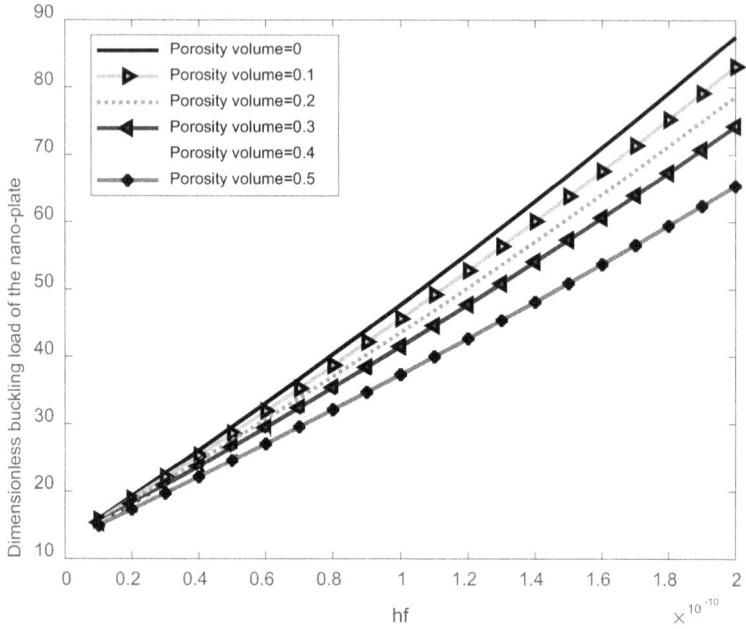

(a) Even distribution

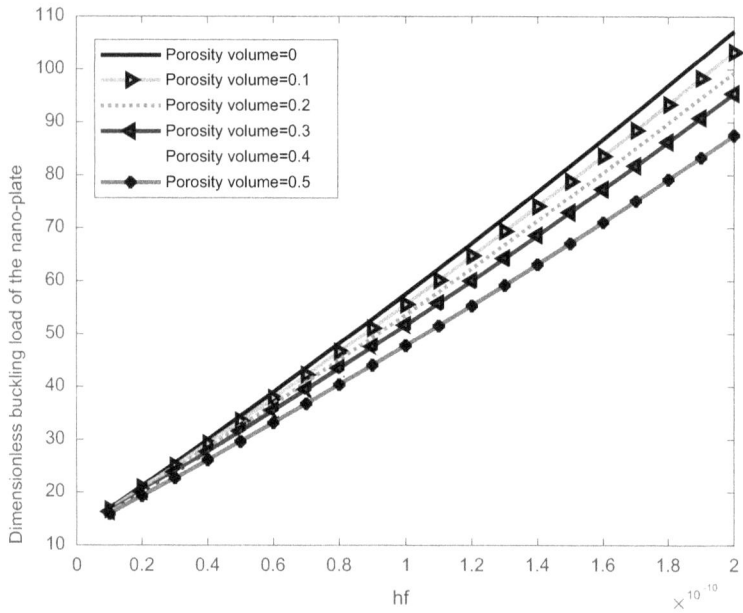

(b) Uneven distribution

FIGURE 2.7　Changes in non-dimensional buckling loads due to facesheet thickness for (a) even porosity and (b) uneven porosity.

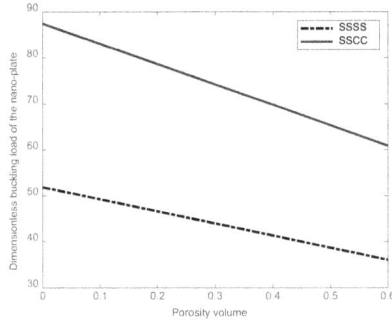

(a) Even distribution

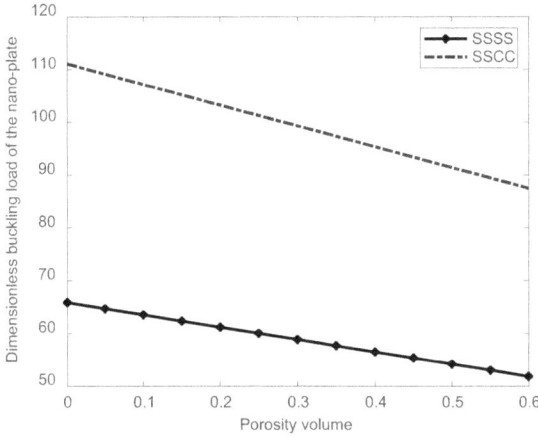

(b) Uneven distribution

FIGURE 2.8 Non-dimensional buckling loads and their relationship to the porosity volume parameter for (a) even porosity and (b) uneven porosity.

the equations. It has been shown that the dimensional buckling load for boundary conditions results in a greater amount of trapped strain than the dimensionless buckling load for simple boundary circumstances. Increasing the thickness of the magnetostrictive layer also reduces the dimensionless buckling load. In addition, it has been discovered that the dimensional buckling loads for simply clamped boundary conditions are much higher than the dimensionless buckling loads for simply-supported boundary conditions.

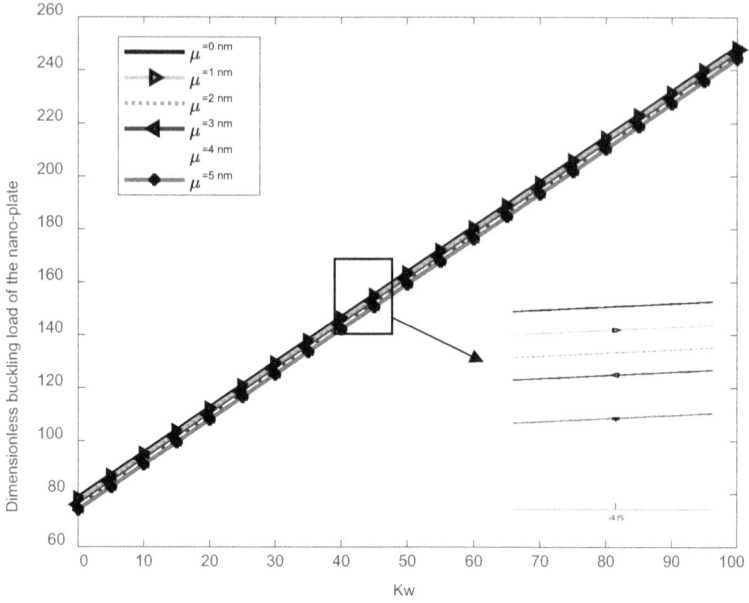

(a) Even distribution

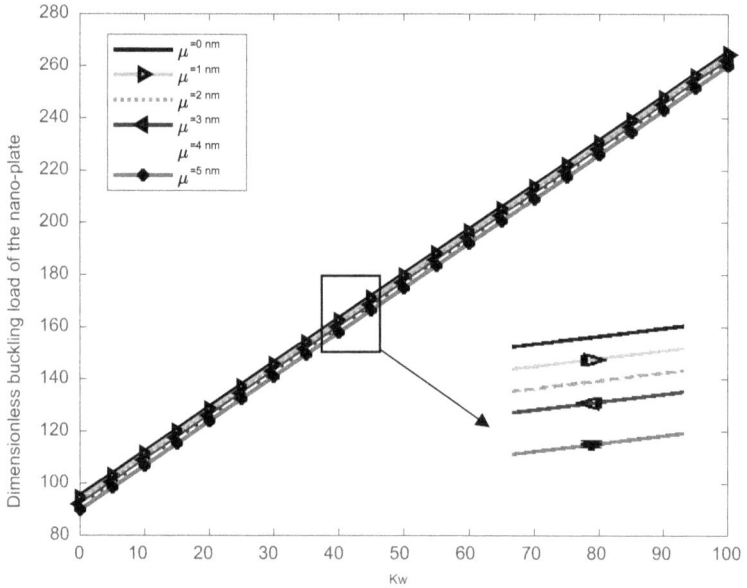

(b) Uneven distribution

FIGURE 2.9   Influence of the kind of foundation media on the buckling loads that are not dimensioned: (a) Even porosity, (b) uneven porosity.

Appendix 2A

$$K_{11} = A_{11}\left(\int_0^b\int_0^a\left(\frac{\partial^3 X_m}{\partial x^3}Y_n\frac{\partial X_m}{\partial x}Y_n\right)dxdy\right) + A_{66}\left(\int_0^b\int_0^a\left(\frac{\partial X_m}{\partial x}Y_n\frac{\partial X_m}{\partial x}\frac{\partial^2 Y_n}{\partial y^2}\right)dxdy\right)$$

$$K_{12} = A_{12}\left(\int_0^b\int_0^a\left(\frac{\partial X_m}{\partial x}\frac{\partial Y_n}{\partial y}X_m\frac{\partial^2 Y_n}{\partial y^2}\right)dxdy\right) + A_{66}\left(\int_0^b\int_0^a\left(\frac{\partial X_m}{\partial x}\frac{\partial Y_n}{\partial y}X_m\frac{\partial^2 Y_n}{\partial y^2}\right)dxdy\right)$$

$$K_{13} = -B_{11}\left(\int_0^b\int_0^a\left(\frac{\partial^3 X_m}{\partial x^3}Y_n X_m Y_n\right)dxdy\right) - B_{12}\left(\int_0^b\int_0^a\left(\frac{\partial X_m}{\partial x}Y_n X_m\frac{\partial^2 Y_n}{\partial y^2}\right)dxdy\right)$$
$$- B_{66}\left(\int_0^b\int_0^a\left(\frac{\partial X_m}{\partial x}Y_n X_m\frac{\partial^2 Y_n}{\partial y^2}\right)dxdy\right)$$

$$K_{14} = S_{11}\left(\int_0^b\int_0^a\left(\frac{\partial^3 X_m}{\partial x^3}Y_n\frac{\partial X_m}{\partial x}Y_n\right)dxdy\right) + S_{66}\left(\int_0^b\int_0^a\left(\frac{\partial X_m}{\partial x}Y_n\frac{\partial X_m}{\partial x}\frac{\partial^2 Y_n}{\partial y^2}\right)dxdy\right)$$

$$K_{15} = S_{12}\left(\int_0^b\int_0^a\left(\frac{\partial X_m}{\partial x}\frac{\partial Y_n}{\partial y}X_m\frac{\partial^2 Y_n}{\partial y^2}\right)dxdy\right) + S_{66}\left(\int_0^b\int_0^a\left(\frac{\partial X_m}{\partial x}\frac{\partial Y_n}{\partial y}X_m\frac{\partial^2 Y_n}{\partial y^2}\right)dxdy\right)$$

$$K_{21} = A_{12}\left(\int_0^b\int_0^a\left(\frac{\partial^2 X_m}{\partial x^2}Y_n\frac{\partial X_m}{\partial x}\frac{\partial Y_n}{\partial y}\right)dxdy\right) + A_{66}\left(\int_0^b\int_0^a\left(\frac{\partial^2 X_m}{\partial x^2}Y_n\frac{\partial X_m}{\partial x}\frac{\partial Y_n}{\partial y}\right)dxdy\right)$$

$$K_{22} = A_{22}\left(\int_0^b\int_0^a\left(X_m\frac{\partial Y_n}{\partial y}X_m\frac{\partial^3 Y_n}{\partial y^3}\right)dxdy\right) + A_{66}\left(\int_0^b\int_0^a\left(\frac{\partial^2 X_m}{\partial x^2}\frac{\partial Y_n}{\partial y}X_m\frac{\partial Y_n}{\partial y}\right)dxdy\right)$$

$$K_{23} = -B_{12}\left(\int_0^b\int_0^a\left(\frac{\partial^2 X_m}{\partial x^2}Y_n X_m\frac{\partial Y_n}{\partial y}\right)dxdy\right) - B_{22}\left(\int_0^b\int_0^a\left(X_m Y_n X_m\frac{\partial^3 Y_n}{\partial y^3}\right)dxdy\right)$$
$$- 2B_{66}\left(\int_0^b\int_0^a\left(\frac{\partial^2 X_m}{\partial x^2}Y_n X_m\frac{\partial Y_n}{\partial y}\right)dxdy\right)$$

$$K_{24} = S_{12}\left(\int_0^b\int_0^a\left(\frac{\partial^2 X_m}{\partial x^2}Y_n\frac{\partial X_m}{\partial x}\frac{\partial Y_n}{\partial y}\right)dxdy\right) + S_{66}\left(\int_0^b\int_0^a\left(\frac{\partial^2 X_m}{\partial x^2}Y_n\frac{\partial X_m}{\partial x}\frac{\partial Y_n}{\partial y}\right)dxdy\right)$$

$$K_{25} = S_{22}\left(\int_0^b\int_0^a\left(X_m\frac{\partial Y_n}{\partial y}X_m\frac{\partial^3 Y_n}{\partial y^3}\right)dxdy\right) + S_{66}\left(\int_0^b\int_0^a\left(\frac{\partial^2 X_m}{\partial x^2}\frac{\partial Y_n}{\partial y}X_m\frac{\partial Y_n}{\partial y}\right)dxdy\right)$$

$$K_{31} = B_{11} \left(\int_0^b \int_0^a \left(\frac{\partial^4 X_m}{\partial x^4} Y_n \frac{\partial X_m}{\partial x} Y_n \right) dxdy \right) + B_{12} \left(\int_0^b \int_0^a \left(\frac{\partial^2 X_m}{\partial x^2} Y_n \frac{\partial X_m}{\partial x} \frac{\partial^2 Y_n}{\partial y^2} \right) dxdy \right)$$

$$+ 2B_{66} \left(\int_0^b \int_0^a \left(\frac{\partial^2 X_m}{\partial x^2} Y_n \frac{\partial X_m}{\partial x} \frac{\partial^2 Y_n}{\partial y^2} \right) dxdy \right)$$

$$K_{32} = B_{12} \left(\int_0^b \int_0^a \left(\frac{\partial^2 X_m}{\partial x^2} \frac{\partial Y_n}{\partial y} X_m \frac{\partial^2 Y_n}{\partial y^2} \right) dxdy \right) + B_{22} \left(\int_0^b \int_0^a \left(X_m \frac{\partial Y_n}{\partial y} X_m \frac{\partial^4 Y_n}{\partial y^4} \right) dxdy \right)$$

$$+ 2B_{66} \left(\int_0^b \int_0^a \left(\frac{\partial^2 X_m}{\partial x^2} \frac{\partial Y_n}{\partial y} X_m \frac{\partial^2 Y_n}{\partial y^2} \right) dxdy \right)$$

$$K_{33} = -D_{11} \left(\int_0^b \int_0^a \left(\frac{\partial^4 X_m}{\partial x^4} Y_n X_m Y_n \right) dxdy \right) - 2D_{12} \left(\int_0^b \int_0^a \left(\frac{\partial^2 X_m}{\partial x^2} Y_n X_m \frac{\partial^2 Y_n}{\partial y^2} \right) dxdy \right)$$

$$- D_{22} \left(\int_0^b \int_0^a \left(X_m Y_n X_m \frac{\partial^4 Y_n}{\partial y^4} \right) dxdy \right) - 4D_{66} \left(\int_0^b \int_0^a \left(\frac{\partial^2 X_m}{\partial x^2} Y_n X_m \frac{\partial^2 Y_n}{\partial y^2} \right) dxdy \right)$$

$$K_{34} = M_{11} \left(\int_0^b \int_0^a \left(\frac{\partial^4 X_m}{\partial x^4} Y_n \frac{\partial X_m}{\partial x} Y_n \right) dxdy \right) + M_{12} \left(\int_0^b \int_0^a \left(\frac{\partial^2 X_m}{\partial x^2} Y_n \frac{\partial X_m}{\partial x} \frac{\partial^2 Y_n}{\partial y^2} \right) dxdy \right)$$

$$+ 2M_{66} \left(\int_0^b \int_0^a \left(\frac{\partial^4 X_m}{\partial x^4} Y_n \frac{\partial X_m}{\partial x} \frac{\partial^2 Y_n}{\partial y^2} \right) dxdy \right)$$

$$K_{35} = M_{12} \left(\int_0^b \int_0^a \left(\frac{\partial^2 X_m}{\partial x^2} \frac{\partial Y_n}{\partial y} X_m \frac{\partial^2 Y_n}{\partial y^2} \right) dxdy \right) + M_{22} \left(\int_0^b \int_0^a \left(X_m \frac{\partial Y_n}{\partial y} X_m \frac{\partial^4 Y_n}{\partial y^4} \right) dxdy \right)$$

$$+ 2M_{66} \left(\int_0^b \int_0^a \left(\frac{\partial^2 X_m}{\partial x^2} \frac{\partial Y_n}{\partial y} X_m \frac{\partial^2 Y_n}{\partial y^2} \right) dxdy \right)$$

$$K_{41} = S_{11} \left(\int_0^b \int_0^a \left(\frac{\partial^3 X_m}{\partial x^3} Y_n \frac{\partial X_m}{\partial x} Y_n \right) dxdy \right) + S_{66} \left(\int_0^b \int_0^a \left(\frac{\partial X_m}{\partial x} Y_n \frac{\partial X_m}{\partial x} \frac{\partial^2 Y_n}{\partial y^2} \right) dxdy \right)$$

$$K_{42} = S_{12} \left(\int_0^b \int_0^a \left(\frac{\partial X_m}{\partial x} \frac{\partial Y_n}{\partial y} X_m \frac{\partial^2 Y_n}{\partial y^2} \right) dxdy \right) + S_{66} \left(\int_0^b \int_0^a \left(\frac{\partial X_m}{\partial x} \frac{\partial Y_n}{\partial y} X_m \frac{\partial^2 Y_n}{\partial y^2} \right) dxdy \right)$$

$$K_{43} = -M_{11}\left(\int_0^b\int_0^a\left(\frac{\partial^3 X_m}{\partial x^3}Y_n X_m Y_n\right)dxdy\right) - M_{12}\left(\int_0^b\int_0^a\left(\frac{\partial X_m}{\partial x}Y_n X_m\frac{\partial^2 Y_n}{\partial y^2}\right)dxdy\right)$$

$$- 2M_{66}\left(\int_0^b\int_0^a\left(\frac{\partial X_m}{\partial x}Y_n X_m\frac{\partial^2 Y_n}{\partial y^2}\right)dxdy\right)$$

$$K_{44} = O_{11}\left(\int_0^b\int_0^a\left(\frac{\partial^3 X_m}{\partial x^3}Y_n\frac{\partial X_m}{\partial x}Y_n\right)dxdy\right) + O_{66}\left(\int_0^b\int_0^a\left(\frac{\partial X_m}{\partial x}Y_n\frac{\partial X_m}{\partial x}\frac{\partial^2 Y_n}{\partial y^2}\right)dxdy\right)$$

$$- L_{55}\left(\int_0^b\int_0^a\left(\frac{\partial X_m}{\partial x}Y_n\right)dxdy\right)$$

$$K_{45} = O_{12}\left(\int_0^b\int_0^a\left(\frac{\partial X_m}{\partial x}\frac{\partial Y_n}{\partial y}X_m\frac{\partial^2 Y_n}{\partial y^2}\right)dxdy\right) + O_{66}\left(\int_0^b\int_0^a\left(\frac{\partial X_m}{\partial x}\frac{\partial Y_n}{\partial y}X_m\frac{\partial^2 Y_n}{\partial y^2}\right)dxdy\right)$$

$$K_{51} = S_{12}\left(\int_0^b\int_0^a\left(\frac{\partial^2 X_m}{\partial x^2}Y_n\frac{\partial X_m}{\partial x}\frac{\partial Y_n}{\partial y}\right)dxdy\right) + S_{66}\left(\int_0^b\int_0^a\left(\frac{\partial^2 X_m}{\partial x^2}Y_n\frac{\partial X_m}{\partial x}\frac{\partial Y_n}{\partial y}\right)dxdy\right)$$

$$K_{52} = S_{22}\left(\int_0^b\int_0^a\left(X_m\frac{\partial Y_n}{\partial y}X_m\frac{\partial^3 Y_n}{\partial y^3}\right)dxdy\right) + S_{66}\left(\int_0^b\int_0^a\left(\frac{\partial^2 X_m}{\partial x^2}\frac{\partial Y_n}{\partial y}X_m\frac{\partial Y_n}{\partial y}\right)dxdy\right)$$

$$K_{53} = -M_{12}\left(\int_0^b\int_0^a\left(\frac{\partial^2 X_m}{\partial x^2}Y_n X_m\frac{\partial Y_n}{\partial y}\right)dxdy\right) - M_{22}\left(\int_0^b\int_0^a\left(X_m Y_n X_m\frac{\partial^3 Y_n}{\partial y^3}\right)dxdy\right)$$

$$- 2M_{66}\left(\int_0^b\int_0^a\left(\frac{\partial^2 X_m}{\partial x^2}Y_n X_m\frac{\partial Y_n}{\partial y}\right)dxdy\right)$$

$$K_{54} = O_{12}\left(\int_0^b\int_0^a\left(\frac{\partial^2 X_m}{\partial x^2}Y_n\frac{\partial X_m}{\partial x}\frac{\partial Y_n}{\partial y}\right)dxdy\right) + O_{66}\left(\int_0^b\int_0^a\left(\frac{\partial^2 X_m}{\partial x^2}Y_n\frac{\partial X_m}{\partial x}\frac{\partial Y_n}{\partial y}\right)dxdy\right)$$

$$K_{55} = O_{22}\left(\int_0^b\int_0^a\left(X_m\frac{\partial Y_n}{\partial y}X_m\frac{\partial^3 Y_n}{\partial y^3}\right)dxdy\right) + O_{66}\left(\int_0^b\int_0^a\left(\frac{\partial^2 X_m}{\partial x^2}\frac{\partial Y_n}{\partial y}X_m\frac{\partial Y_n}{\partial y}\right)dxdy\right)$$

$$- L_{44}\left(\int_0^b\int_0^a\left(X_m\frac{\partial Y_n}{\partial y}\right)dxdy\right)$$

$$C_{11} = C_{12} = C_{14} = C_{15} = 0 \quad C_{13} = \int_{-\frac{h_c}{2}}^{\frac{h_c}{2}} e_{31}(\alpha)K_c C(t)dz$$

$$C_{21} = C_{22} = C_{24} = C_{25} = 0 \quad C_{23} = \int_{-\frac{h_c}{2}}^{\frac{h_c}{2}} e_{32}(\beta) K_c C(t) dz$$

$$C_{31} = \int_{-\frac{h_c}{2}}^{\frac{h_c}{2}} e_{31}(\alpha) K_c C(t) dz \quad C_{32} = \int_{-\frac{h_c}{2}}^{\frac{h_c}{2}} e_{31}(\beta) K_c C(t) dz$$

$$C_{33} = c_d \left(1 + \mu^2 \alpha^2 + \mu^2 \beta^2\right) \quad C_{34} = \int_{-\frac{h_c}{2}}^{\frac{h_c}{2}} z e_{31}(\alpha) K_c C(t) dz$$

$$C_{35} = \int_{-\frac{h_c}{2}}^{\frac{h_c}{2}} z e_{32}(\beta) K_c C(t) dz$$

$$C_{41} = C_{42} = C_{44} = C_{45} = 0 \quad C_{43} = \int_{-\frac{h_c}{2}}^{\frac{h_c}{2}} z e_{31}(\alpha) K_c C(t) dz$$

$$C_{51} = C_{52} = C_{54} = C_{55} = 0 \quad C_{53} = \int_{-\frac{h_c}{2}}^{\frac{h_c}{2}} z e_{32}(\beta) K_c C(t) dz$$

$$A_{11} = \int_{-h_c/2}^{h_c/2} C_{11}^c dz + \int_{-\frac{(hc+h_f)}{2}}^{-\frac{hc}{2}} C_{11}^f dz + \int_{\frac{h_c}{2}}^{\frac{(h_c+h_f)}{2}} C_{11}^f dz$$

$$A_{12} = \int_{-h_c/2}^{h_c/2} C_{12}^c dz + \int_{-\frac{(hc+h_f)}{2}}^{-\frac{hc}{2}} C_{12}^f dz + \int_{\frac{h_c}{2}}^{\frac{(h_c+h_f)}{2}} C_{12}^f dz$$

$$A_{22} = \int_{-h_c/2}^{h_c/2} C_{22}^c dz + \int_{-\frac{(hc+h_f)}{2}}^{-\frac{hc}{2}} C_{22}^f dz + \int_{\frac{h_c}{2}}^{\frac{(h_c+h_f)}{2}} C_{22}^f dz$$

$$A_{44} = \int_{-h_c/2}^{h_c/2} C_{44}^c dz + \int_{-\frac{(hc+h_f)}{2}}^{-\frac{hc}{2}} C_{44}^f dz + \int_{\frac{h_c}{2}}^{\frac{(h_c+h_f)}{2}} C_{44}^f dz$$

$$A_{55} = \int_{-h_c/2}^{h_c/2} C_{55}^c dz + \int_{-\frac{(hc+h_f)}{2}}^{-\frac{hc}{2}} C_{55}^f dz + \int_{\frac{h_c}{2}}^{\frac{(h_c+h_f)}{2}} C_{55}^f dz$$

$$A_{66} = \int_{-h_c/2}^{h_c/2} C_{66}^c dz + \int_{-\frac{(hc+h_f)}{2}}^{-\frac{hc}{2}} C_{66}^f dz + \int_{\frac{h_c}{2}}^{\frac{(h_c+h_f)}{2}} C_{66}^f dz$$

$$B_{11} = \int_{-h_c/2}^{h_c/2} z C_{11}^c dz + \int_{-\frac{(hc+h_f)}{2}}^{\frac{-hc}{2}} z C_{11}^f dz + \int_{\frac{h_c}{2}}^{\frac{(h_c+h_f)}{2}} z C_{11}^f dz$$

$$B_{12} = \int_{-h_c/2}^{h_c/2} z C_{12}^c dz + \int_{-\frac{(hc+h_f)}{2}}^{\frac{-hc}{2}} z C_{12}^f dz + \int_{\frac{h_c}{2}}^{\frac{(h_c+h_f)}{2}} z C_{12}^f dz$$

$$B_{22} = \int_{-h_c/2}^{h_c/2} z C_{22}^c dz + \int_{-\frac{(hc+h_f)}{2}}^{\frac{-hc}{2}} z C_{22}^f dz + \int_{\frac{h_c}{2}}^{\frac{(h_c+h_f)}{2}} z C_{22}^f dz$$

$$B_{44} = \int_{-h_c/2}^{h_c/2} z C_{44}^c dz + \int_{-\frac{(hc+h_f)}{2}}^{\frac{-hc}{2}} z C_{44}^f dz + \int_{\frac{h_c}{2}}^{\frac{(h_c+h_f)}{2}} z C_{44}^f dz$$

$$B_{55} = \int_{-h_c/2}^{h_c/2} z C_{55}^c dz + \int_{-\frac{(hc+h_f)}{2}}^{\frac{-hc}{2}} z C_{22}^f dz + \int_{\frac{h_c}{2}}^{\frac{(h_c+h_f)}{2}} z C_{55}^f dz$$

$$B_{66} = \int_{-h_c/2}^{h_c/2} z C_{66}^c dz + \int_{-\frac{(hc+h_f)}{2}}^{\frac{-hc}{2}} z C_{66}^f dz + \int_{\frac{h_c}{2}}^{\frac{(h_c+h_f)}{2}} z C_{66}^f dz$$

$$D_{11} = \int_{-h_c/2}^{h_c/2} z^2 C_{11}^c dz + \int_{-\frac{(hc+h_f)}{2}}^{\frac{-hc}{2}} z^2 C_{11}^f dz + \int_{\frac{h_c}{2}}^{\frac{(h_c+h_f)}{2}} z^2 C_{11}^f dz$$

$$D_{12} = \int_{-h_c/2}^{h_c/2} z^2 C_{12}^c dz + \int_{-\frac{(hc+h_f)}{2}}^{\frac{-hc}{2}} z^2 C_{12}^f dz + \int_{\frac{h_c}{2}}^{\frac{(h_c+h_f)}{2}} z^2 C_{12}^f dz$$

$$D_{22} = \int_{-h_c/2}^{h_c/2} z^2 C_{22}^c dz + \int_{-\frac{(hc+h_f)}{2}}^{\frac{-hc}{2}} z^2 C_{22}^f dz + \int_{\frac{h_c}{2}}^{\frac{(h_c+h_f)}{2}} z^2 C_{22}^f dz$$

$$D_{44} = \int_{-h_c/2}^{h_c/2} z^2 C_{44}^c dz + \int_{-\frac{(hc+h_f)}{2}}^{\frac{-hc}{2}} z^2 C_{44}^f dz + \int_{\frac{h_c}{2}}^{\frac{(h_c+h_f)}{2}} z^2 C_{44}^f dz$$

$$D_{55} = \int_{-h_c/2}^{h_c/2} z^2 C_{55}^c dz + \int_{-\frac{(hc+h_f)}{2}}^{\frac{-hc}{2}} z^2 C_{55}^f dz + \int_{\frac{h_c}{2}}^{\frac{(h_c+h_f)}{2}} z^2 C_{55}^f dz$$

$$D_{66} = \int_{-h_c/2}^{h_c/2} z^2 C_{66}^c dz + \int_{-\frac{(hc+h_f)}{2}}^{\frac{-hc}{2}} z^2 C_{66}^f dz + \int_{\frac{h_c}{2}}^{\frac{(h_c+h_f)}{2}} z^2 C_{66}^f dz$$

$$S_{11} = \int_{-h_c/2}^{h_c/2} f(z)C_{11}^c dz + \int_{-\frac{(hc+h_f)}{2}}^{\frac{-hc}{2}} f(z)C_{11}^f dz + \int_{\frac{h_c}{2}}^{\frac{(h_c+h_f)}{2}} f(z)C_{11}^f dz$$

$$S_{12} = \int_{-h_c/2}^{h_c/2} f(z)C_{12}^c dz + \int_{-\frac{(hc+h_f)}{2}}^{\frac{-hc}{2}} f(z)C_{12}^f dz + \int_{\frac{h_c}{2}}^{\frac{(h_c+h_f)}{2}} f(z)C_{12}^f dz$$

$$S_{22} = \int_{-h_c/2}^{h_c/2} f(z)C_{22}^c dz + \int_{-\frac{(hc+h_f)}{2}}^{\frac{-hc}{2}} f(z)C_{22}^f dz + \int_{\frac{h_c}{2}}^{\frac{(h_c+h_f)}{2}} f(z)C_{22}^f dz$$

$$S_{44} = \int_{-h_c/2}^{h_c/2} f(z)C_{44}^c dz + \int_{-\frac{(hc+h_f)}{2}}^{\frac{-hc}{2}} f(z)C_{44}^f dz + \int_{\frac{h_c}{2}}^{\frac{(h_c+h_f)}{2}} f(z)C_{44}^f dz$$

$$S_{55} = \int_{-h_c/2}^{h_c/2} f(z)C_{55}^c dz + \int_{-\frac{(hc+h_f)}{2}}^{\frac{-hc}{2}} f(z)C_{55}^f dz + \int_{\frac{h_c}{2}}^{\frac{(h_c+h_f)}{2}} f(z)C_{55}^f dz$$

$$S_{66} = \int_{-h_c/2}^{h_c/2} f(z)C_{66}^c dz + \int_{-\frac{(hc+h_f)}{2}}^{\frac{-hc}{2}} f(z)C_{66}^f dz + \int_{\frac{h_c}{2}}^{\frac{(h_c+h_f)}{2}} f(z)C_{66}^f dz$$

$$M_{11} = \int_{-h_c/2}^{h_c/2} f(z)zC_{11}^c dz + \int_{-\frac{(hc+h_f)}{2}}^{\frac{-hc}{2}} f(z)zC_{11}^f dz + \int_{\frac{h_c}{2}}^{\frac{(h_c+h_f)}{2}} f(z)zC_{11}^f dz$$

$$M_{12} = \int_{-h_c/2}^{h_c/2} f(z)zC_{12}^c dz + \int_{-\frac{(hc+h_f)}{2}}^{\frac{-hc}{2}} f(z)zC_{12}^f dz + \int_{\frac{h_c}{2}}^{\frac{(h_c+h_f)}{2}} f(z)zC_{12}^f dz$$

$$M_{22} = \int_{-h_c/2}^{h_c/2} f(z)zC_{22}^c dz + \int_{-\frac{(hc+h_f)}{2}}^{\frac{-hc}{2}} f(z)zC_{22}^f dz + \int_{\frac{h_c}{2}}^{\frac{(h_c+h_f)}{2}} f(z)zC_{22}^f dz$$

$$M_{44} = \int_{-h_c/2}^{h_c/2} f(z)zC_{44}^c dz + \int_{-\frac{(hc+h_f)}{2}}^{\frac{-hc}{2}} f(z)zC_{44}^f dz + \int_{\frac{h_c}{2}}^{\frac{(h_c+h_f)}{2}} f(z)zC_{44}^f dz$$

$$M_{55} = \int_{-h_c/2}^{h_c/2} f(z)zC_{55}^c dz + \int_{-\frac{(hc+h_f)}{2}}^{\frac{-hc}{2}} f(z)zC_{22}^f dz + \int_{\frac{h_c}{2}}^{\frac{(h_c+h_f)}{2}} f(z)zC_{55}^f dz$$

$$M_{66} = \int_{-h_c/2}^{h_c/2} f(z)zC_{66}^c dz + \int_{-\frac{(hc+h_f)}{2}}^{\frac{-hc}{2}} f(z)zC_{66}^f dz + \int_{\frac{h_c}{2}}^{\frac{(h_c+h_f)}{2}} f(z)zC_{66}^f dz$$

$$O_{11} = \int_{-h_c/2}^{h_c/2} f(z)^2 C_{11}^c dz + \int_{-\frac{(hc+h_f)}{2}}^{\frac{-hc}{2}} f(z)^2 C_{11}^f dz + \int_{\frac{h_c}{2}}^{\frac{(h_c+h_f)}{2}} f(z)^2 C_{11}^f dz$$

$$O_{12} = \int_{-h_c/2}^{h_c/2} f(z)^2 C_{12}^c dz + \int_{-\frac{(hc+h_f)}{2}}^{\frac{-hc}{2}} f(z)^2 C_{12}^f dz + \int_{\frac{h_c}{2}}^{\frac{(h_c+h_f)}{2}} f(z)^2 C_{12}^f dz$$

$$O_{22} = \int_{-h_c/2}^{h_c/2} f(z)^2 C_{22}^c dz + \int_{-\frac{(hc+h_f)}{2}}^{\frac{-hc}{2}} f(z)^2 C_{22}^f dz + \int_{\frac{h_c}{2}}^{\frac{(h_c+h_f)}{2}} f(z)^2 C_{22}^f dz$$

$$O_{44} = \int_{-h_c/2}^{h_c/2} f(z)^2 C_{44}^c dz + \int_{-\frac{(hc+h_f)}{2}}^{\frac{-hc}{2}} f(z)^2 C_{44}^f dz + \int_{\frac{h_c}{2}}^{\frac{(h_c+h_f)}{2}} f(z)^2 C_{44}^f dz$$

$$O_{55} = \int_{-h_c/2}^{h_c/2} f(z)^2 C_{55}^c dz + \int_{-\frac{(hc+h_f)}{2}}^{\frac{-hc}{2}} f(z)^2 C_{55}^f dz + \int_{\frac{h_c}{2}}^{\frac{(h_c+h_f)}{2}} f(z)^2 C_{55}^f dz$$

$$O_{66} = \int_{-h_c/2}^{h_c/2} f(z)^2 C_{66}^c dz + \int_{-\frac{(hc+h_f)}{2}}^{\frac{-hc}{2}} f(z)^2 C_{66}^f dz + \int_{\frac{h_c}{2}}^{\frac{(h_c+h_f)}{2}} f(z)^2 C_{66}^f dz$$

$$L_{44} = \int_{-h_c/2}^{h_c/2} f(z)'^2 C_{44}^c dz + \int_{-\frac{(hc+h_f)}{2}}^{\frac{-hc}{2}} f(z)'^2 C_{44}^f dz + \int_{\frac{h_c}{2}}^{\frac{(h_c+h_f)}{2}} f(z)'^2 C_{44}^f dz$$

$$L_{55} = \int_{-h_c/2}^{h_c/2} f(z)'^2 C_{55}^c dz + \int_{-\frac{(hc+h_f)}{2}}^{\frac{-hc}{2}} f(z)'^2 C_{55}^f dz + \int_{\frac{h_c}{2}}^{\frac{(h_c+h_f)}{2}} f(z)^2 C_{55}^f dz$$

REFERENCES

1. Ebrahimi, F. and M. Mokhtari, Free vibration analysis of a rotating mori-Tanaka-based functionally graded beam via differential transformation method. Arabian Journal for Science and Engineering, 2015. **41**: p. 1–14.
2. Ebrahimi, F. and M. Mokhtari, *Free vibration analysis of a rotating Mori-Tanaka-based functionally graded beam via differential transformation method*. Arabian Journal for Science and Engineering, 2016. **41**: p. 577–590.
3. Ebrahimi, F., P. Haghi, and A.M. Zenkour, *Modelling of thermally affected elastic wave propagation within rotating Mori-Tanaka-based heterogeneous nanostructures*. Microsystem Technologies, 2018. **24**: p. 2683–2693.
4. Ebrahimi, F. and A. Dabbagh, *Vibration analysis of multi-scale hybrid nanocomposite plates based on a Halpin-Tsai homogenization model*. Composites Part B: Engineering, 2019. **173**: p. 106955.
5. Taheri, M. and F. Ebrahimi, *Buckling analysis of CFRP plates: A porosity-dependent study considering the GPLs-reinforced interphase between fiber and matrix*. The European Physical Journal Plus, 2020. **135**: p. 1–19.

6. Ebrahimi, F., et al., *Analysis of propagation characteristics of elastic waves in heterogeneous nanobeams employing a new two-step porosity-dependent homogenization scheme*. Advances in Nano Research, 2019. **7**(2): p. 135.

7. Reddy, J., C. Wang, and S. Kitipornchai, *Axisymmetric bending of functionally graded circular and annular plates*. European Journal of Mechanics-A/Solids, 1999. **18**(2): p. 185–199.

8. Ebrahimi, F. and M.R. Barati, *Electro-magnetic effects on nonlocal dynamic behavior of embedded piezoelectric nanoscale beams*. Journal of Intelligent Material Systems and Structures, 2017. **28**(15): p. 2007–2022.

9. Ebrahimi, F., et al., *Vibration analysis of porous magneto-electro-elastically actuated carbon nanotube-reinforced composite sandwich plate based on a refined plate theory*. Engineering with Computers, 2021. **37**: p. 921–936.

10. Ebrahimi, F. and M.F. Ahari, *Magnetostriction-assisted active control of the multi-layered nanoplates: Effect of the porous functionally graded facesheets on the system's behavior*. Engineering with Computers, 2021: p. 1–15.

11. Touratier, M., *An efficient standard plate theory*. International Journal of Engineering Science, 1991. **29**(8): p. 901–916.

12. Touratier, M., *A generalization of shear deformation theories for axisymmetric multilayered shells*. International Journal of Solids and Structures, 1992. **29**(11): p. 1379–1399.

13. Touratier, M., *A refined theory of laminated shallow shells*. International Journal of Solids and Structures, 1992. **29**(11): p. 1401–1415.

14. Shi, G., *A new simple third-order shear deformation theory of plates*. International Journal of Solids and Structures, 2007. **44**(13): p. 4399–4417.

15. Wang, C., J.N. Reddy, and K. Lee, *Shear deformable beams and plates: Relationships with classical solutions*. 2000: Elsevier.

16. Vidal, P. and O. Polit, *A family of sinus finite elements for the analysis of rectangular laminated beams*. Composite Structures, 2008. **84**(1): p. 56–72.

17. Qaderi, S., F. Ebrahimi, and V. Mahesh, *Free vibration analysis of graphene platelets-reinforced composites plates in thermal environment based on higher-order shear deformation plate theory*. International Journal of Aeronautical and Space Sciences, 2019. **20**: p. 902–912.

18. Eringen, A.C. and D. Edelen, *On nonlocal elasticity*. International Journal of Engineering Science, 1972. **10**(3): p. 233–248.

19. Eringen, A.C. and J. Wegner, *Nonlocal continuum field theories*. Applied Mechanics Reviews, 2003. **56**(2): p. B20–B22.

20. Peddieson, J., G.R. Buchanan, and R.P. McNitt, *Application of nonlocal continuum models to nanotechnology*. International Journal of Engineering Science, 2003. **41**(3–5): p. 305–312.

21. Lim, C., G. Zhang, and J. Reddy, *A higher-order nonlocal elasticity and strain gradient theory and its applications in wave propagation*. Journal of the Mechanics and Physics of Solids, 2015. **78**: p. 298–313.

22. Ebrahimi, F. and A. Dabbagh, *Wave propagation analysis of magnetostrictive sandwich composite nanoplates via nonlocal strain gradient theory*. Proceedings of the Institution of Mechanical Engineers, Part C: Journal of Mechanical Engineering Science, 2018. **232**(22): p. 4180–4192.

23. Ebrahimi, F., et al., *Hygro-thermal effects on wave dispersion responses of magnetostrictive sandwich nanoplates*. Advances in Nano Research, 2019. **7**(3): p. 157.

24. Ebrahimi, F., A. Dabbagh, and T. Rabczuk, *On wave dispersion characteristics of magnetostrictive sandwich nanoplates in thermal environments*. European Journal of Mechanics-A/Solids, 2021. **85**: p. 104130.

25. Ebrahimi, F. and A. Dabbagh, *Thermo-magnetic field effects on the wave propagation behavior of smart magnetostrictive sandwich nanoplates.* The European Physical Journal Plus, 2018. **133**: p. 1–12.

26. Dong, X., J. Ou, and X. Guan, Applications of magnetostrictive materials in civil structures: A review. In *The 6th International Workshop on Advanced Smart Materials and Smart Structures Technology.* 2011.

27. Ebrahimi, F., K. Khosravi, and A. Dabbagh, *A novel spatial-temporal nonlocal strain gradient theorem for wave dispersion characteristics of FGM nanoplates.* Waves in Random and Complex Media, 2021: p. 1–20.

28. Thai, H.-T. and S.-E. Kim, *A simple quasi-3D sinusoidal shear deformation theory for functionally graded plates.* Composite Structures, 2013. **99**: p. 172–180.

29. Rao, S.S., *Vibration of continuous systems.* 2019: John Wiley & Sons.

30. Vinyas, M., et al., *A finite element-based assessment of free vibration behaviour of circular and annular magneto-electro-elastic plates using higher order shear deformation theory.* Journal of Intelligent Material Systems and Structures, 2019. **30**(16): p. 2478–2501.

31. Shariati, A., et al., *On buckling characteristics of polymer composite plates reinforced with graphene platelets.* Engineering with Computers, 2022: p. 1–12.

32. Mohammadi, M., A. Saidi, and E. Jomehzadeh, *A novel analytical approach for the buckling analysis of moderately thick functionally graded rectangular plates with two simply-supported opposite edges.* Proceedings of the Institution of Mechanical Engineers, Part C: Journal of Mechanical Engineering Science, 2010. **224**(9): p. 1831–1841.

33. Bodaghi, M. and A. Saidi, *Levy-type solution for buckling analysis of thick functionally graded rectangular plates based on the higher-order shear deformation plate theory.* Applied Mathematical Modelling, 2010. **34**(11): p. 3659–3673.

34. Thai, H.-T. and D.-H. Choi, *An efficient and simple refined theory for buckling analysis of functionally graded plates.* Applied Mathematical Modelling, 2012. **36**(3): p. 1008–1022.

35. Sobhy, M., *A comprehensive study on FGM nanoplates embedded in an elastic medium.* Composite Structures, 2015. **134**: p. 966–980.

36. Barati, M.R., A.M. Zenkour, and H. Shahverdi, *Thermo-mechanical buckling analysis of embedded nanosize FG plates in thermal environments via an inverse cotangential theory.* Composite Structures, 2016. **141**: p. 203–212.

37. Karami, B., M. Janghorban, and A. Tounsi, *Galerkin's approach for buckling analysis of functionally graded anisotropic nanoplates/different boundary conditions.* Engineering with Computers, 2019. **35**(4): p. 1297–1316.

3   Vibration Analysis of Magnetostrictive Materials and Structures

Background

A mechanical phenomenon known as vibration may be defined as the occurrence of oscillations at a point of equilibrium. This term originated from the Latin word vibrationem ("to shake"). Oscillations might have a regular pattern, like the swinging of a pendulum, or they can be unpredictable, like the movement of a tire over gravel. There are certain situations in which vibration is desired, such as when it is produced by a tuning fork, a reed in a woodwind or harmonica instrument, a mobile phone, or a loudspeaker.

In many contexts, vibration is undesirable because it results in the loss of energy and the production of noise that is not intended. For instance, the vibrational motion that is produced by working motors, electric motors, or any other kind of mechanical device is often not desired. Such vibrations may be created by things like unequal friction or the meshing of gear teeth, as well as imbalances in spinning components. Unwanted vibrations are often reduced to a minimum by careful design.

When a mechanical system is started in motion with an initial input and then allowed to vibrate freely, the result is a phenomenon known as free vibration. This form of vibration may be produced by doing things like striking a tuning fork and allowing it to ring, as well as by bringing a kid onto a swing and letting them go. As a result of the mechanical system vibrating at one or more of its inherent frequencies, the subject becomes immobile.

Vibration Analysis of Magnetostrictive Materials and Structures

The vibrational properties of Magnetostrictive Materials and Structures are the subject of a sufficient numerical and analytical examination in this chapter. The influence of different factors on the vibrational behavior of magnetostrictive material is studied using a variety of theoretical frameworks and geometrical representations. In order to accomplish this objective, the vibrational behavior of beams and plates is discussed in Sections 3.1 and 3.2.

DOI: 10.1201/9781003355427-3

3.1 VIBRATION ANALYSIS OF MAGNETOSTRICTIVE BEAMS

The behavior of beams composed of magnetostrictive material is the subject of discussion in this subsection. In order to circumvent the behavior of magneto-strictive beams, a variety of examples, each of which is exhaustive, are shown, explored, and explained. In this section, illustrative examples will be offered to demonstrate the impact of various variations' parameters.

3.1.1 FORMULATION

Think of a nanobeam as being embedded on a Winkler-Pasternak elastic sub-strate for the sake of this discussion. The nanobeam has the dimensions length L, breadth b, and total thickness H in the x, y, and z directions, respectively. The nano-face sheets of the top include aluminum (Al), and the bottom surface contains aluminum oxide (Al_2O_3), while the core layer is formed of a magnetostrictive material. The nano-face sheets of the top contain aluminum (Al) as the topmost material. In this particular investigation, the Cartesian methodology, which can be seen shown in Figure 3.1, was used.

The displacement field may be described in the following manner by using the presumptions provided by the higher-order shear deformation (Reddy) beam theory [1]:

$$U_x(x,z,t) = u_0(x,t) + z\phi(x) - \alpha z^3\left(\phi + \frac{\partial w_0}{\partial x}\right) \tag{3.1}$$

$$V_y(x,z,t) = 0 \tag{3.2}$$

$$W_z(x,z,t) = w_0(x,t) \tag{3.3}$$

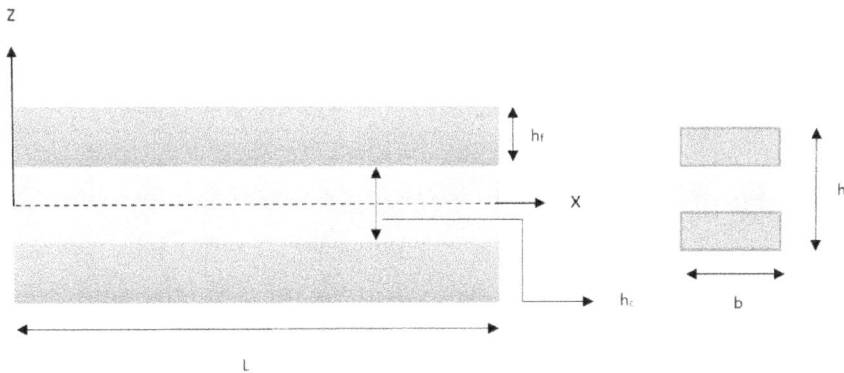

FIGURE 3.1 Sandwich composite Beam arrangement.

where u_0 and w_0 represent the longitudinal and transverse displacements, respectively, and ϕ represents the rotation of the cross section at each point along the neutral axis.

The proposed composite's homogenization is reached as explained in Section 2.1., according to Eqs. (2.9–2.10). The nonlocal theory of Eringen is applied to nonlocal theory, as shown in Eq (2.20).

It is possible to define the relationship between the core layer and stress, strain, and magnetic field as follows [2]:

$$\left(1-\mu^2\nabla^2\right)\begin{Bmatrix}\sigma_{xx}^{c,nl}\\\tau_{xz}^{c,nl}\end{Bmatrix}=\begin{bmatrix}C_{11}^c & 0\\0 & C_{55}^c\end{bmatrix}\begin{Bmatrix}\varepsilon_{xx}^c\\\gamma_{xz}^c\end{Bmatrix}-H_{zz}\begin{Bmatrix}\bar{e}_{31}\\0\end{Bmatrix} \qquad (3.4)$$

where σ_{ij}, ε_{ij} $(i,j=x,y,z)$, respectively, represent the stress and strain constants. In addition to this, H_{zz} represents a magnetic field. In addition to this, it is important to understand that C_{ij} is the abbreviation for the elastic stiffness coefficient [3]:

$$C_{11}^c=\frac{E_c}{1-\vartheta_c^2},\qquad C_{55}^c=\frac{E_c}{2\left(1+\vartheta_c\right)} \qquad (3.5)$$

in which the modulus of elasticity and Poisson's ratio of the core layer are denoted by E_c and v_c, respectively, and $\bar{e}_{ij}$ are the converted magnetostrictive moduli that may be achieved according to the explanation given in Eq. (2.23) [2].

Calculating the magnetic field (H_{zz}) may be done with the help of the Velocity feedback gain relation, which is described in Eqs. (2.24, 2.25) [2]:

Nonzero stresses in the magnetostrictive layer are characterized as the following:

$$\varepsilon_{xx}^c=\epsilon_{xx}^c+z\kappa_{xx}^c+z^3\chi_{xx}^c\qquad \gamma_{xz}^c=\epsilon_{xz}^c+z^2X_{xz}^c \qquad (3.6)$$

in which

$$\epsilon_{xx}^c=\frac{\partial u_0}{\partial x},\ \kappa_{xx}^c=\frac{\partial\phi}{\partial x},\ \chi_{xx}^c=-\alpha\left(\frac{\partial\phi}{\partial x}+\frac{\partial^2 w_0}{\partial x^2}\right)$$

$$\epsilon_{xz}^c=\frac{\partial w_0}{\partial x}+\phi,\ X_{xz}^c=-\beta\left(\frac{\partial w_0}{\partial x}+\phi\right) \qquad (3.7)$$

The link between stress and strain may be described in the following manner with regard to the FG facesheets:

$$\left(1-\mu^2\nabla^2\right)\begin{Bmatrix}\sigma_{xx}^{f,nl}\\\tau_{xz}^{f,nl}\end{Bmatrix}=\begin{bmatrix}C_{11}^f & 0\\0 & C_{55}^f\end{bmatrix}\begin{Bmatrix}\varepsilon_{xx}^f\\\gamma_{xz}^f\end{Bmatrix} \qquad (3.8)$$

elasticity may be defined as having the following characteristics:

$$C_{11}^f = \frac{E(z)}{1-\vartheta_f^2} \quad C_{55}^f = \frac{E(z)}{2(1+\vartheta_f)} \tag{3.9}$$

in which the elastic modulus of the face sheet is referred to as E_f, and the poison's constant of the face sheet is referred to as v_f.

The nonzero stresses of the FG facesheets may be described as the following, if at all possible:

$$\varepsilon_{xx}^f = \epsilon_{xx}^f + z\kappa_{xx}^f + z^3\chi_{xx}^f \quad \gamma_{xz}^f = \epsilon_{xz}^f + z^2 X_{xz}^f \tag{3.10}$$

in which

$$\epsilon_{xx}^f = \frac{\partial u_0}{\partial x}, \kappa_{xx}^f = \frac{\partial \phi}{\partial x}, \chi_{xx}^f = -\alpha\left(\frac{\partial \phi}{\partial x} + \frac{\partial^2 w_0}{\partial x^2}\right)$$

$$\epsilon_{xz}^f = \frac{\partial w_0}{\partial x} + \phi, X_{xz}^f = -\beta\left(\frac{\partial w_0}{\partial x} + \phi\right) \tag{3.11}$$

in which α and β are defined as follows: $\alpha = \dfrac{4}{3h^2}, \beta = \dfrac{4}{h^2}$.

3.1.2 GOVERNING EQUATION

In order to get the equations that regulate the equilibrium, the displacement gradients may be broken down into their component portions, and the coefficients δu_0, δw_0, and $\delta\phi$ can each be set to zero. The first formulation of Hamilton's concept is stated in Eq. (2.32) [4].

An equation for the first variation in strain energy caused by the nanobeam may be written as follows:

$$\delta U^c = \int_v \left(N_{xx}^c\left(\delta\epsilon_{xx}^c\right) + M_{xx}^c\left(\delta\kappa_{xx}^c\right) + G_{xx}^c\left(\delta\chi_{xx}^c\right) + N_{xz}^c\left(\delta\epsilon_{xz}^c\right) + P_{xz}^c\left(\delta X_{xz}^c\right)\right)dV$$

$$\delta U^f = \int_v \left(N_{xx}^f\left(\delta\epsilon_{xx}^f\right) + M_{xx}^f\left(\delta\kappa_{xx}^f\right) + G_{xx}^f\left(\delta\chi_{xx}^f\right) + N_{xz}^f\left(\delta\epsilon_{xz}^f\right) + P_{xz}^f\left(\delta X_{xz}^f\right)\right)dV \tag{3.12}$$

where

$$\left(N_{ij}^c, M_{ij}^c, P_{ij}^c, G_{ij}^c\right) = \int_{-\frac{h_c}{2}}^{\frac{h_c}{2}} \sigma_{ij}^c\left(1, z, z^2, z^3\right)dz, \quad (i, j = x, z)$$

$$\left(N_{ij}^f, M_{ij}^f, P_{ij}^f, G_{ij}^f\right) = \int_{\frac{h_c}{2}}^{\frac{h_c}{2}+h_f} \sigma_{ij}^f\left(1, z, z^2, z^3\right)dz + \int_{-\left(\frac{h_c}{2}+h_f\right)}^{-\frac{h_c}{2}} \sigma_{ij}^f\left(1, z, z^2, z^3\right)dz, \quad (i, j = x, z) \tag{3.13}$$

In the equations that have been presented so far, the terms "core layer" and "face-sheets" are denoted by the superscripts "c" and "f," respectively.

An external force either transfers work as a form of energy into the nanoplate or removes work as a form of energy from the nanoplate. In the system being investigated, "work" is defined as the application of energy by external forces. The first variant of the external task may be phrased as Eq. (2.39) [5, 6].

One possible expression for the first kinetic energy variation of the nanobeam is as follows:

$$
\begin{aligned}
\delta T = \int_v \bigg[&-I_0 \left(\frac{\partial^2 u_0}{\partial t^2} \right) \delta u_0 - I_1 \left(\frac{\partial^2 \phi}{\partial t^2} \right) \delta u_0 + I_3 \alpha \left(\frac{\partial^2 \phi}{\partial t^2} \right) \delta u_0 - I_3 \alpha \left(\frac{\partial^3 w_0}{\partial x \partial t^2} \right) \delta u_0 \\
&-I_1 \left(\frac{\partial^2 u_0}{\partial t^2} \right) \delta \phi - I_2 \left(\frac{\partial^2 \phi}{\partial t^2} \right) \delta \phi - 2 I_4 \alpha \left(\frac{\partial^2 \phi}{\partial t^2} \right) \delta \phi - I_4 \alpha \left(\frac{\partial^3 w_0}{\partial x \partial t^2} \right) \delta \phi + I_3 \alpha \left(\frac{\partial^2 u_0}{\partial t^2} \right) \delta \phi \\
&-I_6 \alpha^2 \left(\frac{\partial^2 \phi}{\partial t^2} \right) \delta \phi - I_6 \alpha^2 \left(\frac{\partial^3 w_0}{\partial x \partial t^2} \right) \delta \phi + I_6 \alpha^2 \left(\frac{\partial^3 \phi}{\partial x \partial t^2} \right) \delta w_0 + I_6 \alpha^2 \left(\frac{\partial^4 w_0}{\partial x^2 \partial t^2} \right) \\
&\delta \phi - I_4 \alpha \left(\frac{\partial^3 \phi}{\partial x \partial t^2} \right) \delta w_0 - I_3 \alpha \left(\frac{\partial^3 u_0}{\partial x \partial t^2} \right) \delta w_0 - I_0 \left(\frac{\partial^2 w_0}{\partial t^2} \right) \delta w_0 \bigg] dA
\end{aligned}
$$

(3.14)

whereby the moment of inertia of the masses is defined as:

$$
I_i = \left(\int_{-h_c/2}^{h_c/2} Y_i \rho_c dz \right) + \left(\int_{-\frac{(h_c + h_f)}{2}}^{-h_c} Y_i \rho_f dz + \int_{h_c}^{\frac{(h_c + h_f)}{2}} Y_i \rho_f dz \right),
$$

$$
Y_i = \left[1, z, z^2, z^3, z^4, z^6 \right], \quad i = (1, 2, 3, 4, 6)
$$

(3.15)

The first strain energy, work generated by applied forces, and kinetic energy variations are now replaced into the Euler-Lagrange equation for the sinusoidal shear deformation by substituting Eqs. (3.12–3.15) into Eq. (2.32). This is done in order to calculate the sinusoidal shear deformation. This gives rise to the three equations of equilibrium, which may be written as follows:

$$
\delta u_0 : \frac{\partial N_{xx}}{\partial x} - e_{31} h_c K_c C(t) \frac{\partial^2 w_0}{\partial x \partial t} = I_0 \left(\frac{\partial^2 u_0}{\partial t^2} \right) + I_1 \left(\frac{\partial^2 \phi}{\partial t^2} \right) - I_3 \alpha \left(\frac{\partial^2 \phi}{\partial t^2} \right) - I_3 \alpha \left(\frac{\partial^3 w_0}{\partial x \partial t^2} \right)
$$

(3.16)

$$
\begin{aligned}
\delta \phi_x : &\frac{\partial M_{xx}}{\partial x} - \alpha \frac{\partial G_{xx}}{\partial x} - N_{xz} + \beta P_{xz} = I_1 \left(\frac{\partial^2 u_0}{\partial t^2} \right) + I_2 \left(\frac{\partial^2 \phi}{\partial t^2} \right) - 2 \alpha I_4 \left(\frac{\partial^2 \phi}{\partial t^2} \right) \\
&- \alpha I_4 \left(\frac{\partial^3 w_0}{\partial x \partial t^2} \right) + \alpha I_3 \left(\frac{\partial^2 u_0}{\partial t^2} \right) + \alpha^2 I_6 \left(\frac{\partial^2 \phi}{\partial t^2} \right) + \alpha^2 I_6 \left(\frac{\partial^2 u_0}{\partial t^2} \right)
\end{aligned}
$$

(3.17)

$$\delta w_0 : \alpha \frac{\partial^2 G_{xx}}{\partial x^2} + \frac{\partial N_{xz}}{\partial x} - \beta \frac{\partial P_{xz}}{\partial x} - k_w \cdot w_0 + k_g \left(\frac{\partial^2 w_0}{\partial x^2} + \frac{\partial^2 w_0}{\partial y^2} \right) - e_{31} h_c K_c C(t) \frac{\partial^2 U_x}{\partial x \partial t}$$

$$= I_0 \left(\frac{\partial^2 w_0}{\partial t^2} \right) - \alpha^2 I_6 \left(\frac{\partial^3 \phi}{\partial x \partial t^2} \right) - \alpha^2 I_6 \left(\frac{\partial^4 w_0}{\partial x^2 \partial t^2} \right) + \alpha I_4 \left(\frac{\partial^3 \phi}{\partial x \partial t^2} \right) + \alpha I_4 \left(\frac{\partial^3 u_0}{\partial x \partial t^2} \right)$$

$$(3.18)$$

3.1.3 SOLUTION

In order for Navier's technique to be usable, the given solutions must satisfy the problem's boundary conditions. The natural frequency may be determined by entering the planned response (Eq. 3.19) into the system's governing equation. According to Navier's solution, under the assumption that the boundary condition is simply maintained, the displacements may be described as follows:

$$\begin{Bmatrix} u_0(x,y,t) \\ \phi_x(x,y,t) \\ w_0(x,y,t) \end{Bmatrix} = \begin{Bmatrix} \sum_{m=1}^{\infty} \sum_{n=1}^{\infty} U_{mn} \cos(ox) e^{i\omega t} \\ \sum_{m=1}^{\infty} \sum_{n=1}^{\infty} \phi_{xmn} \sin(ox) e^{i\omega t} \\ \sum_{m=1}^{\infty} \sum_{n=1}^{\infty} \phi_{xmn} W_{mn} \sin(ox) e^{i\omega t} \end{Bmatrix} \qquad (3.19)$$

in which o is defined as: $o = \dfrac{m\pi}{L}$. It is possible to determine the natural frequencies by solving Eq. (3.19), which may be done by translating it into Eqs. (3.16–3.18) and isolating the differences in u_0, w_0, and ϕ. To represent natural frequencies properly, the determinant of the bellows equation will be adjusted to zero.

$$\left\| [K] - \omega^2 [M] + i\omega [C] \right\| = 0 \qquad (3.20)$$

where K, M, and C are matrices expressing the mass stiffness, and damping matrices are provided in Appendix 3A.

In this section, in order to evaluate the appropriateness and usefulness of the proposed system, comparisons are performed between the results of the indicated system and the references [7, 8] that may be found in the existing body of knowledge. It is important to note that while comparing the results, the magnetostrictive core is not taken into account at all. The results that may be seen in Table 3.1 need to be used in the future as a benchmark for comparisons.

As seen in Table 3.1, when the nonlocal parameter grows, the dimensionless natural frequencies drop. When the nonlocal parameter grows, the nanoplate becomes more flexible, causing this result. One may argue that raising the small-scale parameter decreases the system's energy, resulting in the system's weakened state. As a result, one may claim that the local system's dimensionless natural frequency is greater than that of the nonlocal system. In other words, the total

TABLE 3.1

Natural Frequency Comparison That Disregards Dimensions (b = 1,000 nm, L = 10,000 nm, k = 0)

L/h	μ^2	Ref [7]	Ref [8]	Present	Error %
20	0	9.8797	9.8296	9.83558	0.0608
	1	9.4238	9.3777	9.38343	0.0611
	2	9.0257	8.9829	8.98839	0.0611
	3	8.6741	8.6341	8.63939	0
	4	8.3607	8.3230	8.32811	0.0613
	5	8.0789	8.0433	8.04823	0.0612
50	0	9.8724	9.8631	9.86454	0.0145
	1	9.4172	9.4097	9.41105	0.0143
	2	9.0205	9.0136	9.01486	0.0139
	3	8.6700	8.6636	8.66482	0.0140
	4	8.3575	8.3515	8.35264	0.0136
	5	8.0765	8.0708	8.07193	0.0140
100	0	9.8700	9.8680	9.86871	0.0071
	1	9.4162	9.4143	9.41503	0.0077
	2	9.0197	9.0180	9.01866	0.0073
	3	8.6695	8.6678	8.66848	0.0078
	4	9.3571	8.3555	8.35616	0.0078
	5	8.0762	8.0747	8.07534	0.0079

stiffness is greater than its counterpart due to the addition of the stiffness caused by the pair stress effect to the classical stiffness.

As can be seen, the results display good convergence, and the amount of error occurs is rather small. It should be noted that the nondimensional natural frequency shown in the previous table is achieved in line with the following:

$$\bar{\omega} = \omega L^2 \sqrt{\frac{\rho_a A}{E_a I}} \tag{3.21}$$

3.1.4 RESULTS

In order to investigate the effect of the side-to-thickness ratio (ah c ratio) on the nondimensional natural frequency of the nanobeam, Table 3.2 is prepared for a range of porosity volume values for both an even and an uneven porosity distribution. As can be seen in Table 3.2, the nondimensional natural frequency falls as the degree of porosity in the material rises. On the other hand, the system is said to be stable when the thickness of the core layer increases in tandem with

TABLE 3.2

The Natural Frequency That Is Dimensionless across Aspect Ratios and Porosity Distributions ($\mu^2 = 0\ nm^2$, $K_w = K_g = 0$, $k = 0$, $L = 100\ nm$, $h_f = a/20$, $Kc = 5 \times 10^5$)

Porosity	L / h_c			
Even	5	10	20	30
0	5.82889	13.1553	28.7085	52.8847
0.1	4.73574	10.8321	23.2774	38.2703
0.2	4.44857	9.91991	20.7874	33.1265
0.3	4.15842	9.03527	18.4935	28.7206
0.4	3.86388	8.17315	16.3595	24.8623
Uneven	5	10	20	30
0	5.82889	13.1553	28.7085	52.8847
0.1	4.83431	10.4742	21.8915	36.6443
0.2	3.83717	8.00367	16.2437	25.7714
0.3	2.80746	5.65151	11.3126	17.4511
0.4	1.70249	3.33073	6.79244	10.4644

nondimensional frequencies. This is mainly because, as stated by $\left(\omega = \sqrt{\dfrac{k}{m}} \right)$, the contribution of mass $(m = \rho.L.b.h)$ is outweighed by the influence of the moment of inertia $\left(k \sim I \rightarrow I = \dfrac{1}{12} bh^3 \right)$.

Figure 3.2a,b has also been drafted to facilitate research into the influence that the thickness of the magnetostrictive layer has on the nondimensional natural frequency. The foundation values for Winkler and Pasternak in this figure are intended to be zero ($K_w = K_g = 0$). The gradient index should also be 0. For each of the six possible values of the nonlocal parameter, these graphics show the nondimensional natural frequency for both the even and uneven distributions. As can be seen in this figure, the impact of mass and inertia causes the nondimensional natural frequency to drop as the thickness of the magnetostrictive layer grows. This phenomenon may be attributed to the fact that mass and inertia are inversely proportional. In addition to this, when the nonlocal parameter is raised, the nondimensional natural frequency decreases.

(a) Even distribution
(b) Uneven distribution

Table 3.3 provides an explanation of how to compute the nondimensional natural frequency. In the following table, the values of the dimensionless natural frequency are shown for three different possible combinations of the gradient index

(a)

(b)

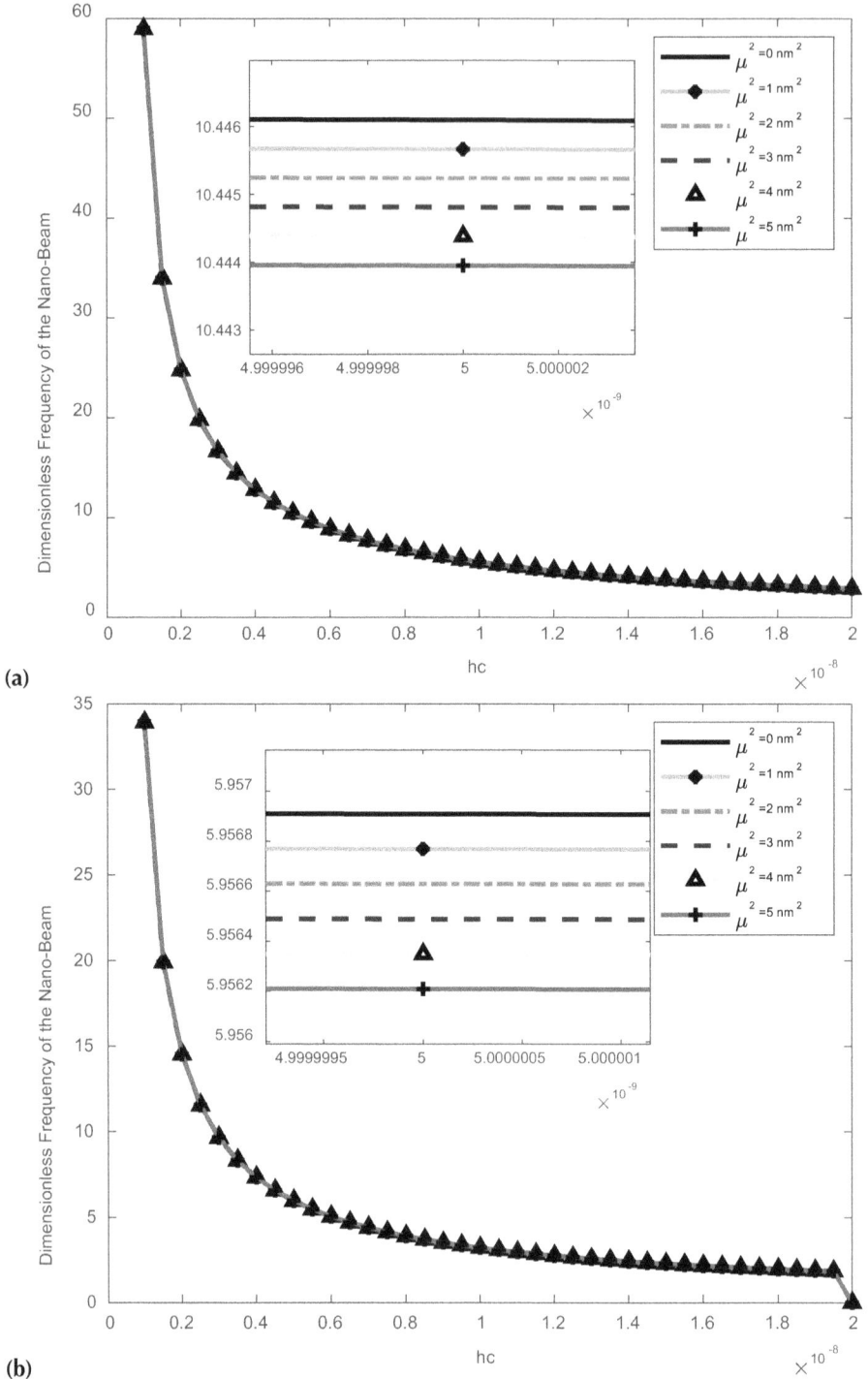

FIGURE 3.2 The magnetostrictive layer thickness affects the natural frequency in the absence of dimensions for (a) even porosity and (b) uneven porosity.

TABLE 3.3

Frequency in the Absence of Dimensions for Different Gradient Indices
($a = 100$ nm, $b = 10$ nm, $hc = a/20$, $h_f = L/40$, $\mu^2 = 1\,nm^2$, $\xi = 0.3$)

Porosity	Feedback gain	$k = 0$	$k = 0.5$	$k = 1$	$k = 1.5$	$k = 5$	$k = 10$
Even	$k_c C(t) = 1 \times 10^5$	65.7937	60.3395	58.7531	58.1704	0.0025203	58.2366
	$k_c C(t) = 5 \times 10^5$	10.4465	8.29733	7.26215	6.66553	0.0440799	5.02004
	$k_c C(t) = 9 \times 10^5$	5.64169	4.49014	3.93582	3.61627	0.156094	2.73434
Uneven	$k_c C(t) = 1 \times 10^5$	23.5508	27.2564	30.2211	15.8013	4.34674	2.18437
	$k_c C(t) = 5 \times 10^5$	5.95705	3.70736	2.64942	2.04708	0.774881	0.412734
	$k_c C(t) = 9 \times 10^5$	3.25649	2.03608	1.45908	1.12935	0.429286	0.228952

and the velocity feedback gain parameters. Due to the fact that aluminum oxide has a higher Young's modulus than aluminum, as can be observed from these data, dimensionless natural frequencies decrease as the gradient index increases; this trend may have a physical explanation. In addition, as can be seen in the table, increasing the velocity feedback gain lowers the nondimensional natural frequency.

Natural frequencies are shown to decrease as the porosity volume % is multiplied (Figure 3.3a,b). This is primarily the result of the fact that the stiffness of the nanoplate will be significantly influenced whenever the porosity volume fraction is increased. As a direct consequence of this, the natural frequency will be lowered.

(a) Even distribution
(b) Uneven distribution

As can be seen in Table 3.4, the distribution of nondimensional natural frequencies is more evenly spread out. This is primarily due to the fact that, in order to have an equal distribution, there must be more cavities; this, in turn, makes the nanobeam more rigid. In addition to this, the fact that $K_w = 0$ and $K_g = 0$ in Table 3.4 suggest that the Winkler and Pasternak foundation value need to be constant.

As a concluding graphic, Figure 3.4 depicts how the Pasternak foundation affects the nondimensional natural frequency. Note that the gradient index has been locked in at 0 throughout the drawing of Figure 3.4. The graph that goes along with this post shows how, as the Pasternak foundation parameter is raised, the dimensionless natural frequency of the nanobeam increases. This phenomenon results from the fact that the nanobeam becomes more rigid when the Pasternak foundation is raised.

In order to monitor the effect of the facesheet thickness on the nondimensional natural frequency, Table 3.5 has been provided for the constant value of the nonlocal parameter ($\mu^2 = 1\,nm^2$). The dimensionless natural frequency rises when

(a)

(b)

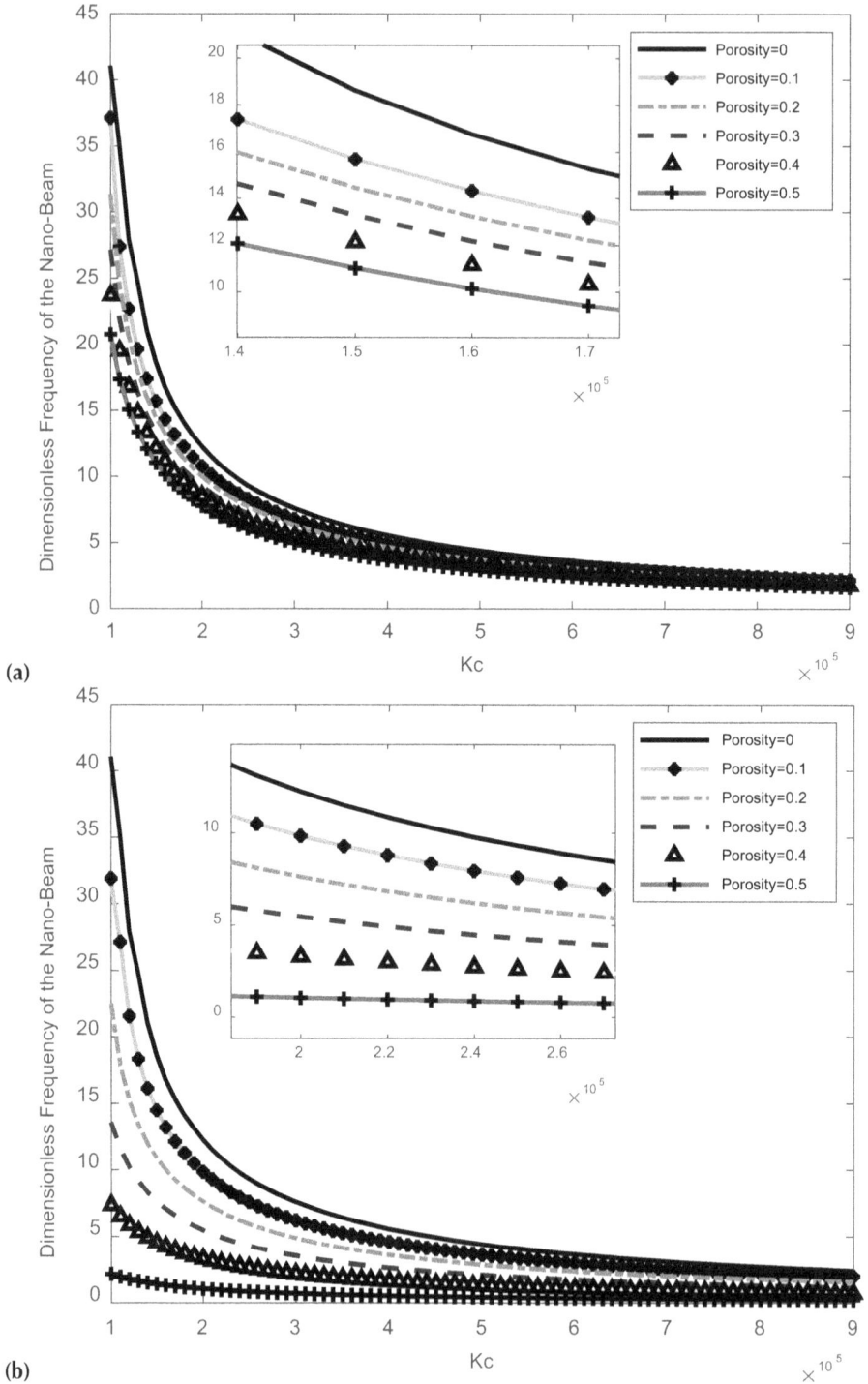

FIGURE 3.3   A study of the frequency-independent effects of velocity feedback gain for (a) even porosity and (b) uneven porosity.

TABLE 3.4

Natural Frequency Comparison in a Dimensionless Space Between the SSSS and SSCC Boundary Conditions $\mu^2 = 1\,nm^2$, $\xi = 0.3$, k = 1, hc = a/6, $h_f = a/40$, a = 100 nm, b = 10 nm, $K_c C(t) = 1 \times 10^5$)

Winkler Foundation	Pasternak Foundation	Even porosity	Uneven porosity
$K_w = 0$	$K_g = 0$	19.3766	6.67446
$K_w = 0$	$K_g = 0.02$	25.3995	12.268
$K_w = 0.001$	$K_g = 0.02$	25.5662	12.4251
$K_w = 0.01$	$K_g = 0.02$	27.0446	13.8144
$K_w = 0$	$K_g = 0.04$	30.7078	17.1997
$K_w = 0.001$	$K_g = 0.04$	30.8668	17.3438
$K_w = 0.01$	$K_g = 0.04$	32.3016	18.6312

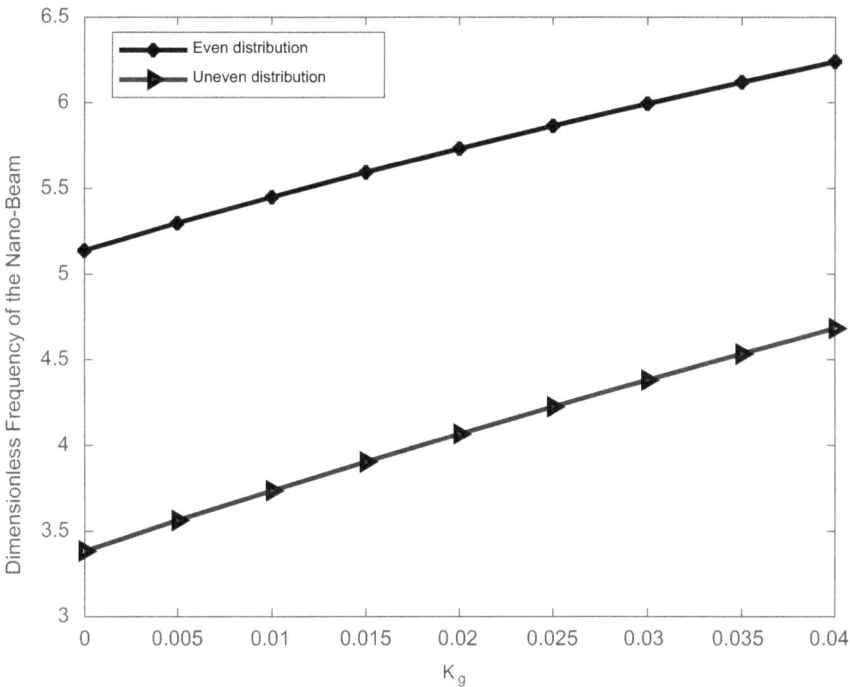

FIGURE 3.4 The Pasternak effect on the natural frequency beyond dimensions.

the face-sheet thickness increases, as seen in Table 3.5. This is due to the fact that an increase in the value of h_f causes an increase in the stiffness of the nanoplate, which in turn, increases the thickness of the nanoplate. As a consequence, higher natural frequencies are produced.

TABLE 3.5

The Natural Frequency in Dimensions for a Square Nanobeam with Varying Porosity and Gradient Index ($\xi = 0.1$, $\mu^2 = 1nm^2$, $K_w = K_g = 0$, $hc = a/6$, a = 100 nm, b = 10 nm, $k = 1$, $K_w = K_g = 0$, $K_c C(t) = 5 \times 10^5$)

	Even porosity	Uneven porosity
$h_f = L/12$	6.48177	6.5841
$h_f = L/15$	5.69961	5.54628
$h_f = L/20$	4.82582	4.48084
$h_f = L/30$	3.82594	3.37976

3.1.5 Concluding Remarks

In this study, the vibrational behavior of a composite magnetostrictive nanobeam that is integrated with functionally graded facesheets is investigated using a higher-order shear deformation beam theory. In addition, the equations that regulate the system may be derived by utilizing Hamilton's principle. The following is a synopsis of some of the remarkable conclusions reached by the present research:

A greater value for the velocity feedback gain will result in a lower value for the non-dimensional natural frequency. To put it another way, the magnetic field has a damping effect on the oscillations.

As the thickness of the FG facesheet increases, the dimensionless natural frequency of the nanobeam also increases.

The dimensionless natural frequency of the nanobeam is shown to decrease as the thickness of the core layer increases.

3.2 VIBRATION ANALYSIS OF MAGNETOSTRICTIVE PLATES

The goal of this study is to learn more about the vibrational properties of a composite sandwich made up of magnetostrictive nanoplates and functionally graded material facesheets. It is predicted that Terfenol-D will be used for the core layer, with functionally graded materials used for the outer and inner plies. Eringen's nonlocal hypothesis is used here to account for the localized effect. Yet, using FSDT, we can determine the nanoplate's kinematic relations. This occurs due to the presence of a Winkler-Pasternak medium underneath the nanoplate. Using Hamilton's principle, we are able to derive the governing equations, which are then solved analytically using the Navier method for the simply supported boundary condition.

3.2.1 FORMULATION

The Mindlin-Oufliand plate theory makes the assumption that there is a linear variation in displacement throughout the thickness of the plate, but that the thickness of the plate does not vary as a result of the deformation. A further presumption is that no account is taken for the normal tension that runs parallel to the thickness. The condition in question is often referred to as the aircraft stress condition. On the other hand, the Reisner static theory makes the assumption that the shear stress is quadratic over the thickness of the plate, but the bending stress is assumed to be linear. Because of this, the through-thickness displacement will no longer be necessarily linear when the plate thickness is deformed. Instead, it will alter as a result of the deformation. As a result, Reisner's static theory does not make advantage of situations involving plane stress. It is common practice to refer to the Mindlin-Oufliand plate theory as the FSDT of plates. Since it presumes a linear displacement over the thickness, the FSDT is incompatible with Reisner's stationary plane theory. In contrast to the assumptions made by the classical theory, the first-order shear theory postulates that planes that were initially perpendicular to an intermediate surface would no longer be in this orientation after the surface has been deformed.

The fundamental presumptions behind the first order shear deformation model:

1- The normal stresses that run along the thickness of the material are disregarded, which is what is meant by the term "plane stress theory."
2- The components of the displacement change in a linear fashion all the way through the thickness of the plate.
3- The plate's thickness remains the same throughout the deformation process.
4- The shear stress along the thickness of the plate is of the second order, while the bending stress is linear along the plate's lengthwise direction.

As shown in Figure 3.5, we use the Cartesian coordinate system, with the origin located at one corner of the plate, and the axes extending along the length, width, and thickness of the core layer.

In this particular piece of writing, the FSDT will be used in order to conceive of the challenge. Shown here is an example of how the displacement field corresponding to the FSDT may be provided [9]:

$$U_x(x,y,z,t) = u_0(x,y,t) + \phi_x(x,y,t) \tag{3.22}$$

$$V_y(x,y,z,t) = v_0(x,y,t) + \phi_y(x,y,t) \tag{3.23}$$

$$W_z(x,y,z,t) = w_0(x,y,t) \tag{3.24}$$

where the full displacement fields in the Cartesian x, y, and z directions are denoted by U_x, V_y, and W_z. The link between the core layer's stresses, strains, and magnetic properties may be expressed in Eq. (2.7) [10].

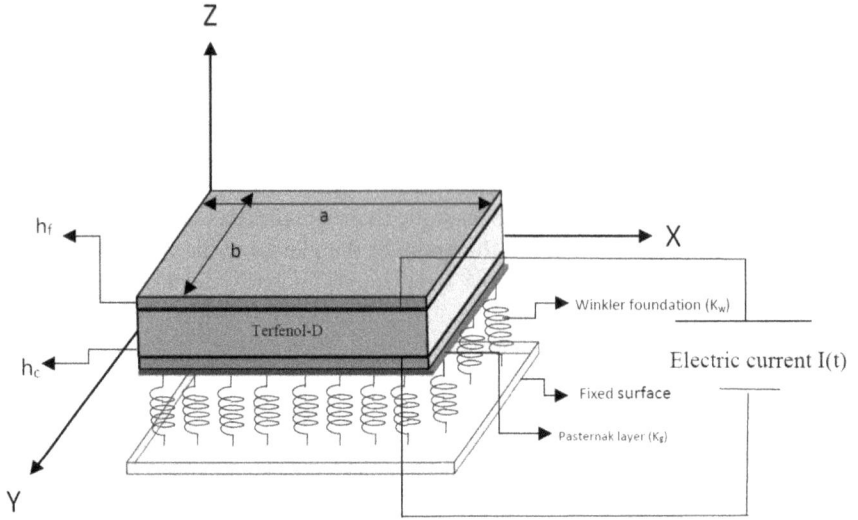

FIGURE 3.5   Sandwich composite plate arrangement [3].

The core layer's nonzero stresses may be characterized as the following terms:

$$
\left\{
\begin{array}{c}
\varepsilon^c_{xx} \\
\varepsilon^c_{yy} \\
\varepsilon^c_{yz} \\
\varepsilon^c_{xz} \\
\varepsilon^c_{xy}
\end{array}
\right\}
=
\left\{
\begin{array}{c}
\epsilon^c_{xx} \\
\epsilon^c_{yy} \\
\epsilon^c_{yz} \\
\epsilon^c_{xz} \\
\epsilon^c_{xy}
\end{array}
\right\}
+ z
\left\{
\begin{array}{c}
\kappa^c_{xx} \\
\kappa^c_{yy} \\
\kappa^c_{yz} \\
\kappa^c_{xz} \\
\kappa^c_{xy}
\end{array}
\right\}
\tag{3.25}
$$

where

$$
\left\{
\begin{array}{c}
\epsilon^c_{xx} \\
\epsilon^c_{yy} \\
\epsilon^c_{yz} \\
\epsilon^c_{xz} \\
\epsilon^c_{xy}
\end{array}
\right\}
=
\left\{
\begin{array}{c}
\dfrac{\partial u_0}{\partial x} \\[2mm]
\dfrac{\partial v_0}{\partial y} \\[2mm]
\left(\phi_y + \dfrac{\partial w_0}{\partial y}\right) \\[2mm]
\left(\phi_x + \dfrac{\partial w_0}{\partial x}\right) \\[2mm]
\left(\dfrac{\partial u_0}{\partial y} + \dfrac{\partial v_0}{\partial x}\right)
\end{array}
\right\}
\qquad
\left\{
\begin{array}{c}
\kappa^c_{xx} \\
\kappa^c_{yy} \\
\kappa^c_{yz} \\
\kappa^c_{xz} \\
\kappa^c_{xy}
\end{array}
\right\}
=
\left\{
\begin{array}{c}
\dfrac{\partial \phi_x}{\partial x} \\[2mm]
\dfrac{\partial \phi_y}{\partial y} \\[2mm]
0 \\[2mm]
0 \\[2mm]
\dfrac{\partial \phi_x}{\partial y} + \dfrac{\partial \phi_y}{\partial x}
\end{array}
\right\}
\tag{3.26}
$$

Relationship between stress and strain for the facesheets is described in Eq. (2.14) [11]. Definitions that may be used for elastic coefficients are stated in Eq. (2.15). The stresses that are greater than zero in the upper and lower layers may be characterized as follows:

$$
\begin{Bmatrix} \varepsilon_{xx}^f \\ \varepsilon_{yy}^f \\ \varepsilon_{yz}^f \\ \varepsilon_{xz}^f \\ \varepsilon_{xy}^f \end{Bmatrix} = \begin{Bmatrix} \epsilon_{xx}^f \\ \epsilon_{yy}^f \\ \epsilon_{yz}^f \\ \epsilon_{xz}^f \\ \epsilon_{xy}^f \end{Bmatrix} + z \begin{Bmatrix} \kappa_{xx}^f \\ \kappa_{yy}^f \\ \kappa_{yz}^f \\ \kappa_{xz}^f \\ \kappa_{xy}^f \end{Bmatrix}
\tag{3.27}
$$

in which

$$
\begin{Bmatrix} \epsilon_{xx}^f \\ \epsilon_{yy}^f \\ \epsilon_{yz}^f \\ \epsilon_{xz}^f \\ \epsilon_{xy}^f \end{Bmatrix} = \begin{Bmatrix} \dfrac{\partial u_0}{\partial x} \\ \dfrac{\partial v_0}{\partial y} \\ \left(\phi_y + \dfrac{\partial w_0}{\partial y} \right) \\ \left(\phi_x + \dfrac{\partial w_0}{\partial x} \right) \\ \left(\dfrac{\partial u_0}{\partial y} + \dfrac{\partial v_0}{\partial x} \right) \end{Bmatrix}
\qquad
\begin{Bmatrix} \kappa_{xx}^f \\ \kappa_{yy}^f \\ \kappa_{yz}^f \\ \kappa_{xz}^f \\ \kappa_{xy}^f \end{Bmatrix} = \begin{Bmatrix} \dfrac{\partial \phi_x}{\partial x} \\ \dfrac{\partial \phi_y}{\partial y} \\ 0 \\ 0 \\ \dfrac{\partial \phi_x}{\partial y} + \dfrac{\partial \phi_y}{\partial x} \end{Bmatrix}
\tag{3.28}
$$

3.2.2 Governing Equation

In this section, the motion equations will be derived by the use of the energy technique. In order to do this, the Hamilton principle will be used. The essence of this idea might be stated as explained in Eq. (2.18) [4].

The variation in the strain energy between the facesheets and the core layer may be expressed as the following equation:

$$
\delta U^c = \int_0^a \int_0^b \Big(N_{xx}^c \left(\delta \epsilon_{xx}^c \right) + M_{xx}^c \left(\delta \kappa_{xx}^c \right) + N_{yy}^c \left(\delta \epsilon_{yy}^c \right) + M_{yy}^c \left(\delta \kappa_{yy}^c \right) + N_{xz}^c \left(\delta \epsilon_{xz}^c \right)
$$
$$
+ M_{xz}^c \left(\delta \kappa_{xz}^c \right) + N_{yz}^c \left(\delta \epsilon_{yz}^c \right) + M_{yz}^c \left(\delta \kappa_{yz}^c \right) + N_{xy}^c \left(\delta \epsilon_{xy}^c \right) + M_{xy}^c \left(\delta \kappa_{xy}^c \right) \Big) dA
\tag{3.29}
$$

$$
\delta U^f = \int_0^a \int_0^b \Big(N_{xx}^f \left(\delta \epsilon_{xx}^f \right) + M_{xx}^f \left(\delta \kappa_{xx}^f \right) + N_{yy}^f \left(\delta \epsilon_{yy}^f \right) + M_{yy}^f \left(\delta \kappa_{yy}^f \right) + N_{xz}^f \left(\delta \epsilon_{xz}^f \right)
$$
$$
+ M_{xz}^f \left(\delta \kappa_{xz}^f \right) + N_{yz}^f \left(\delta \epsilon_{yz}^f \right) + M_{yz}^f \left(\delta \kappa_{yz}^f \right) + N_{xy}^f \left(\delta \epsilon_{xy}^f \right) + M_{xy}^f \left(\delta \kappa_{xy}^f \right) \Big) dA
\tag{3.32}
$$

where

$$\left(N_{ij}^c, M_{ij}^c\right) = \int_{-\frac{h_c}{2}}^{\frac{h_c}{2}} \sigma_{ij}^c \left(1,z\right) dz, \quad (i,j = x,y,z) \tag{3.31}$$

$$\left(N_{ij}^f, M_{ij}^f\right) = \int_{\frac{h_c}{2}}^{\frac{h_c}{2}+h_f} \sigma_{ij}^f \left(1,z\right) dz + \int_{-\left(\frac{h_c}{2}+h_f\right)}^{-\frac{h_c}{2}} \sigma_{ij}^f \left(1,z\right) dz, \quad (i,j = x,y,z) \tag{3.32}$$

In the equations that have been presented thus far, the letters 'c' and 'f' in the superscripts refer, respectively, to the core layer and the facesheets. Additionally, the forces and moments are denoted by the notations $N_{i,j}$ and $M_{i,j}$, respectively. Nonlocal connections of forces and bending moments may be expressed in the following ways:

$$(1-\mu^2\nabla^2)\begin{bmatrix} N_{xx} \\ N_{yy} \\ N_{yz} \\ N_{xz} \\ N_{xy} \end{bmatrix} = \begin{bmatrix} A_{11} & A_{12} & 0 & 0 & 0 \\ A_{21} & A_{22} & 0 & 0 & 0 \\ 0 & 0 & A_{44} & 0 & 0 \\ 0 & 0 & 0 & A_{55} & 0 \\ 0 & 0 & 0 & 0 & A_{66} \end{bmatrix} \begin{pmatrix} \begin{bmatrix} \dfrac{\partial u_0}{\partial x} \\ \dfrac{\partial v_0}{\partial y} \\ \varnothing y + \dfrac{\partial w_0}{\partial y} \\ \varnothing x + \dfrac{\partial w_0}{\partial x} \\ \dfrac{\partial u_0}{\partial y} + \dfrac{\partial v_0}{\partial x} \end{bmatrix} \\ + \begin{bmatrix} B_{11} & B_{12} & 0 & 0 & 0 \\ B_{21} & B_{22} & 0 & 0 & 0 \\ 0 & 0 & B_{44} & 0 & 0 \\ 0 & 0 & 0 & B_{55} & 0 \\ 0 & 0 & 0 & 0 & B_{66} \end{bmatrix} \begin{bmatrix} \dfrac{\partial \phi_x}{\partial x} \\ \dfrac{\partial \phi_y}{\partial y} \\ 0 \\ 0 \\ \dfrac{\partial \phi_x}{\partial x} + \dfrac{\partial \phi_y}{\partial x} \end{bmatrix} \end{pmatrix} \tag{3.33}$$

$$(1-\mu^2\nabla^2)\begin{bmatrix} M_{xx} \\ M_{yy} \\ M_{yz} \\ M_{xz} \\ M_{xy} \end{bmatrix} = \left(\begin{bmatrix} B_{11} & B_{12} & 0 & 0 & 0 \\ B_{21} & B_{22} & 0 & 0 & 0 \\ 0 & 0 & B_{44} & 0 & 0 \\ 0 & 0 & 0 & B_{55} & 0 \\ 0 & 0 & 0 & 0 & B_{66} \end{bmatrix} \begin{bmatrix} \dfrac{\partial u_0}{\partial x} \\ \dfrac{\partial v_0}{\partial y} \\ \varnothing y + \dfrac{\partial w_0}{\partial y} \\ \varnothing x + \dfrac{\partial w_0}{\partial x} \\ \dfrac{\partial u_0}{\partial y} + \dfrac{\partial v_0}{\partial x} \end{bmatrix} \right.$$

$$\left. + \begin{bmatrix} D_{11} & D_{12} & 0 & 0 & 0 \\ D_{21} & D_{22} & 0 & 0 & 0 \\ 0 & 0 & D_{44} & 0 & 0 \\ 0 & 0 & 0 & D_{55} & 0 \\ 0 & 0 & 0 & 0 & D_{66} \end{bmatrix} \begin{bmatrix} \dfrac{\partial \phi_x}{\partial x} \\ \dfrac{\partial \phi_y}{\partial y} \\ 0 \\ 0 \\ \dfrac{\partial \phi_x}{\partial x} + \dfrac{\partial \phi_y}{\partial x} \end{bmatrix} \right) \qquad (3.34)$$

where

$$\begin{bmatrix} A_{11} & B_{11} & D_{11} \\ A_{22} & B_{22} & D_{22} \\ A_{12} & B_{12} & D_{12} \\ A_{44} & B_{44} & D_{44} \\ A_{55} & B_{55} & D_{55} \\ A_{66} & B_{66} & D_{66} \end{bmatrix}$$

$$= \int_{-\frac{hc}{2}}^{\frac{hc}{2}} [1 \quad z \quad z^2] \begin{bmatrix} C_{11}^c \\ C_{22}^c \\ C_{12}^c \\ C_{44}^c \\ C_{55}^c \\ C_{66}^c \end{bmatrix} dz + \int_{\frac{hc}{2}}^{\frac{h_c}{2}+hf} [1 \quad z \quad z^2] \begin{bmatrix} C_{11}^f \\ C_{22}^f \\ C_{12}^f \\ C_{44}^f \\ C_{55}^f \\ C_{66}^f \end{bmatrix} dz + \int_{-\left(\frac{h_c}{2}+hf\right)}^{-\frac{hc}{2}} [1 \quad z \quad z^2] \begin{bmatrix} C_{11}^f \\ C_{22}^f \\ C_{12}^f \\ C_{44}^f \\ C_{55}^f \\ C_{66}^f \end{bmatrix} dz$$

.
$$(3.35)$$

The variation in the work that is done external may be stated as explained in Eq. (2.25). The variation in kinetic energy may be represented mathematically as follows:

$$
\delta T = \frac{1}{2} \int \left[I_0 \left(\frac{\partial^2 u_0}{\partial t^2} \right) \delta u_0 + I_1 \left(\frac{\partial^2 u_0}{\partial t^2} \right) \delta \phi_x + I_1 \left(\frac{\partial^2 \phi_x}{\partial t^2} \right) \delta u_0 + I_2 \left(\frac{\partial^2 \phi_x}{\partial t^2} \right) \delta \phi_x \right.
$$

$$
+ I_0 \left(\frac{\partial^2 v_0}{\partial t^2} \right) \delta v_0 + I_1 \left(\frac{\partial^2 v_0}{\partial t^2} \right) \delta \phi_y + I_1 \left(\frac{\partial^2 \phi_y}{\partial t^2} \right) \delta v_0 + I_2 \left(\frac{\partial^2 \phi_y}{\partial t^2} \right) \delta \phi_y
$$

$$
\left. + I_0 \left(\frac{\partial^2 w_0}{\partial t^2} \right) \delta w_0 \right] dA
$$

(3.36)

Thus, according definition, the mass moments of inertia are as follows:

$$
I_0 = \int_{-\frac{h_c}{2}}^{\frac{h_c}{2}} \rho_c dz + \left(\int_{-\frac{(h_c+h_f)}{2}}^{\frac{-h_c}{2}} \rho_f(z) dz + \int_{\frac{h_c}{2}}^{\frac{(h_c+h_f)}{2}} \rho_f(z) dz \right)
$$

(3.37)

$$
I_1 = \int_{-h_c/2}^{h_c/2} z\rho_c dz + \left(\int_{-\frac{(hc+h_f)}{2}}^{\frac{-hc}{2}} z\rho_f(z) dz + \int_{\frac{h_c}{2}}^{\frac{(h_c+h_f)}{2}} z\rho_f(z) dz \right)
$$

(3.38)

$$
I_2 = \int_{-h_c/2}^{h_c/2} z^2 \rho_c dz + \left(\int_{-\frac{(h_c+h_f)}{2}}^{\frac{-h_c}{2}} z^2 \rho_f(z) dz + \int_{\frac{hc}{2}}^{\frac{(h_c+h_f)}{2}} z^2 \rho_f(z) dz \right)
$$

(3.39)

When Eqs. (3.29) through (3.39) are plugged into Eq. (2.18), the following results are obtained:

$$
\delta u_0 : \frac{\partial N_{xx}}{\partial x} + \frac{\partial N_{xy}}{\partial y} - e_{31} h_c K_c C(t) \frac{\partial^2 w_0}{\partial x \partial t} = I_0 \frac{\partial^2 u}{\partial t^2} + I_1 \frac{\partial^2 \phi_x}{\partial t^2}
$$

(3.40)

$$
\delta v_0 : \frac{\partial N_{xx}}{\partial y} + \frac{\partial N_{xy}}{\partial x} - e_{32} h_c K_c C(t) \frac{\partial^2 w_0}{\partial y \partial t} = I_0 \frac{\partial^2 v}{\partial t^2} + I_1 \frac{\partial^2 \phi_y}{\partial t^2}
$$

(3.41)

$$
\delta w_0 : \frac{\partial N_{xz}}{\partial x} + \frac{\partial N_{yz}}{\partial y} - k_w . w + k_g \left(\frac{\partial^2 w}{\partial x^2} + \frac{\partial^2 w}{\partial y^2} \right) - e_{31} h_c K_c C(t) \frac{\partial^2 U_x}{\partial x \partial t}
$$

$$
- e_{32} h_c K_c C(t) \frac{\partial^2 V_y}{\partial y \partial t} = I_0 \frac{\partial^2 w}{\partial t^2}
$$

(3.42)

$$\delta\varnothing_x : \frac{\partial M_{xx}}{\partial x} + \frac{\partial M_{xy}}{\partial y} - N_{xz} = I_1 \frac{\partial^2 u}{\partial t^2} + I_2 \frac{\partial^2 \phi_x}{\partial t^2} \tag{3.43}$$

$$\delta\varnothing_y : \frac{\partial M_{yy}}{\partial y} + \frac{\partial M_{xy}}{\partial x} - N_{yz} = I_1 \frac{\partial^2 v}{\partial t^2} + I_2 \frac{\partial^2 \phi_y}{\partial t^2} \tag{3.44}$$

3.2.3 SOLUTION

In order to find a solution for the governing equation, an analytical solution developed by Navier was used. The suggested solutions to Navier's issue have to be in accordance with the problem's boundary conditions for the solution to work. In order to proceed, the natural frequency may be calculated by inserting the suggested response into Eq. (3.45), which is the equation that governs the system. Navier's method can definitely solve partial differential equations with excellent accuracy, convergence, and performance, and it has been effectively used to determine a wide range of issues in vibration analysis. In addition, the technique has been used to successfully determine the number of other problems in the field of vibration analysis.

Taking into account the boundary condition of being simply supported, the following is how the displacements may be described in accordance with Navier's solution:

$$\begin{Bmatrix} u_0(x,y,t) \\ v_0(x,y,t) \\ w_0(x,y,t) \\ \phi_x(x,y,t) \\ \phi_y(x,y,t) \end{Bmatrix} = \begin{Bmatrix} \sum_{m=1}^{\infty}\sum_{n=1}^{\infty} U_{mn}\cos(\alpha x)\sin(\beta y)e^{i\omega t} \\ \sum_{m=1}^{\infty}\sum_{n=1}^{\infty} V_{mn}\sin(\alpha x)\cos(\beta y)e^{i\omega t} \\ \sum_{m=1}^{\infty}\sum_{n=1}^{\infty} W_{mn}\sin(\alpha x)\sin(\beta y)e^{i\omega t} \\ \sum_{m=1}^{\infty}\sum_{n=1}^{\infty} \phi_{xmn}\cos(\alpha x)\sin(\beta y)e^{i\omega t} \\ \sum_{m=1}^{\infty}\sum_{n=1}^{\infty} \phi_{ymn}\sin(\alpha x)\cos(\beta y)e^{i\omega t} \end{Bmatrix} \tag{3.45}$$

The amplitude of the vibration is denoted by the variables U_{mn}, V_{mn}, W_{mn}, ϕ_{xmn} and ϕ_{ymn}, respectively. Furthermore, is the symbol for the fundamental frequency. In addition to this, and are defined as $\beta = \frac{n\pi}{y}$ $\alpha = \frac{m\pi}{a}$.

The natural frequencies may be determined by substituting Eq. (3.45) for Eqs. (3.40) through (3.44), and then isolating the variations of $u_0, v_0, w_0, \phi_x, \phi_y$. Next, by solving the resulting equation, one can get the natural frequencies. In order for the bellow equation to accurately represent the natural frequencies, the determinant variable will have to be set to zero.

$$\left\| [K] - \omega^2 [M] + i\omega[C] \right\| = 0 \tag{3.46}$$

where the stiffness, mass, and damping are represented by matrices K, M, and C, respectively; these matrices are defined in Appendix 3B.

3.2.4 RESULTS

Imagine a nanoplate consisting of magnetostrictive material and FG material combined together. The governing equation will be constructed using the FSDT that Mindlin [9] presented, the Powe-law relation [10, 12, 13], and Eringen's non-local theory [6]. A comparison is made before moving on in order to verify the postulated mathematical connections. The findings are contrasted with those that were published in Refs. [14–20], all of which used the Kirchhoff-Love plate theory to produce natural frequencies. They used a nano core that was isotropic and elastic, and it had the following characteristics:

$$Ec = 30 \times 109pa, \, \textit{v}c = 0.25, \, hf = 0, \, a = b = 10 \text{ m}$$

In addition, there is a comparison for the magnetostrictive nanoplate taking into account the feedback gain with respect to relevant Refs. [14–20] included in Table 3.6. It is clear from looking at Table 3.6 that there is a high level of concordance between the findings of this study and the findings of the references that were previously mentioned. It is important to keep in mind that the dimensionless natural frequency may be calculated using the following relation:

$$\Omega = \frac{\omega h_c \sqrt{\rho_c}}{\sqrt{E_c}} \tag{3.47}$$

Table 3.7 shows the results of an additional validation that was carried out in order to assess the effectiveness of the FG nano facesheets. To do this, the

---

TABLE 3.6

A Look at How the Natural Frequencies Change Over a Range of Core Layer Thicknesses

	$K_c C(t) = 0$							$K_c C(t) = 10^3$			$K_c C(t) = 10^4$	
h_c	Present	[14]	[15]	[16]	[17] FSDT	[17] 2D	[21]	[14]	[19]	Present	[20]	Present
0.5	0.01458	0.01480	0.01480	0.01464	-	-	-	-	-	-	–	–
1	0.05693	0.05769	0.05769	0.05673	0.06382	0.05777	0.05769	-	-	-	0.3841	0.385257
2	0.2092	0.2112	0.2112	0.2055	0.2334	0.2121	-	1.046	1.000	1.0466	–	–

TABLE 3.7

Analysis of the Reference Data Versus the Natural Frequency

mxn	k=0. [22]	k=0 Present	Error %	k=0.5 [22]	k=0.5 Present	Error %	k=1 [22]	k=1 Present	Error %	k=2 [22]	k=2 Present	Error %
1×1	3.691	3.687	0.09	3.366	3.3625	0.1	3.217	3.2149	0.09	3.129	3.1285	0.03
1×2	5.832	5.8234	0.1	5.323	5.3142	0.1	5.088	5.0813	0.1	4.943	4.9421	0.02
2×1	11.96	11.928	0.3	10.94	10.907	0.3	10.46	10.431	0.2	10.13	10.131	0.05
2×2	13.92	13.872	0.3	12.74	12.693	0.4	12.18	12.139	0.3	11.79	11.785	0.07
2×3	17.09	17.023	0.4	15.66	15.590	0.4	14.97	14.912	0.4	14.48	14.467	0.09
3×2	26.05	25.889	0.6	23.94	23.767	0.7	22.87	22.739	0.6	22.05	22.024	0.1
3×3	28.87	28.675	0.6	26.55	26.343	0.7	25.37	25.205	0.6	24.44	24.401	0.1

natural frequencies of the FGM rectangular plate obtained in Reference for varying amounts of gradient index (k) are compared with the results of this work. Importantly, the natural frequencies in Ref. [22] are computed using the second-order shear deformation theory. This fact must be emphasized. The purpose of Table 3.7 is to compare the dimensionless natural frequencies observed in this investigation to those reported in published works. As shown in Table 3.7, the suggested system functions well, and the percentage of errors is very low. The formula for the error is $Error \% = \dfrac{|Present - Reference|}{Reference} \times 100$. In addition to this, the dimensionless natural frequency may be calculated using the following relation:

$$\Omega = \frac{\omega a^2 \sqrt{\rho_c}}{h\sqrt{E_c}} \tag{3.48}$$

In this equation, the natural frequency is represented by the symbol ω, whereas the dimensionless natural frequency is represented by the symbol Ω. In the following tables, recommended values are taken into account unless it is specifically specified differently. a = b = 40 nm, hc = 3 nm, h_f = 0.2 nm.

In addition to this, the dimensionless forms that make up the Winkler and Pasternak basis are thought of as the following:

$$K_w = \frac{k_w a^4}{D_{11}} \quad K_g = \frac{k_g a^2}{D_{11}} \tag{3.49}$$

Table 3.8, which was made to keep tracks on the nanoplate's first three mode forms, displays the results of an analysis of the nonlocal parameter's effect on the natural frequencies. Table 3.8 demonstrates that when the value of the nonlocal parameter is raised, the natural frequencies decrease. This is essentially

TABLE 3.8

Frequencies of a Square Nanoplate's Natural Dimensions for Various Nonlocal Parameters ($k=2$, $\xi = 0.1$, $K_w=K_g=0$)

Natural Frequency	Feedback gain	$\mu^2 = 0\,\mathrm{mn}^2$	$\mu^2 = 1\,\mathrm{mn}^2$	$\mu^2 = 2\,\mathrm{mn}^2$	$\mu^2 = 3\,\mathrm{mn}^2$	$\mu^2 = 4\,\mathrm{mn}^2$	$\mu^2 = 0\,\mathrm{mn}^2$
Ω_{11}	$k_c C(t)=1\times10^5$	0.347194	0.346739	0.346285	0.345834	0.345384	0.344936
Ω_{11}	$k_c C(t)=5\times10^5$	0.077878	0.077873	0.077867	0.077862	0.077857	0.077852
Ω_{11}	$k_c C(t)=9\times10^5$	0.043427	0.043426	0.043425	0.043424	0.043423	0.043423
Ω_{22}	$k_c C(t)=1\times10^5$	1.279324	1.272448	1.265686	1.259036	1.252492	1.246054
Ω_{22}	$k_c C(t)=5\times10^5$	0.288361	0.288280	0.288199	0.288118	0.288038	0.287957
Ω_{22}	$k_c C(t)=9\times10^5$	0.160834	0.160820	0.160806	0.160792	0.160778	0.160764
Ω_{33}	$k_c C(t)=1\times10^5$	2.574191	2.542521	2.51067	2.482750	2.454496	2.427241
Ω_{33}	$k_c C(t)=5\times10^5$	0.583020	0.582633	0.582246	0.581860	0.581475	0.581091
Ω_{33}	$k_c C(t)=9\times10^5$	0.325251	0.325184	0.325116	0.325049	0.324982	0.324914

because an increase in the value of the nonlocal parameter leads to a more malleable nanoplate. It is arguable that the local system's inherent frequencies, which are independent of dimensions, are greater than those of the nonlocal system. Increasing the feedback gain causes a decrease in the system's dimensionless natural frequencies, as shown in Table 3.8, allowing us to more precisely control the system's oscillation frequency.

Natural frequencies of the sandwich magnetostrictive composite nanoplate linked with FGM facesheets are analyzed by varying the gradient index and feedback gain (see Tables 3.9 and 3.10). Please feel free to make use of the tables provided here. From these numbers, it is evident that raising the gradient index causes a decrease in natural frequency. Taking into consideration the fact that aluminum oxide has a larger Young's modulus than aluminum might provide a physical justification for this trend. A further observation from this table is that the natural frequency tends to drop as the feedback gain increases.

In addition, the influence of the elastic foundation on the dimensionless natural frequencies is analyzed in Table 3.11 for a variety of various Winkler and Pasternak coefficient values. It has been noticed that the dimensionless natural frequencies rise as the elastic foundation value rises. This is due to the fact that the dimensionless natural frequencies rise as a result of the beneficial influence that the foundation's stiff coefficients have on the nanoplate's overall stiffness.

TABLE 3.9

Natural Frequencies of a Square Nanoplate at Various Gradient Strengths and With Uniform Porosity, Expressed in Terms of Dimensions. ($\xi = 0.1$, $\mu^2 = 0$ nm^2, K$_w$ = 0, K$_g$ = 0)

Natural frequency	Feedback gain	$k = 0$	$k = 0.5$	$k = 1$	$k = 1.5$	$k = \infty$
Ω_{11}	$k_c C(t) = 1 \times 10^5$	0.57681	0.46487	0.40676	0.37119	0.22588
Ω_{11}	$k_c C(t) = 5 \times 10^5$	0.13670	0.10726	0.09255	0.08374	0.04919
Ω_{12}	$k_c C(t) = 1 \times 10^5$	1.36458	1.10543	0.97043	0.88760	0.54683
Ω_{12}	$k_c C(t) = 5 \times 10^5$	0.32465	0.25591	0.22148	0.20081	0.11933
Ω_{22}	$k_c C(t) = 1 \times 10^5$	2.07793	1.69035	1.48794	1.36354	0.84896
Ω_{22}	$k_c C(t) = 5 \times 10^5$	0.49581	0.39234	0.34041	0.30919	0.18556

TABLE 3.10

Natural Frequencies of a Square Nanoplate at Various Gradient Strengths and Porosity Levels, Expressed in Units Independent of Dimension ($\xi = 0.1$, $\mu^2 = 0$ nm^2, K$_w$ = 0, K$_g$ = 0)

Natural frequency	Feedback gain	$k = 0$	$k = 0.5$	$k = 1$	$k = 1.5$	$k = \infty$
Ω_{11}	$k_c C(t) = 1 \times 10^5$	0.536058	0.421369	0.361704	0.325138	0.175202
Ω_{11}	$k_c C(t) = 5 \times 10^5$	0.125431	0.095862	0.081079	0.072217	0.037418
Ω_{12}	$k_c C(t) = 1 \times 10^5$	1.272861	1.006484	0.867252	0.781631	0.427152
Ω_{12}	$k_c C(t) = 5 \times 10^5$	0.299106	0.229824	0.195053	0.174150	0.091406
Ω_{22}	$k_c C(t) = 1 \times 10^5$	1.944331	1.544970	1.335496	1.206339	0.667427
Ω_{22}	$k_c C(t) = 5 \times 10^5$	0.458421	0.353832	0.301187	0.269469	0.143070

In addition, the influence of the elastic foundation on the dimensionless natural frequencies is analyzed in Table 3.12 for a variety of various Winkler and Pasternak coefficient values. It has been noted that when the elastic basis value rises, the dimensionless natural frequencies rise along with it. This is due to the fact that the dimensionless natural frequencies rise when the foundation's stiff coefficients are taken into account, increasing the nanoplate's overall stiffness.

TABLE 3.11

The Natural Frequencies, on a Dimensionless Scale, of a Square Nanoplate With Varying Elastic Base Conditions ($k=1.5$, $\xi=0.1$, $K_c=1\times10^5$)

Winkler foundation	Pasternak foundation	Even porosity	Uneven porosity
$K_w = 0$	$K_g = 0$	0.371194	0.325138
$K_w = 0$	$K_g = 0.02$	1.100962	1.055930
$K_w = 0.001$	$K_g = 0.02$	1.116922	1.071915
$K_w = 0.01$	$K_g = 0.02$	1.250426	1.205638
$K_w = 0$	$K_g = 0.04$	1.495523	1.451207
$K_w = 0.001$	$K_g = 0.04$	1.506767	1.462476
$K_w = 0.01$	$K_g = 0.04$	1.603677	1.559602

TABLE 3.12

The Natural Frequency of the Nanoplate, in Terms of Space Dimensions, for a Variety of Gradient Index and Porosity Volume Fraction Values, Assuming Uniform Porosity ($\mu^2 = 0\,\text{nm}^2$, $Kc = 1\times10^5$, $K_w = 0$, $K_g = 0$)

Porosity volume fraction	$k = 0$	$k = 4$	$k = \infty$
$\xi=0$	0.599224	0.325721	0.255136
$\xi=0.1$	0.576812	0.298524	0.225887
$\xi=0.2$	0.554066	0.270578	0.195363
$\xi=0.3$	0.530975	0.241731	0.163033
$\xi=0.4$	0.507525	0.211746	0.127771
$\xi=0.5$	0.483702	0.180214	0.086419
$\xi=0.6$	0.459489	0.146343	0.008696

Table 3.13 provides a simple overview of how the thickness of the core layer affects the dynamic behaviors of the continuous system for both the uniform and uneven porosity distributions. Naturally, the system will become more stable as the thickness of the core layer increases, since more non-dimensional frequencies mean a more robust system. As shown by the formula $\omega = \sqrt{\dfrac{k}{m}}$, the moment of inertia $\left(k \sim I \rightarrow I = \dfrac{1}{12}bh^3\right)$ has a greater effect than the mass $(m = \rho.a.b.h)$,

TABLE 3.13

Variations in the Nanoplate's Frequency (in a Dimensionless Form) as the Thickness of Its Central Layer Is Varied ($k = 1.5$, $\xi = 0.2$, $\mu^2 = 1 \text{nm}^2$, Kc = 1×10^5, $h_f = 0.2$ nm, $K_w = 0$, $K_g = 0$)

	Even porosity	Uneven porosity
$hc = 1$	0.240745	0.149794
$hc = 1.5$	0.268367	0.173979
$hc = 2$	0.294858	0.199100
$hc = 3$	0.345003	0.249226

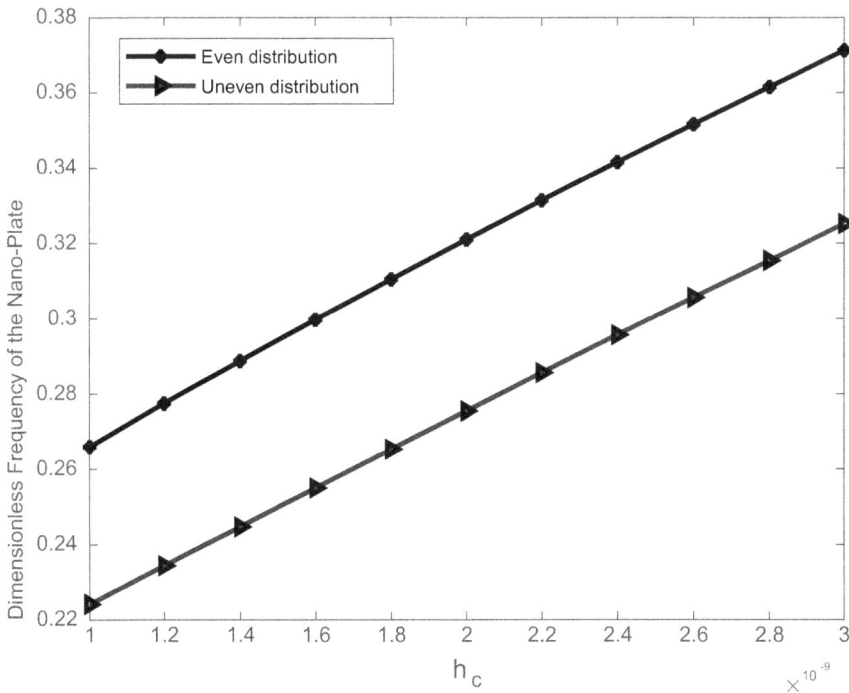

FIGURE 3.6 The relationship between the thickness of the core layer and the dimensionless natural frequencies of a composite magnetostrictive nanoplate with ferroelectric ferrite (FGM) facesheets.

which is why inertia plays a more significant role. The underlying cause of this phenomenon is this.

Plotting Figure 3.6 allows one to investigate how the thickness of the magnetostrictive layer influences the natural frequency of the nanoplate. It is clear from looking at Figure 3.6 that increasing the thickness of the smart magnetostrictive core

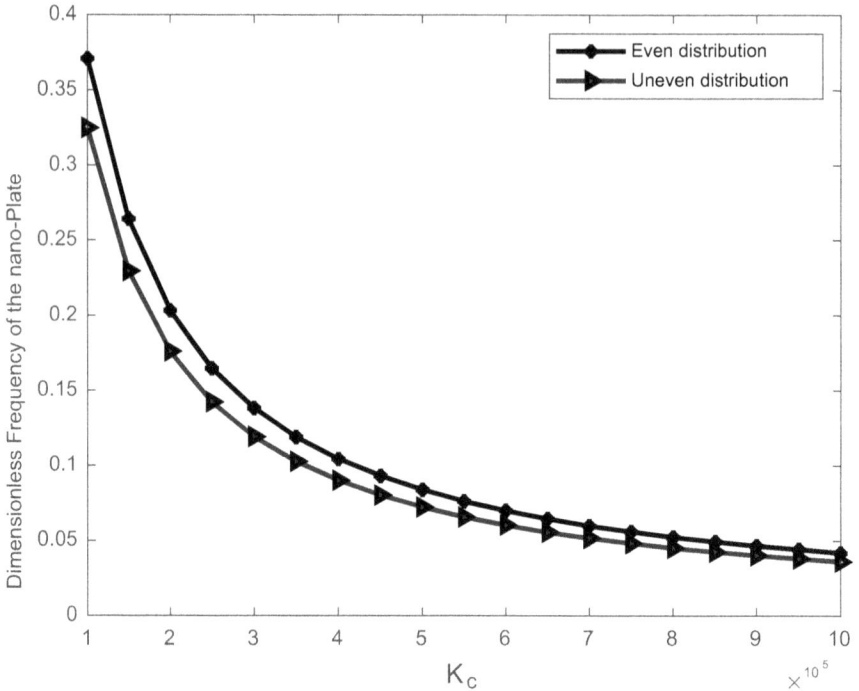

FIGURE 3.7 Composite magnetostrictive nanoplates combined with (FGM) facesheets exhibit dimensionless natural frequencies and positive feedback gain.

results in an increase in the dimensionless natural frequencies. This observation was supported in the paragraphs that came before it from a physics point of view.

In addition, Figure 3.7 illustrates the influence that the feedback gain parameter has on the dimensionless natural frequency. This can be seen by looking at the diagram. The dampening impact that the magnetostriction phenomenon has on the behavior of the system is shown in Figure 3.7, and it can be seen that the natural frequencies of the system dramatically drop as the value of the feedback gain is increased. It cannot be denied that the velocity feedback gain plays a significant part in determining the system's level of steadiness.

The influence of the Pasternak spring on the system's stability and predictability is seen in Figure 3.8. As can be seen in Figure 3.8, the composite magnetostrictive nanoplate associated with FGM facesheets exhibits an increase in its dimensionless natural frequencies as the Pasternak coefficient increases. A pair of Figure 3.8 plots, one for a uniform distribution of pores and the other for a nonuniform distribution, have been made.

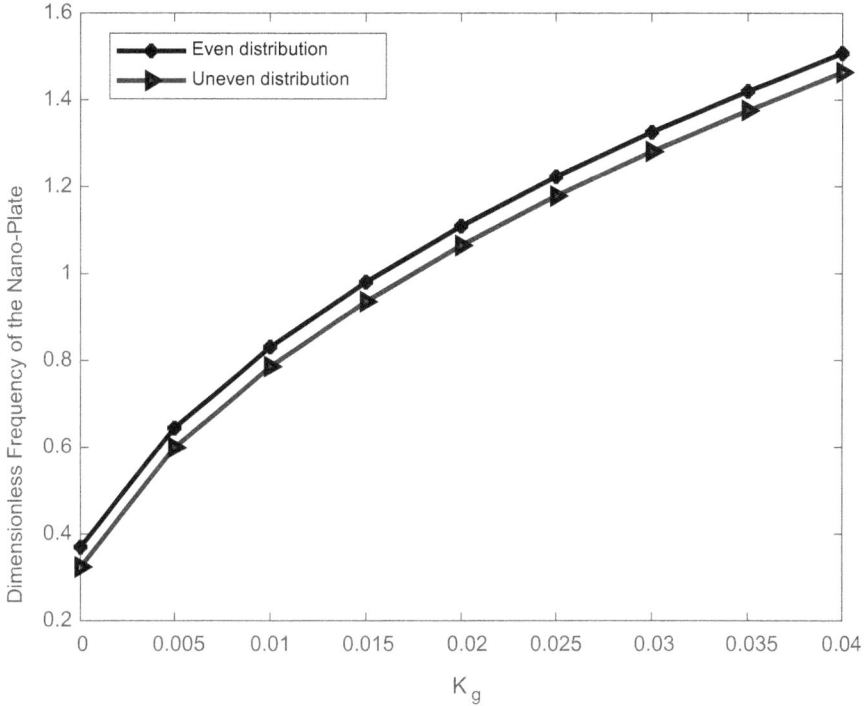

FIGURE 3.8 The composite magnetostrictive nanoplate's natural frequencies when combined with FGM facesheets are compared to those of the Pasternak model.

In Figure 3.9, we explore how changing the Winkler parameter affects the dimensionless natural frequency. Both equal and uneven distributions are of interest to this study. This graph makes it quite evident that increasing the Winkler foundation's value likewise increases the natural frequency. It's possible to get the conclusion that the Pasternak foundation is more effective than the Winkler foundation after comparing Figures 3.8 and 3.9.

To follow the effect of the gradient index on the non-dimensional frequency, the values for the nonlocal parameter $\left(\mu^2 = 0\,\text{nm}^2\right)$ have been held constant in the scheme of Figure 3.10a,b. In order to better observe the impact, this was done. All these calculations are based on the assumption that the Winkler and Pasternak basis maintains its initial value, such that $K_w = 0$ and Kg = 0. In Figure 3.10a,b, we can observe that when the gradient index value increases, the non-dimensional frequency decreases. This is due to the fact that an increase in the gradient index reduces the total stiffness of the employed facesheets. Moreover, an almost exponential attenuation is seen in the inherent frequencies of the composite magnetostrictive

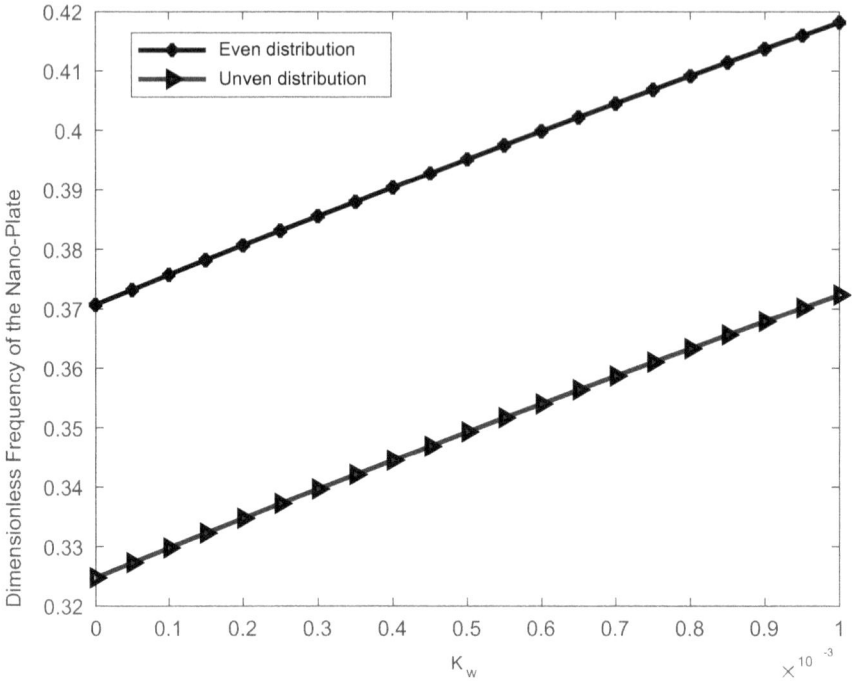

FIGURE 3.9 Composite magnetostrictive nanoplates with (FGM) facesheets versus Winkler gating materials in terms of their inherent frequencies.

nanoplate that has been federated with FGM facesheets for both even and uneven distribution. Hence, if one wants to create a reliable real-world application, they must account for the negative impact that porosity has on the dynamic behavior of the system.

The effect of the porosity volume fraction on the dimensionless natural frequencies is shown graphically in Figure 3.11, which is presented at the very end of the article. It is evident that pores have a negative effect on the stiffness of the facesheets, as shown by the fact that the dimensionless natural frequency for the even distribution decreases as the porosity volume percentage increases. To make sense of Figure 3.11, assume that the gradient index, k, is held constant at 1.5 and that the nanoplate, a, is square with sides of 40 nm on each side. This point has to be highlighted.

3.2.5 CONCLUDING REMARKS

The FSDT was used in this study to analyze the vibrational behavior of a composite magnetostrictive nanoplate that was integrated with functionally graded facesheets. This analysis was carried out in order to better understand the results of

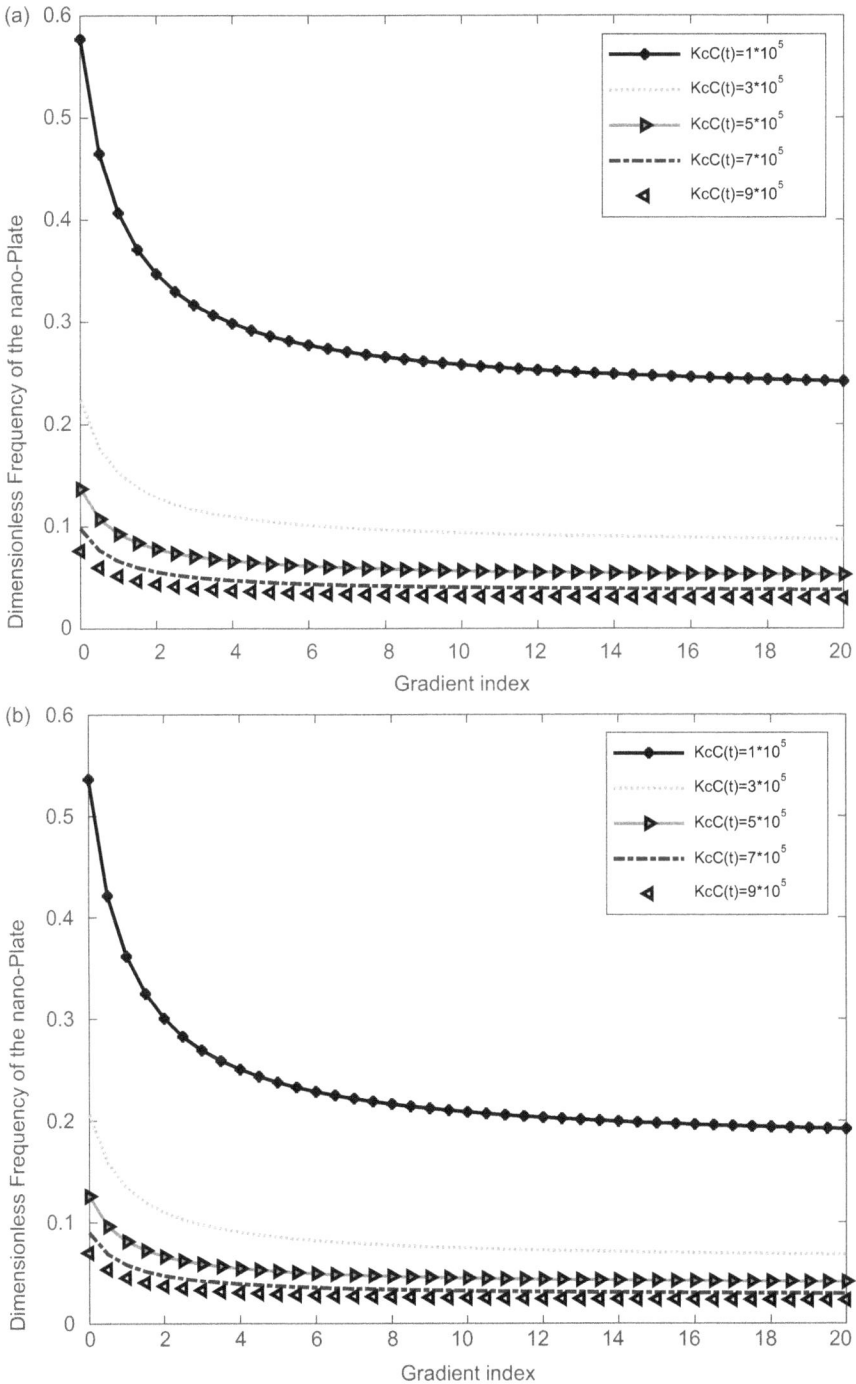

FIGURE 3.10 Determined for a range of feedback gain levels, the effect of the gradient index on the nanoplate's dimensionless natural frequency: (a) Even, (b) Uneven distribution $(K_w = 0, K_g = 0)$.

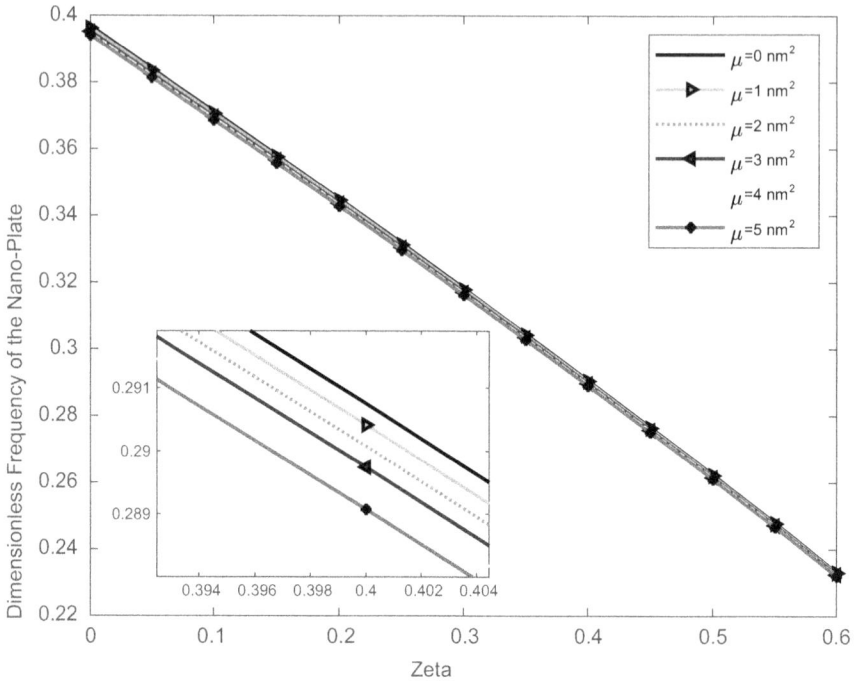

FIGURE 3.11 The frequency of a nanoplate without regard to its dimensions and its influence on achieving uniform porosity ($k = 1.5$, $K_c = 1\times10^5$, $K_w = 0$, $K_g = 0$).

this study. In order to acquire the governing equations, Hamilton's principle was put to use, and then those equations, once obtained, were solved with the assistance of Navier's analytical solution. The following are some points that stood out to the researchers based on the findings of the current study:

- If you disregard the impacts of the possible pores, you may end up with an inaccurate calculation of the frequency of the intelligent system, which may lead to unanticipated events such as resonance at lower frequencies.
- By making intelligent use of the magnetostriction phenomena, the frequency of the oscillations may be easily adjusted to the desired level.
- The natural frequencies of the nanoplate see a significant drop when the value of the feedback gain is made larger.

Appendix 3A

$$K_{11} = -A_{11}\left(o^2\right)$$

$$K_{12} = -B_{11}\left(o^2\right) + \alpha S_{11}\left(o^2\right)$$

$$K_{13} = \alpha S_{11}\left(o^3\right)$$

$$K_{21} = -B_{11}\left(o^2\right) + \alpha S_{11}\left(o^2\right)$$

$$K_{22} = -D_{11}\left(o^2\right) + 2\alpha M_{11}\left(o^2\right) - \alpha^2 L_{11}\left(o^2\right) - A_{55} + 2\beta D_{55} - \beta^2 M_{55}$$

$$K_{23} = \alpha M_{11}\left(o^3\right) - \alpha^2 L_{11}\left(o^3\right) + 2\beta D_{55}\left(o\right) - \beta^2 M_{55}\left(o\right) - A_{55}\left(o\right)$$

$$K_{31} = \alpha S_{11}\left(o^3\right)$$

$$K_{32} = \alpha M_{11}\left(o^3\right) - \alpha^2 L_{11}\left(o^3\right) + 2\beta D_{55}\left(o\right) - \beta^2 M_{55}\left(o\right) - A_{55}\left(o\right)$$

$$K_{33} = -\alpha^2 L_{11}\left(o^4\right) - \beta^2 M_{55}\left(o^2\right) + 2\beta D_{55}\left(o^2\right) - A_{55}\left(o^2\right) - K_w - K_g\left(o^2\right)$$
$$\quad - K_w\left(\mu^2 o^2\right) - K_g\left(\mu^2 o^4\right)$$

$$M_{11} = I_0\left(1 + \mu^2 o^2\right)$$

$$M_{12} = -\left(I_1 - I_3\alpha + \mu^2 I_1 o^2 + I_3\alpha\mu^2 o^2\right)$$

$$M_{13} = -\left(-I_3\alpha(o) - \mu^2 I_3\alpha\left(o^3\right)\right)$$

$$M_{21} = -\left(I_1 - I_3\alpha + \mu^2 I_1 o^2 + I_3\alpha\mu^2 o^2\right)$$

$$M_{22} = -\left(I_2 - 2I_4\alpha + I_6\alpha^2 + I_2\mu^2 o^2 - 2I_4\alpha o^2 + I_6\mu^2\alpha^2 o^2\right)$$

$$M_{23} = -\left(I_6\alpha^2 o - I_4\alpha(o) + I_6\alpha^2\mu^2 o^3 - I_4\alpha\mu^2 o^3\right)$$

$$M_{31} = -\left(-I_3\alpha(o) - \mu^2 I_3\alpha\left(o^3\right)\right)$$

$$M_{32} = -\left(I_6\alpha^2 o - I_4\alpha(o) + I_6\alpha^2\mu^2 o^3 - I_4\alpha\mu^2 o^3\right)$$

$$M_{33} = -\left(I_0 + I_6 o^2\alpha^2 + I_0\mu^2 o^2 + I_6\mu^2 o^4\alpha^2\right)$$

$$A_{11} = \int_{-h_c/2}^{h_c/2} C_{11}^c dz + \int_{-\frac{(hc+h_f)}{2}}^{\frac{-hc}{2}} C_{11}^f dz + \int_{\frac{h_c}{2}}^{\frac{(h_c+h_f)}{2}} C_{11}^f dz$$

$$A_{12} = \int_{-h_c/2}^{h_c/2} C_{12}^c dz + \int_{-\frac{(hc+h_f)}{2}}^{\frac{-hc}{2}} C_{12}^f dz + \int_{\frac{h_c}{2}}^{\frac{(h_c+h_f)}{2}} C_{12}^f dz$$

$$A_{22} = \int_{-h_c/2}^{h_c/2} C_{22}^c dz + \int_{-\frac{(hc+h_f)}{2}}^{\frac{-hc}{2}} C_{22}^f dz + \int_{\frac{h_c}{2}}^{\frac{(h_c+h_f)}{2}} C_{22}^f dz$$

$$A_{44} = \int_{-h_c/2}^{h_c/2} C_{44}^c dz + \int_{-\frac{(hc+h_f)}{2}}^{\frac{-hc}{2}} C_{44}^f dz + \int_{\frac{h_c}{2}}^{\frac{(h_c+h_f)}{2}} C_{44}^f dz$$

$$A_{55} = \int_{-h_c/2}^{h_c/2} C_{55}^c dz + \int_{-\frac{(hc+h_f)}{2}}^{\frac{-hc}{2}} C_{55}^f dz + \int_{\frac{h_c}{2}}^{\frac{(h_c+h_f)}{2}} C_{55}^f dz$$

$$A_{66} = \int_{-h_c/2}^{h_c/2} C_{66}^c dz + \int_{-\frac{(hc+h_f)}{2}}^{\frac{-hc}{2}} C_{66}^f dz + \int_{\frac{h_c}{2}}^{\frac{(h_c+h_f)}{2}} C_{66}^f dz$$

$$B_{11} = \int_{-h_c/2}^{h_c/2} z C_{11}^c dz + \int_{-\frac{(hc+h_f)}{2}}^{\frac{-hc}{2}} z C_{11}^f dz + \int_{\frac{h_c}{2}}^{\frac{(h_c+h_f)}{2}} z C_{11}^f dz$$

$$B_{12} = \int_{-h_c/2}^{h_c/2} z C_{12}^c dz + \int_{-\frac{(hc+h_f)}{2}}^{\frac{-hc}{2}} z C_{12}^f dz + \int_{\frac{h_c}{2}}^{\frac{(h_c+h_f)}{2}} z C_{12}^f dz$$

$$B_{22} = \int_{-h_c/2}^{h_c/2} z C_{22}^c dz + \int_{-\frac{(hc+h_f)}{2}}^{\frac{-hc}{2}} z C_{22}^f dz + \int_{\frac{h_c}{2}}^{\frac{(h_c+h_f)}{2}} z C_{22}^f dz$$

$$B_{44} = \int_{-h_c/2}^{h_c/2} z C_{44}^c dz + \int_{-\frac{(hc+h_f)}{2}}^{\frac{-hc}{2}} z C_{44}^f dz + \int_{\frac{h_c}{2}}^{\frac{(h_c+h_f)}{2}} z C_{44}^f dz$$

$$B_{55} = \int_{-h_c/2}^{h_c/2} z C_{55}^c dz + \int_{-\frac{(hc+h_f)}{2}}^{\frac{-hc}{2}} z C_{22}^f dz + \int_{\frac{h_c}{2}}^{\frac{(h_c+h_f)}{2}} z C_{55}^f dz$$

$$B_{66} = \int_{-h_c/2}^{h_c/2} z C_{66}^c dz + \int_{-\frac{(hc+h_f)}{2}}^{\frac{-hc}{2}} z C_{66}^f dz + \int_{\frac{h_c}{2}}^{\frac{(h_c+h_f)}{2}} z C_{66}^f dz$$

$$D_{11} = \int_{-h_c/2}^{h_c/2} z^2 C_{11}^c dz + \int_{-\frac{(hc+h_f)}{2}}^{\frac{-hc}{2}} z^2 C_{11}^f dz + \int_{\frac{h_c}{2}}^{\frac{(h_c+h_f)}{2}} z^2 C_{11}^f dz$$

$$D_{12} = \int_{-h_c/2}^{h_c/2} z^2 C_{12}^c dz + \int_{-\frac{(hc+h_f)}{2}}^{\frac{-hc}{2}} z^2 C_{12}^f dz + \int_{\frac{h_c}{2}}^{\frac{(h_c+h_f)}{2}} z^2 C_{12}^f dz$$

$$D_{22} = \int_{-h_c/2}^{h_c/2} z^2 C_{22}^c dz + \int_{-\frac{(hc+h_f)}{2}}^{\frac{-hc}{2}} z^2 C_{22}^f dz + \int_{\frac{h_c}{2}}^{\frac{(h_c+h_f)}{2}} z^2 C_{22}^f dz$$

$$D_{44} = \int_{-h_c/2}^{h_c/2} z^2 C_{44}^c dz + \int_{-\frac{(hc+h_f)}{2}}^{\frac{-hc}{2}} z^2 C_{44}^f dz + \int_{\frac{h_c}{2}}^{\frac{(h_c+h_f)}{2}} z^2 C_{44}^f dz$$

$$D_{55} = \int_{-h_c/2}^{h_c/2} z^2 C_{55}^c dz + \int_{-\frac{(hc+h_f)}{2}}^{\frac{-hc}{2}} z^2 C_{55}^f dz + \int_{\frac{h_c}{2}}^{\frac{(h_c+h_f)}{2}} z^2 C_{55}^f dz$$

$$D_{66} = \int_{-h_c/2}^{h_c/2} z^2 C_{66}^c dz + \int_{-\frac{(hc+h_f)}{2}}^{\frac{-hc}{2}} z^2 C_{66}^f dz + \int_{\frac{h_c}{2}}^{\frac{(h_c+h_f)}{2}} z^2 C_{66}^f dz$$

$$S_{11} = \int_{-h_c/2}^{h_c/2} z^3 C_{11}^c dz + \int_{-\frac{(hc+h_f)}{2}}^{\frac{-hc}{2}} z^3 C_{11}^f dz + \int_{\frac{h_c}{2}}^{\frac{(h_c+h_f)}{2}} z^3 C_{11}^f dz$$

$$S_{12} = \int_{-h_c/2}^{h_c/2} z^3 C_{12}^c dz + \int_{-\frac{(hc+h_f)}{2}}^{\frac{-hc}{2}} z^3 C_{12}^f dz + \int_{\frac{h_c}{2}}^{\frac{(h_c+h_f)}{2}} z^3 C_{12}^f dz$$

$$S_{22} = \int_{-h_c/2}^{h_c/2} z^3 C_{22}^c dz + \int_{-\frac{(hc+h_f)}{2}}^{\frac{-hc}{2}} z^3 C_{22}^f dz + \int_{\frac{h_c}{2}}^{\frac{(h_c+h_f)}{2}} z^3 C_{22}^f dz$$

$$S_{44} = \int_{-h_c/2}^{h_c/2} z^3 C_{44}^c dz + \int_{-\frac{(hc+h_f)}{2}}^{\frac{-hc}{2}} z^3 C_{44}^f dz + \int_{\frac{h_c}{2}}^{\frac{(h_c+h_f)}{2}} z^3 C_{44}^f dz$$

$$S_{55} = \int_{-h_c/2}^{h_c/2} z^3 C_{55}^c dz + \int_{-\frac{(hc+h_f)}{2}}^{\frac{-hc}{2}} z^3 C_{55}^f dz + \int_{\frac{h_c}{2}}^{\frac{(h_c+h_f)}{2}} z^3 C_{55}^f dz$$

$$S_{66} = \int_{-h_c/2}^{h_c/2} z^3 C_{66}^c dz + \int_{-\frac{(hc+h_f)}{2}}^{\frac{-hc}{2}} z^3 C_{66}^f dz + \int_{\frac{h_c}{2}}^{\frac{(h_c+h_f)}{2}} z^3 C_{66}^f dz$$

$$M_{11} = \int_{-h_c/2}^{h_c/2} z^4 C_{11}^c dz + \int_{-\frac{(hc+h_f)}{2}}^{\frac{-hc}{2}} z^4 C_{11}^f dz + \int_{\frac{h_c}{2}}^{\frac{(h_c+h_f)}{2}} z^4 C_{11}^f dz$$

$$M_{12} = \int_{-h_c/2}^{h_c/2} z^4 C_{12}^c dz + \int_{-\frac{(hc+h_f)}{2}}^{\frac{-hc}{2}} z^4 C_{12}^f dz + \int_{\frac{h_c}{2}}^{\frac{(h_c+h_f)}{2}} z^4 C_{12}^f dz$$

$$M_{22} = \int_{-h_c/2}^{h_c/2} z^4 C_{22}^c dz + \int_{-\frac{(hc+h_f)}{2}}^{\frac{-hc}{2}} z^4 C_{22}^f dz + \int_{\frac{h_c}{2}}^{\frac{(h_c+h_f)}{2}} z^4 C_{22}^f dz$$

$$M_{44} = \int_{-h_c/2}^{h_c/2} z^4 C_{44}^c dz + \int_{-\frac{(hc+h_f)}{2}}^{-hc} z^4 C_{44}^f dz + \int_{\frac{h_c}{2}}^{\frac{(h_c+h_f)}{2}} z^4 C_{44}^f dz$$

$$M_{55} = \int_{-h_c/2}^{h_c/2} z^4 C_{55}^c dz + \int_{-\frac{(hc+h_f)}{2}}^{-hc} z^4 C_{22}^f dz + \int_{\frac{h_c}{2}}^{\frac{(h_c+h_f)}{2}} z^4 C_{55}^f dz$$

$$M_{66} = \int_{-h_c/2}^{h_c/2} z^4 C_{66}^c dz + \int_{-\frac{(hc+h_f)}{2}}^{-hc} z^4 C_{66}^f dz + \int_{\frac{h_c}{2}}^{\frac{(h_c+h_f)}{2}} z^4 C_{66}^f dz$$

$$L_{11} = \int_{-h_c/2}^{h_c/2} z^6 C_{11}^c dz + \int_{-\frac{(hc+h_f)}{2}}^{-hc} z^6 C_{11}^f dz + \int_{\frac{h_c}{2}}^{\frac{(h_c+h_f)}{2}} z^6 C_{11}^f dz$$

$$L_{12} = \int_{-h_c/2}^{h_c/2} z^6 C_{12}^c dz + \int_{-\frac{(hc+h_f)}{2}}^{-hc} z^6 C_{12}^f dz + \int_{\frac{h_c}{2}}^{\frac{(h_c+h_f)}{2}} z^6 C_{12}^f dz$$

$$L_{22} = \int_{-h_c/2}^{h_c/2} z^6 C_{22}^c dz + \int_{-\frac{(hc+h_f)}{2}}^{-hc} z^6 C_{22}^f dz + \int_{\frac{h_c}{2}}^{\frac{(h_c+h_f)}{2}} z^6 C_{22}^f dz$$

$$L_{44} = \int_{-h_c/2}^{h_c/2} z^6 C_{44}^c dz + \int_{-\frac{(hc+h_f)}{2}}^{-hc} z^6 C_{44}^f dz + \int_{\frac{h_c}{2}}^{\frac{(h_c+h_f)}{2}} z^6 C_{44}^f dz$$

$$L_{55} = \int_{-h_c/2}^{h_c/2} z^6 C_{55}^c dz + \int_{-\frac{(hc+h_f)}{2}}^{-hc} z^6 C_{55}^f dz + \int_{\frac{h_c}{2}}^{\frac{(h_c+h_f)}{2}} z^6 C_{55}^f dz$$

$$L_{66} = \int_{-h_c/2}^{h_c/2} z^6 C_{66}^c dz + \int_{-\frac{(hc+h_f)}{2}}^{-hc} z^6 C_{66}^f dz + \int_{\frac{h_c}{2}}^{\frac{(h_c+h_f)}{2}} z^6 C_{66}^f dz$$

Appendix 3B

$$K_{11} = -A_{11}\left(\alpha^2\right) - A_{66}\left(\beta^2\right) \qquad K_{12} = -A_{12}\left(\alpha\beta\right) - A_{66}\left(\alpha\beta\right)$$

$$K_{13} = 0 \qquad K_{14} = -B_{11}\left(\alpha^2\right) - B_{66}\left(\beta^2\right)$$

$$K_{15} = -B_{12}\left(\alpha\beta\right) - B_{66}\left(\alpha\beta\right)$$

$$K_{21} = -A_{21}\left(\alpha\beta\right) - A_{66}\left(\alpha\beta\right) \qquad K_{22} = -A_{22}\left(\alpha^2\right) - A_{66}\left(\beta^2\right)$$

$$K_{23} = 0 \qquad K_{24} = -B_{21}\left(\alpha\beta\right) - B_{66}\left(\alpha\beta\right)$$

$$K_{25} = -B_{22}\left(\alpha^2\right) - B_{66}\left(\beta^2\right)$$

$$K_{31} = 0 \qquad K_{32} = 0$$

$$K_{33} = -A_{55}\left(\alpha^2\right) - A_{44}\left(\beta^2\right) - K_w - K_g\left(\alpha^2\right) - K_g\left(\beta^2\right) - K_w\left(\mu^2\alpha^2\right) - K_w\left(\mu^2\beta^2\right)$$
$$\qquad - K_g\left(\mu^2\alpha^4\right) - K_g\left(\mu^2\beta^4\right) - 2K_g\left(\mu^2\alpha^2\beta^2\right)$$

$$K_{34} = -A_{55}\left(\alpha\right) \qquad K_{35} = -A_{44}\left(\beta\right)$$

$$K_{41} = -B_{11}\left(\alpha^2\right) - B_{66}\left(\beta^2\right) \qquad K_{42} = -B_{21}\left(\alpha\beta\right) - B_{66}\left(\alpha\beta\right)$$

$$K_{43} = -A_{55}\left(\alpha\right) \qquad K_{44} = -D_{11}\left(\alpha^2\right) - D_{66}\left(\beta^2\right) - A_{55}$$

$$K_{45} = -D_{12}\left(\alpha\beta\right) - D_{66}\left(\alpha\beta\right)$$

$$K_{51} = -B_{12}\left(\alpha\beta\right) - B_{66}\left(\alpha\beta\right) \qquad K_{52} = -B_{22}\left(\alpha^2\right) - B_{66}\left(\beta^2\right)$$

$$K_{53} = -A_{44}\left(\beta\right) \qquad K_{54} = -D_{12}\left(\alpha\beta\right) - D_{66}\left(\alpha\beta\right)$$

$$K_{55} = -D_{22}\left(\beta^2\right) - D_{66}\left(\alpha^2\right) - A_{44}$$

$$M_{11} = I_0\left(1 + \mu^2\alpha^2 + \mu^2\beta^2\right) \qquad M_{14} = I_1\left(1 + \mu^2\alpha^2 + \mu^2\beta^2\right)$$

$$M_{22} = I_0\left(1 + \mu^2\alpha^2 + \mu^2\beta^2\right) \qquad M_{25} = I_1\left(1 + \mu^2\alpha^2 + \mu^2\beta^2\right)$$

$$M_{33} = I_0\left(1 + \mu^2\alpha^2 + \mu^2\beta^2\right) \qquad M_{41} = I_1\left(1 + \mu^2\alpha^2 + \mu^2\beta^2\right)$$

$$M_{44} = I_2\left(1 + \mu^2\alpha^2 + \mu^2\beta^2\right) \qquad M_{52} = I_1\left(1 + \mu^2\alpha^2 + \mu^2\beta^2\right)$$

$$M_{55} = I_2\left(1 + \mu^2\alpha^2 + \mu^2\beta^2\right)$$

$$C_{11} = C_{12} = C_{14} = C_{15} = 0 \qquad C_{13} = \int_{-\frac{h_c}{2}}^{\frac{h_c}{2}} e_{31}\left(\alpha\right) K_c C\left(t\right) dz$$

$$C_{21} = C_{22} = C_{24} = C_{25} = 0 \qquad C_{23} = \int_{-\frac{h_c}{2}}^{\frac{h_c}{2}} e_{32}\left(\beta\right) K_c C\left(t\right) dz$$

$$C_{31} = \int_{-\frac{h_c}{2}}^{\frac{h_c}{2}} e_{31}\left(\alpha\right) K_c C\left(t\right) dz \qquad C_{32} = \int_{-\frac{h_c}{2}}^{\frac{h_c}{2}} e_{31}\left(\beta\right) K_c C\left(t\right) dz$$

$$C_{33} = 0 \qquad C_{34} = \int_{-\frac{h_c}{2}}^{\frac{h_c}{2}} z e_{31}\left(\alpha\right) K_c C\left(t\right) dz$$

$$C_{35} = \int_{-\frac{h_c}{2}}^{\frac{h_c}{2}} z e_{32}\left(\beta\right) K_c C\left(t\right) dz$$

$$C_{41} = C_{42} = C_{44} = C_{45} = 0 \qquad C_{43} = \int_{-\frac{h_c}{2}}^{\frac{h_c}{2}} z e_{31}(\alpha) K_c C(t) dz$$

$$C_{51} = C_{52} = C_{54} = C_{55} = 0 \qquad C_{53} = \int_{-\frac{h_c}{2}}^{\frac{h_c}{2}} z e_{32}(\beta) K_c C(t) dz$$

$$A_{11} = \int_{-h_c/2}^{h_c/2} C_{11}^c dz + \int_{-(hc+h_f)}^{-\frac{hc}{2}} C_{11}^f dz + \int_{\frac{h_c}{2}}^{\frac{(h_c+h_f)}{2}} C_{11}^f dz$$

$$A_{12} = \int_{-h_c/2}^{h_c/2} C_{12}^c dz + \int_{-(hc+h_f)}^{-\frac{hc}{2}} C_{12}^f dz + \int_{\frac{h_c}{2}}^{\frac{(h_c+h_f)}{2}} C_{12}^f dz$$

$$A_{22} = \int_{-h_c/2}^{h_c/2} C_{22}^c dz + \int_{-(hc+h_f)}^{-\frac{hc}{2}} C_{22}^f dz + \int_{\frac{h_c}{2}}^{\frac{(h_c+h_f)}{2}} C_{22}^f dz$$

$$A_{44} = \int_{-h_c/2}^{h_c/2} C_{44}^c dz + \int_{-(hc+h_f)}^{-\frac{hc}{2}} C_{44}^f dz + \int_{\frac{h_c}{2}}^{\frac{(h_c+h_f)}{2}} C_{44}^f dz$$

$$A_{55} = \int_{-h_c/2}^{h_c/2} C_{55}^c dz + \int_{-(hc+h_f)}^{-\frac{hc}{2}} C_{55}^f dz + \int_{\frac{h_c}{2}}^{\frac{(h_c+h_f)}{2}} C_{55}^f dz$$

$$A_{66} = \int_{-h_c/2}^{h_c/2} C_{66}^c dz + \int_{-(hc+h_f)}^{-\frac{hc}{2}} C_{66}^f dz + \int_{\frac{h_c}{2}}^{\frac{(h_c+h_f)}{2}} C_{66}^f dz$$

$$B_{11} = \int_{-h_c/2}^{h_c/2} z C_{11}^c dz + \int_{-(hc+h_f)}^{-\frac{hc}{2}} z C_{11}^f dz + \int_{\frac{h_c}{2}}^{\frac{(h_c+h_f)}{2}} z C_{11}^f dz$$

$$B_{12} = \int_{-h_c/2}^{h_c/2} z C_{12}^c dz + \int_{-(hc+h_f)}^{-\frac{hc}{2}} z C_{12}^f dz + \int_{\frac{h_c}{2}}^{\frac{(h_c+h_f)}{2}} z C_{12}^f dz$$

$$B_{22} = \int_{-h_c/2}^{h_c/2} z C_{22}^c dz + \int_{-(hc+h_f)}^{-\frac{hc}{2}} z C_{22}^f dz + \int_{\frac{h_c}{2}}^{\frac{(h_c+h_f)}{2}} z C_{22}^f dz$$

$$B_{44} = \int_{-h_c/2}^{h_c/2} z C_{44}^c dz + \int_{-(hc+h_f)}^{-\frac{hc}{2}} z C_{44}^f dz + \int_{\frac{h_c}{2}}^{\frac{(h_c+h_f)}{2}} z C_{44}^f dz$$

$$B_{55} = \int_{-h_c/2}^{h_c/2} z C_{55}^c dz + \int_{-(hc+h_f)}^{-\frac{hc}{2}} z C_{22}^f dz + \int_{\frac{h_c}{2}}^{\frac{(h_c+h_f)}{2}} z C_{55}^f dz$$

$$B_{66} = \int_{-h_c/2}^{h_c/2} z C_{66}^c dz + \int_{-\frac{(hc+h_f)}{2}}^{\frac{-hc}{2}} z C_{66}^f dz + \int_{\frac{h_c}{2}}^{\frac{(h_c+h_f)}{2}} z C_{66}^f dz$$

$$D_{11} = \int_{-h_c/2}^{h_c/2} z^2 C_{11}^c dz + \int_{-\frac{(hc+h_f)}{2}}^{\frac{-hc}{2}} z^2 C_{11}^f dz + \int_{\frac{h_c}{2}}^{\frac{(h_c+h_f)}{2}} z^2 C_{11}^f dz$$

$$D_{12} = \int_{-h_c/2}^{h_c/2} z^2 C_{12}^c dz + \int_{-\frac{(hc+h_f)}{2}}^{\frac{-hc}{2}} z^2 C_{12}^f dz + \int_{\frac{h_c}{2}}^{\frac{(h_c+h_f)}{2}} z^2 C_{12}^f dz$$

$$D_{22} = \int_{-h_c/2}^{h_c/2} z^2 C_{22}^c dz + \int_{-\frac{(hc+h_f)}{2}}^{\frac{-hc}{2}} z^2 C_{22}^f dz + \int_{\frac{h_c}{2}}^{\frac{(h_c+h_f)}{2}} z^2 C_{22}^f dz$$

$$D_{44} = \int_{-h_c/2}^{h_c/2} z^2 C_{44}^c dz + \int_{-\frac{(hc+h_f)}{2}}^{\frac{-hc}{2}} z^2 C_{44}^f dz + \int_{\frac{h_c}{2}}^{\frac{(h_c+h_f)}{2}} z^2 C_{44}^f dz$$

$$D_{55} = \int_{-h_c/2}^{h_c/2} z^2 C_{55}^c dz + \int_{-\frac{(hc+h_f)}{2}}^{\frac{-hc}{2}} z^2 C_{55}^f dz + \int_{\frac{h_c}{2}}^{\frac{(h_c+h_f)}{2}} z^2 C_{55}^f dz$$

$$D_{66} = \int_{-h_c/2}^{h_c/2} z^2 C_{66}^c dz + \int_{-\frac{(hc+h_f)}{2}}^{\frac{-hc}{2}} z^2 C_{66}^f dz + \int_{\frac{h_c}{2}}^{\frac{(h_c+h_f)}{2}} z^2 C_{66}^f dz$$

REFERENCES

1. Ebrahimi, F. and A. Jafari, *A higher-order thermomechanical vibration analysis of temperature-dependent FGM beams with porosities.* Journal of Engineering, 2016. **2016**: p. 20.
2. Ebrahimi, F., A. Dabbagh, and T. Rabczuk, *On wave dispersion characteristics of magnetostrictive sandwich nanoplates in thermal environments.* European Journal of Mechanics-A/Solids, 2021. **85**: p. 104130.
3. Ebrahimi, F. and M.F. Ahari, *Magnetostriction-assisted active control of the multi-layered nanoplates: effect of the porous functionally graded facesheets on the system's behavior.* Engineering with Computers, 2021: p. 1–15.
4. Rao, S.S., *Vibration of continuous systems.* 2019: John Wiley & Sons.
5. Ebrahimi, F., M.-S. Shafiee, and M.F. Ahari, *Buckling analysis of single and double-layer annular graphene sheets in thermal environment.* Engineering with Computers, 2022: p. 1–15.
6. Ebrahimi, F., M.-S. Shafiei, and M.F. Ahari, *Vibration analysis of single and multi-walled circular graphene sheets in thermal environment using GDQM.* Waves in Random and Complex Media, 2022: p. 1–40.
7. Eltaher, M., S.A. Emam, and F. Mahmoud, *Free vibration analysis of functionally graded size-dependent nanobeams.* Applied Mathematics and Computation, 2012. **218**(14): p. 7406–7420.

8. Rahmani, O. and O. Pedram, *Analysis and modeling the size effect on vibration of functionally graded nanobeams based on nonlocal Timoshenko beam theory.* International Journal of Engineering Science, 2014. **77**: p. 55–70.

9. Mindlin, R., *Influence of rotatory inertia and shear on flexural motions of isotropic, elastic plates.* 1951.

10. Reddy, J., C. Wang, and S. Kitipornchai, *Axisymmetric bending of functionally graded circular and annular plates.* European Journal of Mechanics-A/Solids, 1999. **18**(2): p. 185–199.

11. Mahinzare, M., H. Akhavan, and M. Ghadiri, *A nonlocal strain gradient theory for rotating thermo-mechanical characteristics on magnetically actuated viscoelastic functionally graded nanoshell.* Journal of Intelligent Material Systems and Structures, 2020. **31**(12): p. 1511–1523.

12. Ebrahimi, F. and M.R. Barati, *Electro-magnetic effects on nonlocal dynamic behavior of embedded piezoelectric nanoscale beams.* Journal of Intelligent Material Systems and Structures, 2017. **28**(15): p. 2007–2022.

13. Ebrahimi, F., et al., *Vibration analysis of porous magneto-electro-elastically actuated carbon nanotube-reinforced composite sandwich plate based on a refined plate theory.* Engineering with Computers, 2021. **37**: p. 921–936.

14. Ghorbanpour Arani, A., H. Khani Arani, and Z. Khoddami Maraghi, *Size-dependent in vibration analysis of magnetostrictive sandwich composite micro-plate in magnetic field using modified couple stress theory.* Journal of Sandwich Structures & Materials, 2019. **21**(2): p. 580–603.

15. Hosseini-Hashemi, S., et al., *Free vibration of functionally graded rectangular plates using first-order shear deformation plate theory.* Applied Mathematical Modelling, 2010. **34**(5): p. 1276–1291.

16. Zhao, X., Y.Y. Lee, and K.M. Liew, *Free vibration analysis of functionally graded plates using the element-free kp-Ritz method.* Journal of Sound and Vibration, 2009. **319**(3): p. 918–939.

17. Matsunaga, H., *Free vibration and stability of functionally graded plates according to a 2-D higher-order deformation theory.* Composite Structures, 2008. **82**(4): p. 499–512.

18. Rezaei, A. and A. Saidi, *Exact solution for free vibration of thick rectangular plates made of porous materials.* Composite Structures, 2015. **134**: p. 1051–1060.

19. Ghorbanpour Arani, A., Z. Khoddami Maraghi, and H. Khani Arani, *Orthotropic patterns of Pasternak foundation in smart vibration analysis of magnetostrictive nanoplate.* Proceedings of the Institution of Mechanical Engineers, Part C: Journal of Mechanical Engineering Science, 2016. **230**(4): p. 559–572.

20. Khani Arani, H., M. Shariyat, and A. Mohammadian, *Vibration analysis of magnetostrictive nano-plate by using modified couple stress and nonlocal elasticity theories.* International Journal of Materials and Metallurgical Engineering, 2020. **14**(9): p. 229–234.

21. Rezaei, A.S. and A.R. Saidi, *Exact solution for free vibration of thick rectangular plates made of porous materials.* Composite Structures, 2015. **134**: p. 1051–1060.

22. Jha, D.K., T. Kant, and R.K. Singh, *Higher order shear and normal deformation theory for natural frequency of functionally graded rectangular plates.* Nuclear Engineering and Design, 2012. **250**: p. 8–13.

4 Wave Propagation of Magnetostrictive Materials and Structures

Background

The fundamental idea behind the wave propagation phenomena in solids is that any externally created disturbance will move from a desired place on a substrate to other contiguous sites in the continuum as it moves from that ideal point. If I'm being completely forthright with you, the phenomenon of wave propagation is directly associated with the interatomic interactions of very small particles such as atoms and molecules. However, in common problems in the field of mechanical engineering, either solid or fluid mechanics, the system that is selected and assumed to be the host of the dispersion is considered to be a continuous system. This is the case regardless of whether the mechanics being studied are solid or fluid. As a result of this assumption, the relations of continuum mechanics are employed in order to mathematically define the phenomenon that was discussed before. The stated assumption may be made more clear by examining the basic principles of continuous systems, in particular those that pertain to solid mechanics. This is the subject matter that will be covered in this chapter.

The motion of elastic waves is sufficiently enough to transmit a significant quantity of energy across very great distances in the medium in which they are traveling. In addition, it is essential to keep in mind that wave propagation is a dynamic process that, in order to take place, calls for a certain collection of prerequisites. To put it another way, waves are unable to go through a stiff body like a wall. Hence, the presence of inertia is the primary condition that must be fulfilled for the propagation of waves in a medium that is desired.

When seen from the perspective of a mechanical engineer, the phenomena of wave propagation in an elastic solid are very significant. The extensive use of wave propagation in real-world situations is the primary driver behind this reality. A wave is introduced as an input into a system in many different applications, and several bits of information are gleaned from the waves' reflection or growth in the medium as they travel through it. For example, the propagation of elastic waves is one of the most effective means of locating flaws and voids in a medium that has certain desired characteristics. This process may be finished by transmitting

DOI: 10.1201/9781003355427-4

an elastic wave into a medium and then calculating the differences in the characteristics of the transmitted and reflected branches of the wave. In other words, nondestructive tests (NDTs) are used in many applications to discover defects developed in the process of manufacture; but, in other instances, such exams are not appropriate. Because of this, a technique based on wave propagation is applied in order to locate and identify the flaws.

Wave Propagation of Magnetostrictive Materials and Structures

The wave propagation in magnetostrictive materials and structures are investigated in this chapter, along with a broad variety of analytical and numerical solution approaches. The discussion will begin with an examination of the wave propagation in magnetostrictive materials, followed by an explanation of many different cases.

4.1 WAVE PROPAGATION OF MAGNETOSTRICTIVE PLATES

In this section, wave propagation of magnetostrictive will be discussed through various examples for plate structures. Also, the effect of hygrothermal environment will be explained through figures and tables.

4.1.1 FORMULATION

Examine the use of a classical plate model in order to define the wave propagation issue of MSNPs while taking into account the effects of scale. The MSNPs are exposed to a thermal loading. When it comes to magnetostrictive materials, the relationship between stress, strain, and magnetic field may be stated as Eq. (2.21). Moreover, C_{ij} represents the elastic stiffness coefficient, which may be expressed as Eq. (2.22).

Understanding the reciprocal magnetoelastic interaction found in magnetostrictive materials is intriguing in this light. Materials of this kind may be resilient to both stretching and compressing after being subjected to a magnetic field. According to the text, this phenomenon is caused by the reorientation of the magnetic moments of individual atoms. To further the goal of capturing the magnetization effect, a feedback control system is employed as the basis for an effort to account for the intensity of the magnetic field in MSCNP. Using this framework, we may write the magnetostrictive coefficients as Eq. (2.23).

To take into consideration the magnetization effects of the MSCNP, the current study makes use of a unidirectional magnetic field along the z axis, which is then applied to the MSCNP. Following is an expression that may be used to describe the strength of the magnetic field as Eq. (2.10).

In point of fact, the value of the coil constant may be expressed as Eq. (2.25). On the other hand, we are going to discuss the stress-strain relations of the top layer and the bottom layer. In the same way as in the central core, the following equations are used in this case for the top and lower facesheets as Eq. (2.28, 2.29).

The current study models the equations of motion for the plates in accordance with the classical plate theory, which results in the following form for the equations [1]:

$$u_x(x,y,z,t) = u(x,y,t) - z\frac{\partial w(x,y,t)}{\partial x} \tag{4.1}$$

$$u_y(x,y,z,t) = v(x,y,t) - z\frac{\partial w(x,y,t)}{\partial y} \tag{4.2}$$

$$u_z(x,y,z,t) = w(x,y,t) \tag{4.3}$$

whereby the variables u, v and w correspond to the displacement components in the x, y, and z directions, respectively. Hence, the stresses of the core plate that are not zero may be characterized as the following:

$$\begin{Bmatrix} \varepsilon_{xx}^{core} \\ \varepsilon_{yy}^{core} \\ \varepsilon_{xy}^{core} \end{Bmatrix} = \begin{Bmatrix} \varepsilon_{xx}^{core} \\ \varepsilon_{yy}^{core} \\ \varepsilon_{xy}^{core} \end{Bmatrix} + z \begin{Bmatrix} \kappa_{xx}^{core} \\ \kappa_{yy}^{core} \\ \kappa_{xy}^{core} \end{Bmatrix} \tag{4.4}$$

where

$$\begin{Bmatrix} \varepsilon_{xx}^{core} \\ \varepsilon_{yy}^{core} \\ \varepsilon_{xy}^{core} \end{Bmatrix} = \begin{Bmatrix} \dfrac{\partial u}{\partial x} \\ \dfrac{\partial v}{\partial y} \\ \dfrac{\partial u}{\partial y} + \dfrac{\partial v}{\partial x} \end{Bmatrix}, \begin{Bmatrix} \kappa_{xx}^{core} \\ \kappa_{yy}^{core} \\ \kappa_{xy}^{core} \end{Bmatrix} = \begin{Bmatrix} -\dfrac{\partial^2 w}{\partial x^2} \\ -\dfrac{\partial^2 w}{\partial y^2} \\ -2\dfrac{\partial^2 w}{\partial x \partial y} \end{Bmatrix} \tag{4.5}$$

Furthermore, the same relations may be constructed for the face sheets, which are as follows:

$$\begin{Bmatrix} \varepsilon_{xx}^{face} \\ \varepsilon_{yy}^{face} \\ \varepsilon_{xy}^{face} \end{Bmatrix} = \begin{Bmatrix} \varepsilon_{xx}^{face} \\ \varepsilon_{yy}^{face} \\ \varepsilon_{xy}^{face} \end{Bmatrix} + z \begin{Bmatrix} \kappa_{xx}^{face} \\ \kappa_{yy}^{face} \\ \kappa_{xy}^{face} \end{Bmatrix} \tag{4.6}$$

where

$$
\left\{\begin{array}{l} \varepsilon_{xx}^{face} \\ \varepsilon_{yy}^{face} \\ \varepsilon_{xy}^{face} \end{array}\right\} = \left\{\begin{array}{l} \dfrac{\partial u}{\partial x} \\ \dfrac{\partial v}{\partial y} \\ \dfrac{\partial u}{\partial y} + \dfrac{\partial v}{\partial x} \end{array}\right\}, \left\{\begin{array}{l} \kappa_{xx}^{face} \\ \kappa_{yy}^{face} \\ \kappa_{xy}^{face} \end{array}\right\} = \left\{\begin{array}{l} -\dfrac{\partial^2 w}{\partial x^2} \\ -\dfrac{\partial^2 w}{\partial y^2} \\ -2\dfrac{\partial^2 w}{\partial x \partial y} \end{array}\right\} \tag{4.7}
$$

In light of this, the Euler-Lagrange equations of the MSP may be written as follows, in accordance with Hamilton's principle [2]:

$$
\int_0^t \delta\left(U - T + V\right) dt = 0 \tag{4.8}
$$

where U, T, and V represent the amount of work done by external forces, the amount of energy that is stretched, and the kinetic energy, respectively. The difference in strain energy between the core and face sheets may now be expressed as Eqs. (2.33–2.35). Moreover, the variation of kinetic energy may be characterized by Eqs. (2.42–2.43).

Now, we can write down the variation of the external work done by external forces in the following form:

$$
\delta V = \int_0^L \left(-k_w \delta w + k_p \left(\frac{\partial^2 \delta w}{\partial x^2} + \frac{\partial^2 \delta w}{\partial y^2} \right) - c_d \frac{\partial w}{\partial t} - N^T \left(\frac{\partial^2 \delta w}{\partial x^2} + \frac{\partial^2 \delta w}{\partial y^2} \right) \right) dx \tag{4.9}
$$

where k_w represents the Winkler coefficient, k_p represents the Pasternak coefficient, and C_d represents the damping coefficient of the elastic medium. Moreover, the thermal loading that is being applied to the system is denoted by the letter N^T. The Euler-Lagrange equations of MSCPs may be stated as follows after Eqs. (2.33), (2.34), (2.42), and (4.9) have been replaced in Eq. (4.8) and the coefficients of $\delta u, \delta v$ and δw have been set to zero:

$$
\frac{\partial N_{xx}}{\partial x} + \frac{\partial N_{xy}}{\partial y} = I_0 \frac{\partial^2 u}{\partial t^2} + \frac{e_{31} K_c C(t) h_c}{E_c} \frac{\partial^2 w}{\partial x \partial t} \tag{4.10}
$$

$$
\frac{\partial N_{xy}}{\partial x} + \frac{\partial N_{yy}}{\partial y} = I_0 \frac{\partial^2 v}{\partial t^2} + \frac{e_{32} K_c C(t) h_c}{E_c} \frac{\partial^2 w}{\partial y \partial t} \tag{4.11}
$$

$$
\frac{\partial^2 M_{xx}}{\partial x^2} + 2\frac{\partial^2 M_{xy}}{\partial x \partial y} + \frac{\partial^2 M_{yy}}{\partial y^2} - N^T \nabla^2 w = I_0 \frac{\partial^2 w}{\partial t^2} - I_2 \nabla^2 \frac{\partial^2 w}{\partial t^2}
$$
$$
+ k_w w - k_p \nabla^2 w + c_d \frac{\partial w}{\partial t} + \frac{e_{31} K_c C(t) h_c}{E_c} \frac{\partial^2 u}{\partial x \partial t} + \frac{e_{32} K_c C(t) h_c}{E_c} \frac{\partial^2 v}{\partial y \partial t} \tag{4.12}
$$

According to nonlocal strain gradient elasticity theory, an effective theory that is robust enough to take into account the scale impacts should include two factors, one that is nonlocal and the other that is based on length scales. So, the fundamental formulation of this theory may be stated as follows:

$$\sigma_{ij} = \sigma_{ij}^{(0)} - \frac{d\sigma_{ij}^{(1)}}{dx} \tag{4.13}$$

where $\sigma_{ij}^{(0)}$ and $\sigma_{ij}^{(1)}$ are higher order stresses that are connected to the strain (ε_{xx}) and strain gradient $(\varepsilon_{xx,x})$, respectively. The classical stress is denoted by $\sigma_{ij}^{(0)}$, while the higher order stress is denoted by $\sigma_{ij}^{(1)}$. These strains may be stated in the following format:

$$\sigma_{ij}^{(0)} = \int_0^L C_{ijkl}\,\alpha_0\left(x,x',e_0a\right)\varepsilon_{kl}'\left(x'\right)dx' \tag{4.14}$$

$$\sigma_{ij}^{(1)} = l^2 \int_0^L C_{ijkl}\,\alpha_1\left(x,x',e_1a\right)\varepsilon_{kl,x}'\left(x'\right)dx' \tag{4.15}$$

The coefficient of elasticity is indicated by C_{ijkl}, and the parameters e_0a and e_1a are introduced to account for the effects of nonlocality. Additionally, the letter l represents the strain gradient effects. The constitutive relation of the NSGT may be written as follows once the nonlocal kernel functions $\alpha_0\left(x,x',e_0a\right)$ and $\alpha_1\left(x,x',e_1a\right)$ meet the specified conditions:

$$\left(1-(e_1a)^2\nabla^2\right)\left(1-(e_0a)^2\nabla^2\right)\sigma_{ij} = C_{ijkl}\left(1-(e_1a)^2\nabla^2\right)\varepsilon_{kl}$$
$$-C_{ijkl}l^2\left(1-(e_0a)^2\nabla^2\right)\nabla^2\varepsilon_{kl} \tag{4.16}$$

where symbol ∇^2 represents the Laplacian operator. Considering that $e_1 = e_0 = e$, the general constitutive connection derived in Eq. (4.16), with reference to the effect of temperature gradients, may be expressed as follows:

$$\left(1-\mu^2\nabla^2\right)\sigma_{ij} = C_{ijkl}\left(1-\lambda^2\nabla^2\right)\left(\varepsilon_{kl}-\alpha_{ij}T\right) \tag{4.17}$$

where the nonlocal parameter is denoted by $\mu = ea$, and the length-scale parameter is denoted by $\lambda = l$. In addition, α_{ij} represents the coefficient of thermal expansion, and T is the temperature expressed in Kelvin degrees. Now that the initial nonlocal connections of normal forces and bending moments have been established, we can get the final nonlocal relations by integrating Eq. (4.17) across the cross-sectional area of the plate as follows:

$$\left(1-\mu^2\nabla^2\right)\begin{bmatrix} M_{xx} \\ M_{yy} \\ M_{xy} \end{bmatrix} =$$

$$\left(1-\lambda^2\nabla^2\right)\left(\begin{bmatrix} B_{11} & B_{12} & 0 \\ B_{21} & B_{22} & 0 \\ 0 & 0 & B_{66} \end{bmatrix}\begin{bmatrix} \dfrac{\partial u}{\partial x} \\ \dfrac{\partial v}{\partial y} \\ \dfrac{\partial u}{\partial y}+\dfrac{\partial v}{\partial x} \end{bmatrix} + \begin{bmatrix} D_{11} & D_{12} & 0 \\ D_{21} & D_{22} & 0 \\ 0 & 0 & D_{66} \end{bmatrix}\begin{bmatrix} -\dfrac{\partial^2 w}{\partial x^2} \\ -\dfrac{\partial^2 w}{\partial y^2} \\ -2\dfrac{\partial^2 w}{\partial x\partial y} \end{bmatrix}\right) \qquad (4.18)$$

$$\left(1-\mu^2\nabla^2\right)\begin{bmatrix} M_{xx} \\ M_{yy} \\ M_{xy} \end{bmatrix} =$$

$$\left(1-\lambda^2\nabla^2\right)\left(\begin{bmatrix} B_{11} & B_{12} & 0 \\ B_{21} & B_{22} & 0 \\ 0 & 0 & B_{66} \end{bmatrix}\begin{bmatrix} \dfrac{\partial u}{\partial x} \\ \dfrac{\partial v}{\partial y} \\ \dfrac{\partial u}{\partial y}+\dfrac{\partial v}{\partial x} \end{bmatrix} + \begin{bmatrix} D_{11} & D_{12} & 0 \\ D_{21} & D_{22} & 0 \\ 0 & 0 & D_{66} \end{bmatrix}\begin{bmatrix} -\dfrac{\partial^2 w}{\partial x^2} \\ -\dfrac{\partial^2 w}{\partial y^2} \\ -2\dfrac{\partial^2 w}{\partial x\partial y} \end{bmatrix}\right) \qquad (4.19)$$

where

$$\begin{bmatrix} A_{11} & B_{11} & D_{11} \\ A_{22} & B_{22} & D_{22} \\ A_{12} & B_{12} & D_{12} \\ A_{66} & B_{66} & D_{66} \end{bmatrix} = \int_{-\frac{h_c}{2}}^{\frac{h_c}{2}}\int_0^b \begin{bmatrix} 1 & z & z^2 \end{bmatrix}\begin{bmatrix} C_{11} \\ C_{22} \\ C_{12} \\ C_{66} \end{bmatrix} dy\,dz$$

$$+\int_{-\frac{(h_c+h_f)}{2}}^{-\frac{h_c}{2}}\int_0^b \begin{bmatrix} 1 & z & z^2 \end{bmatrix}\begin{bmatrix} Q_{11} \\ Q_{22} \\ Q_{12} \\ Q_{66} \end{bmatrix} dy\,dz + \int_{\frac{h_c}{2}}^{\frac{(h_c+h_f)}{2}}\int_0^b \begin{bmatrix} 1 & z & z^2 \end{bmatrix}\begin{bmatrix} Q_{11} \\ Q_{22} \\ Q_{12} \\ Q_{66} \end{bmatrix} dy\,dz \qquad (4.20)$$

$$S_{21} = S_{12},\left(S = A,B,D\right)$$

4.1.1.1 Governing Equation

Now, the final governing equations of MSNPs may be obtained by inserting Eqs. (4.18) and (4.19) in Eqs. (4.10) to (4.12):

$$
\left(1-\lambda^2\nabla^2\right)
\begin{pmatrix}
A_{11}\dfrac{\partial^2 u}{\partial x^2}+\left(A_{12}+A_{66}\right)\dfrac{\partial^2 v}{\partial x\partial y}+A_{66}\dfrac{\partial^2 u}{\partial y^2}-B_{11}\dfrac{\partial^3 w}{\partial x^3}- \\[2mm]
\left(B_{12}+2B_{66}\right)\dfrac{\partial^3 w}{\partial x\partial y^2}
\end{pmatrix}
$$
$$
+\left(1-\mu^2\nabla^2\right)\left(-I_0\dfrac{\partial^2 u}{\partial t^2}-\dfrac{e_{31}K_cC(t)h_c}{E_c}\dfrac{\partial^2 w}{\partial x\partial t}\right)=0
\tag{4.21}
$$

$$
\left(1-\lambda^2\nabla^2\right)
\begin{pmatrix}
A_{22}\dfrac{\partial^2 v}{\partial y^2}+\left(A_{12}+A_{66}\right)\dfrac{\partial^2 u}{\partial x\partial y}+A_{66}\dfrac{\partial^2 v}{\partial x^2}-B_{22}\dfrac{\partial^3 w}{\partial y^3}- \\[2mm]
\left(B_{12}+2B_{66}\right)\dfrac{\partial^3 w}{\partial x^2\partial y}
\end{pmatrix}
$$
$$
+\left(1-\mu^2\nabla^2\right)\left(-I_0\dfrac{\partial^2 v}{\partial t^2}-\dfrac{e_{32}K_cC(t)h_c}{E_c}\dfrac{\partial^2 w}{\partial y\partial t}\right)=0
\tag{4.22}
$$

$$
\left(1-\lambda^2\nabla^2\right)
\begin{pmatrix}
B_{11}\dfrac{\partial^3 u}{\partial x^3}+\left(B_{12}+2B_{66}\right)\dfrac{\partial^3 u}{\partial x\partial y^2}+B_{22}\dfrac{\partial^3 v}{\partial y^3}+\left(B_{12}+2B_{66}\right)\dfrac{\partial^3 v}{\partial x^2\partial y} \\[2mm]
-D_{11}\dfrac{\partial^4 w}{\partial x^4}-2\left(D_{12}+2D_{66}\right)\dfrac{\partial^4 w}{\partial x^2\partial y^2}+D_{22}\dfrac{\partial^4 w}{\partial y^4}
\end{pmatrix}
$$
$$
+\left(1-\mu^2\nabla^2\right)
\begin{pmatrix}
-I_0\dfrac{\partial^2 w}{\partial t^2}+I_2\nabla^2\dfrac{\partial^2 w}{\partial t^2}-k_w w+k_p\nabla^2 w-c_d\dfrac{\partial w}{\partial t}- \\[2mm]
N^T\nabla^2 w+\dfrac{e_{31}K_cC(t)h_c}{E_c}\dfrac{\partial^2 u}{\partial x\partial t}+\dfrac{e_{32}K_cC(t)h_c}{E_c}\dfrac{\partial^2 v}{\partial y\partial t}
\end{pmatrix}=0
\tag{4.23}
$$

4.1.2 SOLUTION

In this instance, the governing equations are solved using a technique known as an analytical solution approach. The following is how the displacements are expected to look [3]:

$$
\begin{Bmatrix} u \\ v \\ w \end{Bmatrix}=
\begin{Bmatrix}
U\exp\left[i\left(\beta_1 x+\beta_2 y-\omega t\right)\right] \\
V\exp\left[i\left(\beta_1 x+\beta_2 y-\omega t\right)\right] \\
W\exp\left[i\left(\beta_1 x+\beta_2 y-\omega t\right)\right]
\end{Bmatrix}
\tag{4.24}
$$

where U, V, and W represent the amplitudes of the waves, β_1 and β_2 represent the wave numbers in the x and y directions, respectively, and ω represents the circular frequency of the scattered waves. By using Eq. (4.24) as a substitute for the variables $u, v,$ and w in Eqs. (4.21) through (4.23), the following equation may be obtained:

$$\left([K]-\omega^2[M]\right)\begin{bmatrix} U \\ V \\ W \end{bmatrix} = \begin{bmatrix} 0 \\ 0 \\ 0 \end{bmatrix} \tag{4.25}$$

After working out this equation, you will get the resultant value for the circular frequency. The phase velocity may be calculated as follows after this value has been divided by the number of waves:

$$c_p = \frac{\omega}{\beta} \tag{4.26}$$

4.1.3 RESULTS

In this section, we are going to look at the wave dispersion characteristics of an MSNP from a numerical perspective in order to have a better understanding of the influence that the different parameters have. In this particular piece of writing, it is assumed that the center core is constructed out of Terfenol-D, and that both the top and bottom face sheets are constructed out of alumina. In addition, the results of the validation are shown in Table 4.1, which demonstrate the precision and effectiveness of the proposed methodology.

Figure 4.1 depicts, given a variety of nonlocal and length scale factors, how the variance of wave frequency versus wave number changes over time. The purpose of this is to highlight the significant role that these coefficients play in determining how MSNPs respond to dispersion. It is possible to work out that the wave frequency amounts may be lowered whenever the nonlocality is raised from zero to a suitable nonzero value in the case of NE ($\mu = 0$). This is something that can be found out by working backward from the frequency amounts. As a result, the stiffness-softening impact can be clearly identified in this figure, and it manifests itself as a decrease in wave frequency when a nonlocal parameter is introduced. On the other hand, an increase in the magnitude of wave frequency might be reported as length scale parameter is meant to be intensified in a given amount of nonzero nonlocal parameter ($\mu = $ constant). This can be the case when there is a certain nonzero quantity of nonlocal parameter. So, the length scale parameter has the potential to enhance the value of the dispersion replies provided by MSNPs after it has been stimulated to a greater extent.

The phase velocity of MSNPs is shown in Figure 4.2, which displays the combined impact of raising the temperature and increasing the amount of velocity feedback gained. The graphic indicates that there is a reduction in phase velocity

TABLE 4.1

First Dimensionless Fundamental Frequencies of Functionally Graded Nanoplates Compared With Regard to Nonlocality, Plate Thickness, and Aspect Ratio ($n = 5$, $a = 10$ nm).

a/b	μ (nm^2)	$a/h = 10$		$a/h = 20$	
		Present	Natarajan et al. [4]	Present	Natarajan et al. [4]
1	0	0.0461	0.0441	0.0116	0.0113
	1	0.0422	0.0403	0.0106	0.0103
	2	0.0391	0.0374	0.0098	0.0096
	4	0.0345	0.0330	0.0086	0.0085
2	0	0.1137	0.1055	0.0288	0.0279
	1	0.0930	0.0863	0.0235	0.0229
	2	0.0806	0.0748	0.0204	0.0198
	4	0.0659	0.0612	0.0167	0.0162

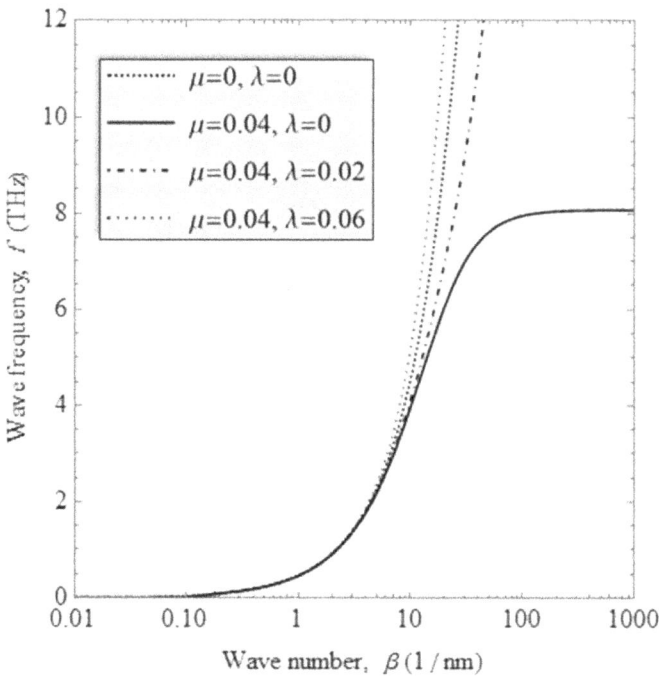

FIGURE 4.1 The MSNP wave frequency is impacted by nonlocal and length scale characteristics together.

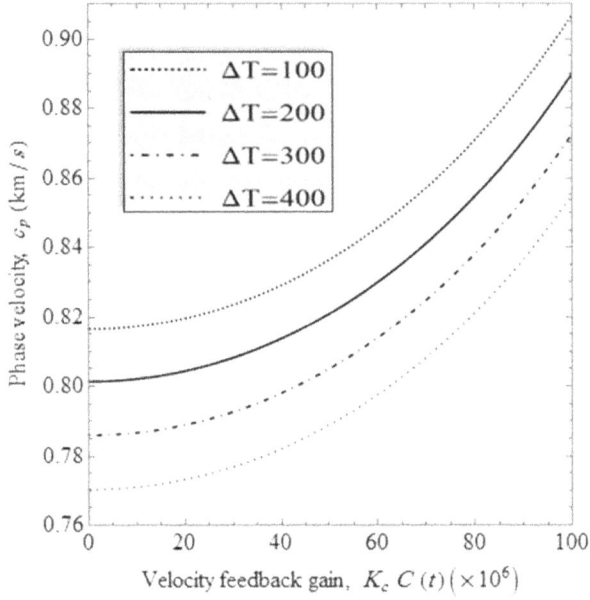

(a) $\beta = 0.05 \times 10^9$

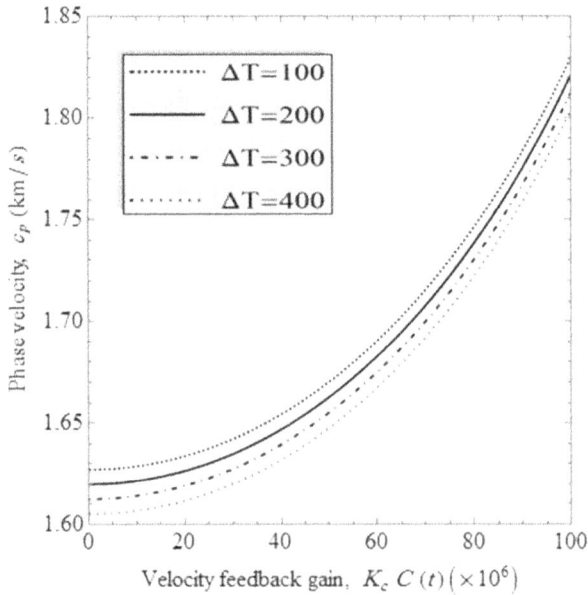

(b) $\beta = 0.1 \times 10^9$

FIGURE 4.2 The relationship between the feedback gain and phase velocity as a function of temperature increase ($k_w = 0.1$, $k_p = 0.5$, $C_d = 0.1$, $\lambda < \mu$). (a) $\beta = 0.05 \times 10^9$; (b) $\beta = 0.1 \times 10^9$.

whenever there is an increase in temperature. In addition, the influence of an increase in temperature may be observed more clearly in lower wave numbers. To put it another way, when the temperature is raised by the same amount, the phase velocity might be impacted more by the wave number of the smaller waves than by the wave numbers of the larger waves. Changes in the velocity feedback gain, which is an impact that is rising in magnitude, are responsible for a further influence. With the addition of velocity feedback gain, amplification of the phase velocity of MSNPs is really possible. As a result, the results may be summed up as follows: the impact of temperature increase decreased, but the effect of velocity feedback gain strengthened.

In addition, the effects of temperature rise, foundation parameters, and wave number are addressed in Figure 4.3, which is a diagram of the variations in phase velocity versus temperature increase for various Winkler and Pasternak coefficients and wave numbers. It is easy to see that when the temperature rises, the values of the phase velocity will eventually decrease. In actuality, a delay in this damping tendency may be seen in conditions with greater wave numbers. In order to assign zero to the phase velocity amounts when a larger wave number is desired, a bigger temperature increase is required, as shown by the result. In the meanwhile, it is essential to note that the foundation parameters have an amplifying effect on the phase velocity values of MSNPs. Higher phase velocities are the consequence of an increase in the total number of Winkler and Pasternak coefficients, as shown. If higher foundation parameters are desired, it is obvious that greater temperature increases are necessary to completely damp the phase velocity of the system. Even if greater foundation parameters are not sought, this is the case.

Figure 4.4 depicts the effect that the damping coefficient has on the phase velocity of MSNPs by showing the fluctuations of phase velocity versus temperature rise for a variety of various damping coefficient value combinations. It can be seen that the phase velocity slows down as the temperature rises, which is the same as what was shown in the preceding figure. This tendency toward lessening may be influenced by a number of variables, one of which is the damping coefficient of a visco-Pasternak medium. When a larger damping coefficient is used, it is obvious that the phase velocity may be slowed down with just a little increase in temperature. Adding a temperature increase is not a feasible method for dampening the dispersion responses of MSNPs in large wave numbers, as can be seen from the diagram. This is due to the fact that in these instances, an incredible temperature rise is necessary, which is essentially an impossible method.

In conclusion, Figure 4.5 depicts the fluctuation of phase velocity against damping coefficient for a variety of temperature rises and foundation parameters. It should come as no surprise that one may deduce from this image that all of the aforementioned patterns can be seen. As the damping coefficient is increased, the phase velocity of MSNPs decreases, as was previously anticipated. On the other side, the size of the foundation parameters' influence on the wave propagation responses increases as that influence is increased. When larger temperature rise

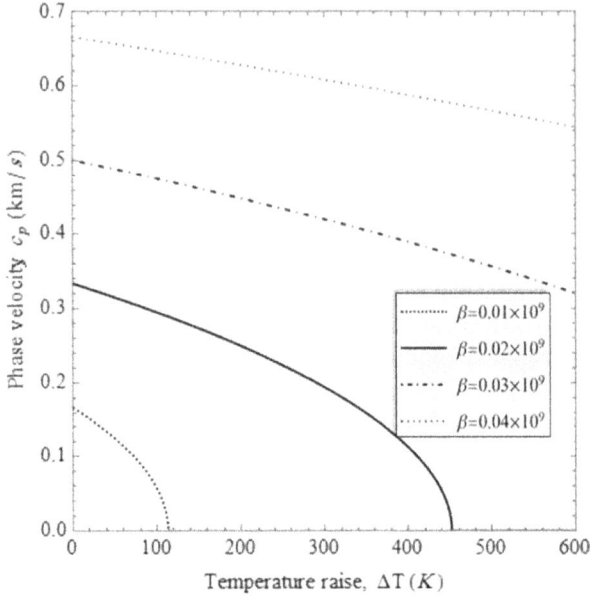

(a) $k_w = 0$, $k_p = 0$

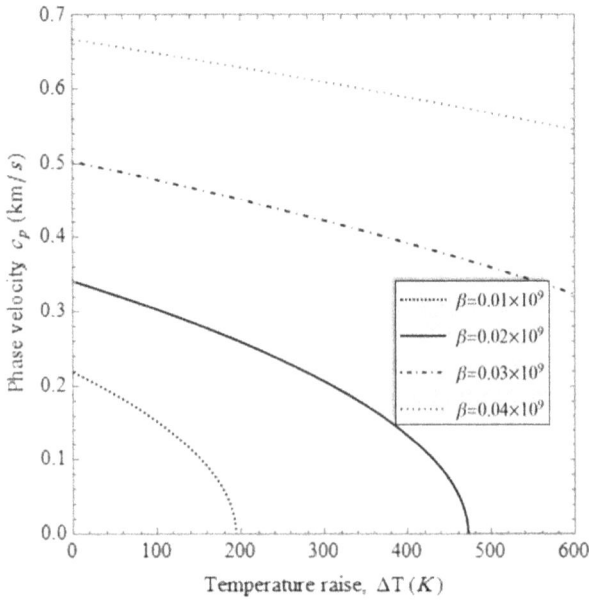

(b) $k_w = 0.2$, $k_p = 0$

FIGURE 4.3　Changes in phase velocity as a function of temperature increase for a range of base parameter and wave number values ($C_d = 0$, $\lambda = \mu$, $K_c C(t) = 10^6$). (a) $k_w = 0$, $k_p = 0$; (b) $k_w = 0.2$, $k_p = 0$; (c) $k_w = 0.2$, $k_p = 1$; (d) $k_w = 0.5$, $k_p = 2$.

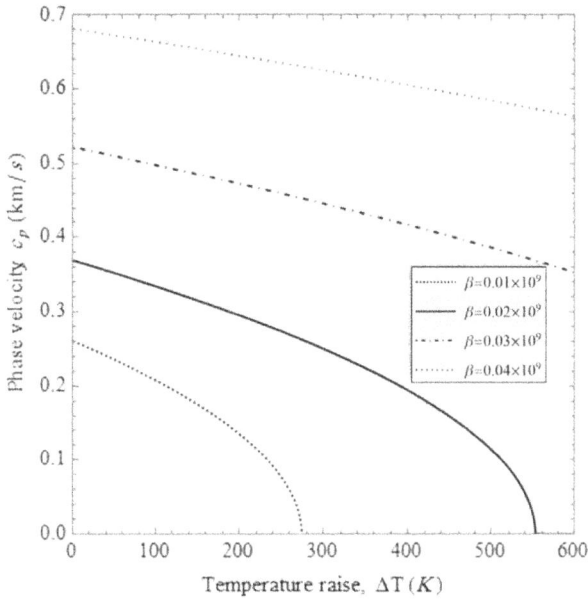

(c) $k_w = 0.2$, $k_p = 1$

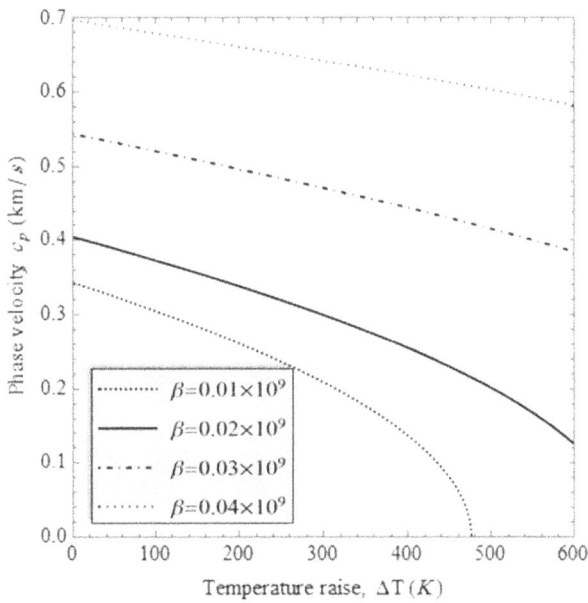

(d) $k_w = 0.5$, $k_p = 2$

FIGURE 4.3 *(Continued)*

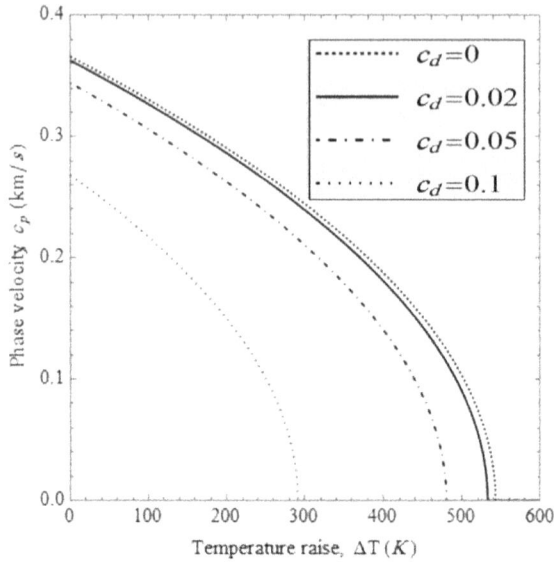

(a) $\beta = 0.02 \times 10^9$

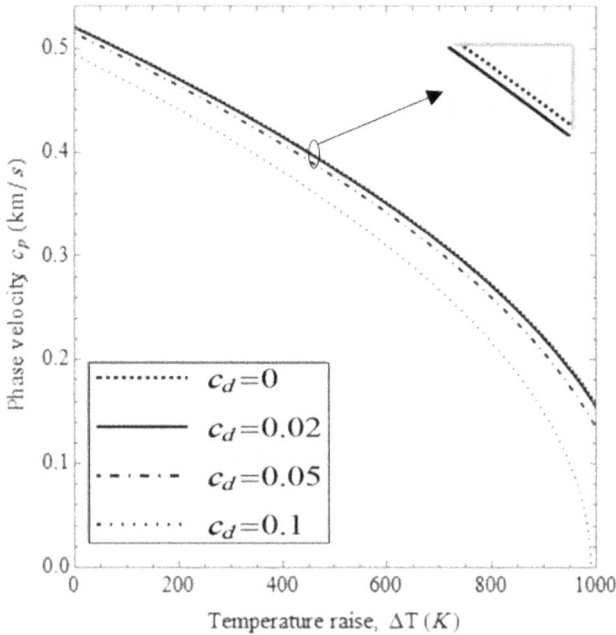

(b) $\beta = 0.03 \times 10^9$

FIGURE 4.4 The effect of varying damping coefficients and wave numbers on the relationship between phase velocity and temperature increase ($k_w = 0.1$, $k_p = 1$, $\lambda = \mu$). (a) $\beta = 0.02 \times 10^9$; (b) $\beta = 0.03 \times 10^9$.

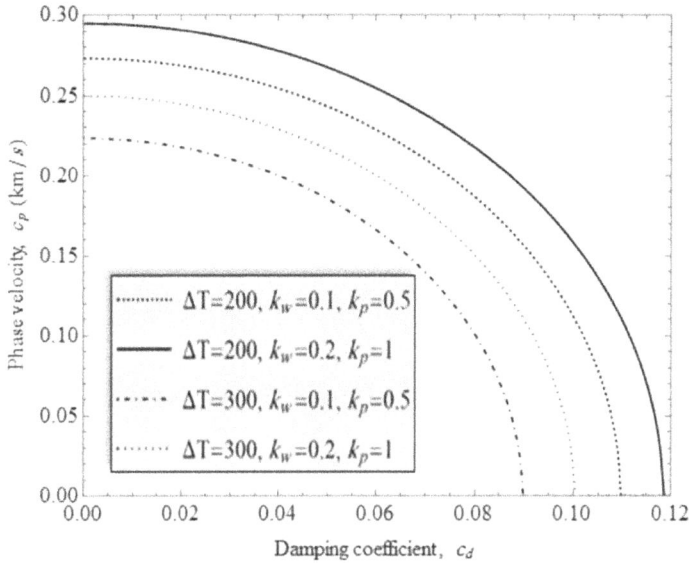

FIGURE 4.5 Changes in the phase velocity of MSNPs caused by varying initial conditions and elevated temperatures $(\beta = 0.02 \times 10^9)$.

values are applied, it is obvious that the phase velocity values may be reduced, leading to a lower overall value.

The size impacts on the wave frequency of MSCNPs with regard to different fiber angles are displayed in Figure 4.6, which may be seen next. As the number of waves rises, it is easy to notice that the frequency of the waves also increases, at least initially, in every circumstance. Nevertheless, if the nonlocal parameter becomes nonzero and the effects of strain gradient stress field are disregarded (by setting to zero $\lambda = 0$), the wave frequency will reach a value that is finite. After then, the wave frequency does not alter in any way. In the other direction, in the situations of NSGT ($\lambda \neq 0$), the wave frequency tends to increase extremely quickly toward infinity. In a similar vein, it is important to point out that once the nonlocal and length scale parameters are both set to zero, the wave frequency once again trends toward infinity. This is an important point to make. In addition, it is obvious that the wave dispersion curves are practically identical in fiber angles 0 and 90, but a significant drop may be reported whenever the fiber angle is believed to be 45. This is the case since 45 is the angle that lies between 0 and 90.

In addition, the scale effects are discussed once more by showing in Figure 4.7 the fluctuation of phase velocity versus wave number for various fiber angles. The figure indicates that once the fiber angles are meant to be either 0 or 90 degrees, there is no difference in the phase velocity values. In addition, when the fiber angle is changed to 45 degrees, a different circumstance, it is possible to see a

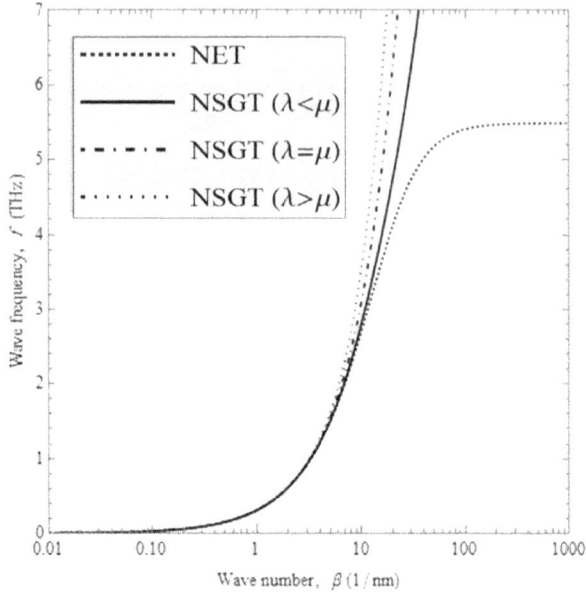

(a) $\theta = 0$

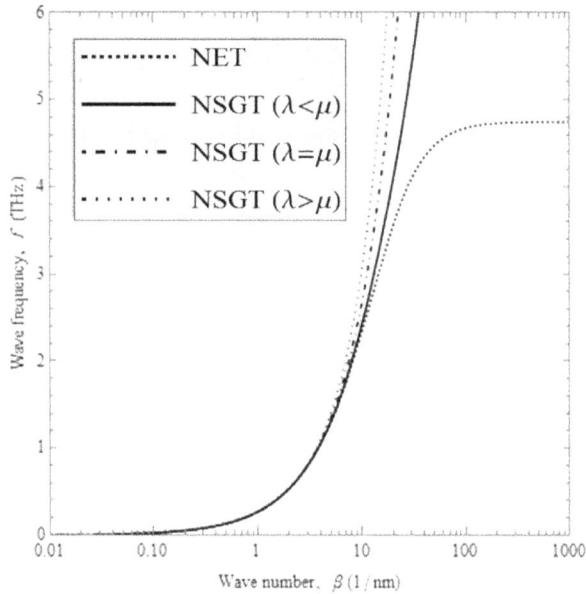

(b) $\theta = 45$

FIGURE 4.6   The relationship between wave frequency and wave number varies depending on the continuum theory and the fiber's angle of incidence. $(k_w = k_p = 100, c_d = 0)$. (a) $\theta = 0$; (b) $\theta = 45$; (c) $\theta = 90$.

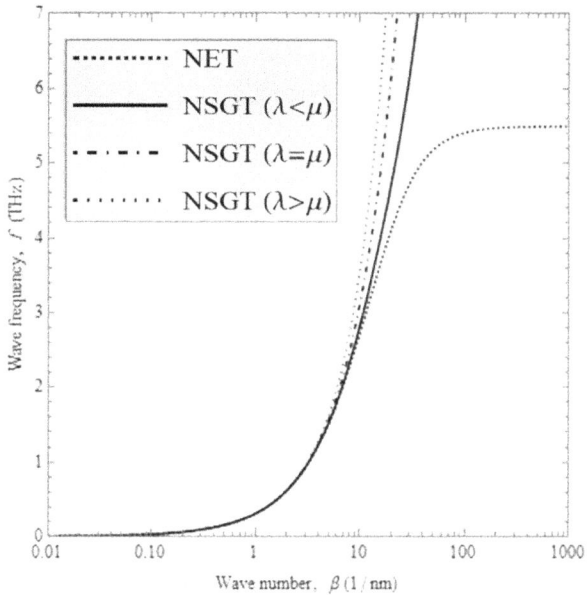

(c) θ = 90

FIGURE 4.6 (*Continued*)

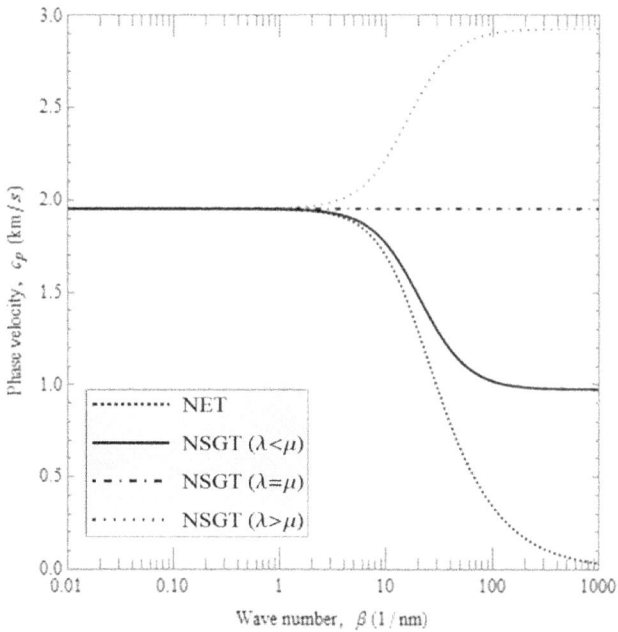

(a) θ = 0

FIGURE 4.7 Different continuum theories and fiber angles result in a distinct relationship between phase velocity and wave number. $(k_w = k_p = 100, c_d = 0)$. (a) θ = 0; (b) θ = 45; (c) θ = 90.

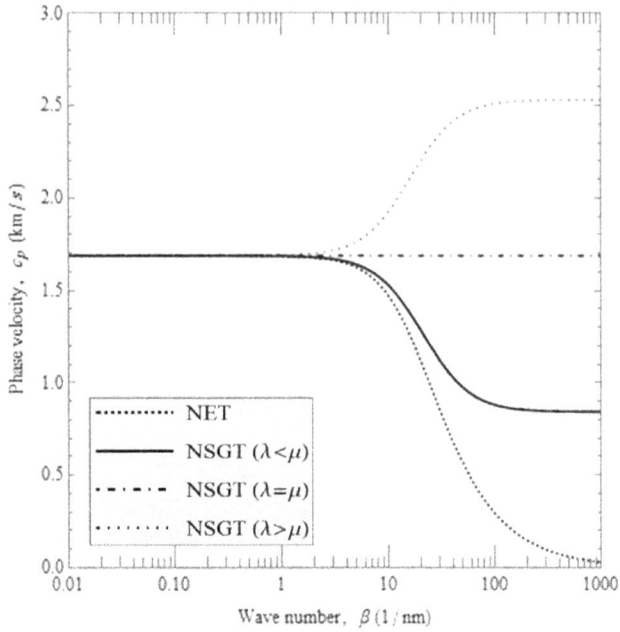

(b) $\theta = 45$

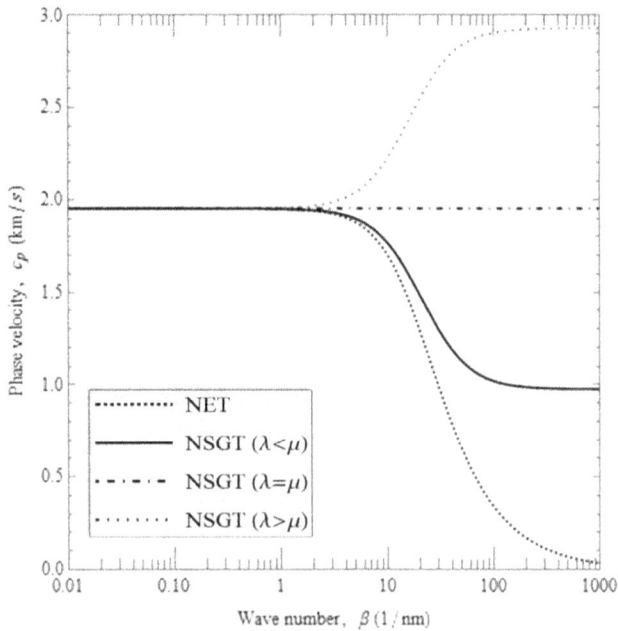

(c) $\theta = 90$

FIGURE 4.7 (*Continued*)

reduction in the total number of wave responses. On the other hand, there is no evidence of any size dependence whatsoever. In this picture, just as in the one before it, the nonlocal parameter has a diminishing effect on the phase velocity values, however the length scale parameter has the ability to enhance the wave responses of MSCNPs.

In addition, a review of the effects of velocity feedback gain on the phase velocity of MSCNPs in a chosen wave number is shown in Figure 4.8. It is possible to draw the conclusion that an increase in the total amount of velocity feedback gain has the potential to encourage the phase velocity of MSCNPs to increase. In addition to that, the scale effects have been shown inside this figure as well. In point of fact, according to our expectations, the phase velocity may be increased if the length scale parameter is included in the calculation.

Figure 4.9 is depicted at the end to highlight the influences of foundation parameters on the wave dispersion answers of MSCNPs by drawing the variations of phase velocity versus damping coefficient for a variety of wave numbers. This is done so that the reader can see how the foundation parameters affect the wave dispersion answers. According to the illustration, the phase velocity of each successive wave number decreases step by step until it reaches zero as the damping coefficient rises. In addition, it is abundantly obvious that the values of phase velocity get dampened in larger damping coefficients whenever increasing wave numbers are used. When included to the equation, the Winkler and Pasternak coefficients each have the potential to cause an increase in the phase velocity quantities that are being calculated. It is important to point out that

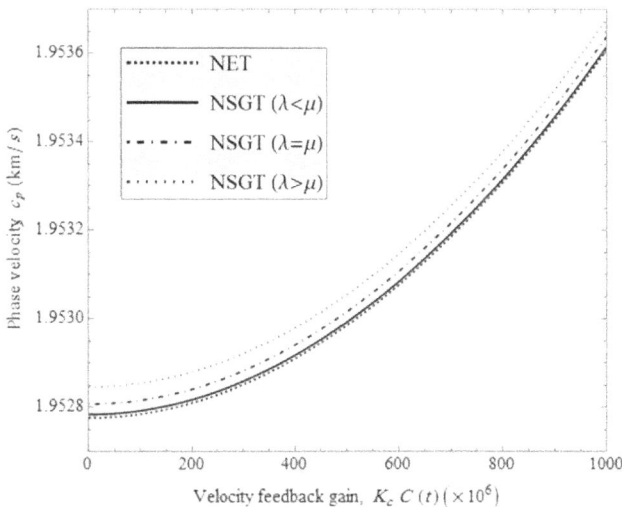

FIGURE 4.8 Nonlocal continuum theories' phase velocities and the impact of velocity feedback gain $(\beta = 0.1 \times 10^9)$.

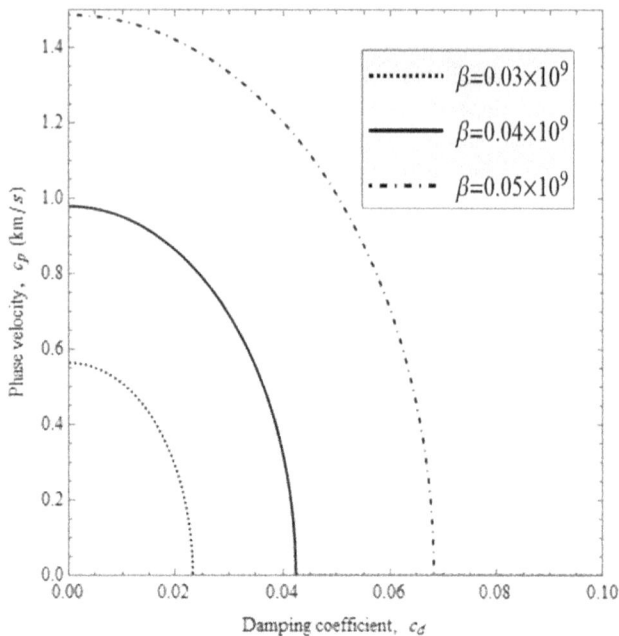

(a) $k_w = 0$, $k_p = 0$

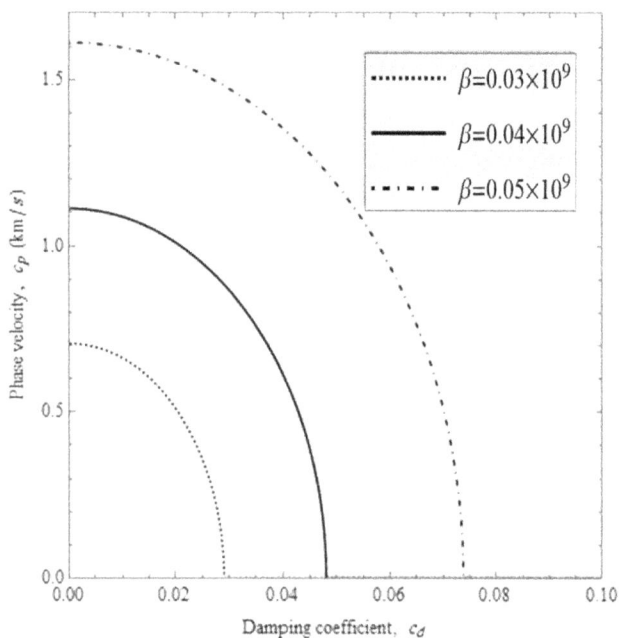

(b) $k_w = 4$, $k_p = 0$

FIGURE 4.9 The effect of initial conditions on the phase velocity of multi-scale complex network particles (MSCNPs) at various wave numbers ($\lambda < \mu$). (a) $k_w = 0$, $k_p = 0$; (b) $k_w = 4$, $k_p = 0$; (c) $k_w = 4$, $k_p = 8$; (d) $k_w = 12$, $k_p = 12$.

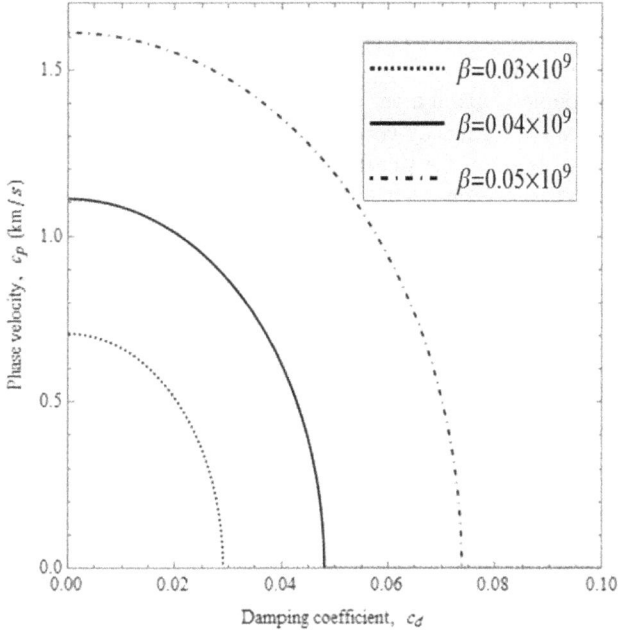

(c) $k_w = 4$, $k_p = 8$

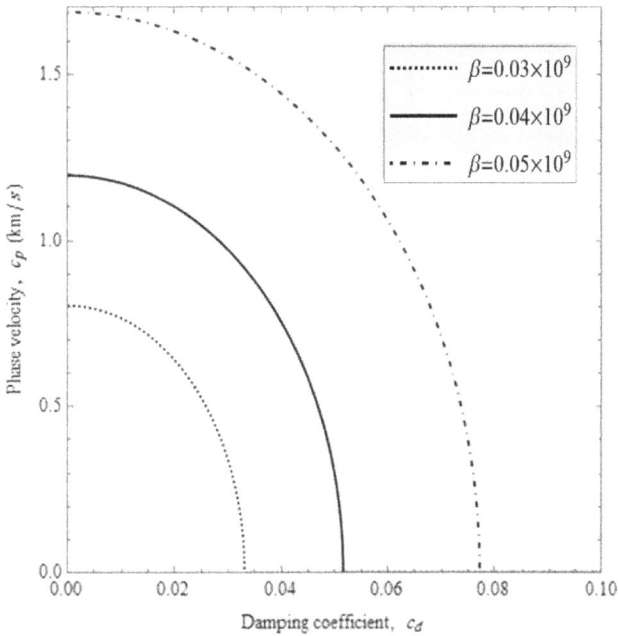

(d) $k_w = 12$, $k_p = 12$

FIGURE 4.9   *(Continued)*

stronger damping coefficients are necessary to fully damp the phase velocities when greater Winkler or Pasternak coefficients are used. This is something that should be kept in mind.

Figure 4.10 depicts, given a variety of nonlocal and length scale factors, how the variance of wave frequency versus wave number changes over time. The purpose of this is to highlight the significant role that these coefficients play in determining how MSNPs respond to dispersion. It is possible to work out that the wave frequency amounts may be lowered whenever the nonlocality is raised from zero to a suitable nonzero value in the case of NE ($\lambda = 0$). This is something that can be found out by working backward from the frequency amounts. As a result, the stiffness-softening impact can be clearly identified in this figure, and it manifests itself as a decrease in wave frequency when a nonlocal parameter is introduced. On the other hand, an increase in the magnitude of wave frequency might be reported as length scale parameter is meant to be intensified in a given amount of nonzero nonlocal parameter (μ = constant). This can be the case when there is a certain nonzero quantity of nonlocal parameter. So, the length scale parameter has the potential to enhance the value of the dispersion replies provided by MSNPs after it has been stimulated to a greater extent.

Following that, Figure 4.11 is drawn with the primary intention of demonstrating the influence of the moisture expansion coefficient of Terfenol-D on the phase

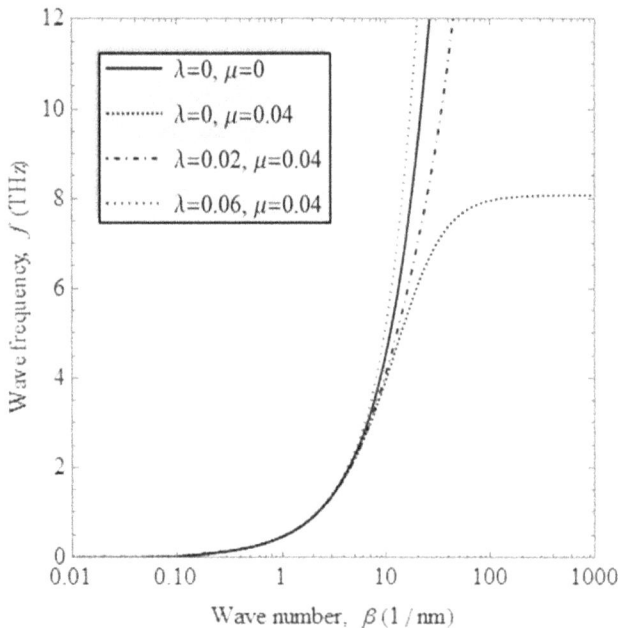

FIGURE 4.10 Changes in MSNPs' wave frequencies as a function of nonlocality and scale ($k_w = 0.01$, $k_p = 0.05$, $C_d = 0$, $\Delta T = 100$, $\Delta C = 0$).

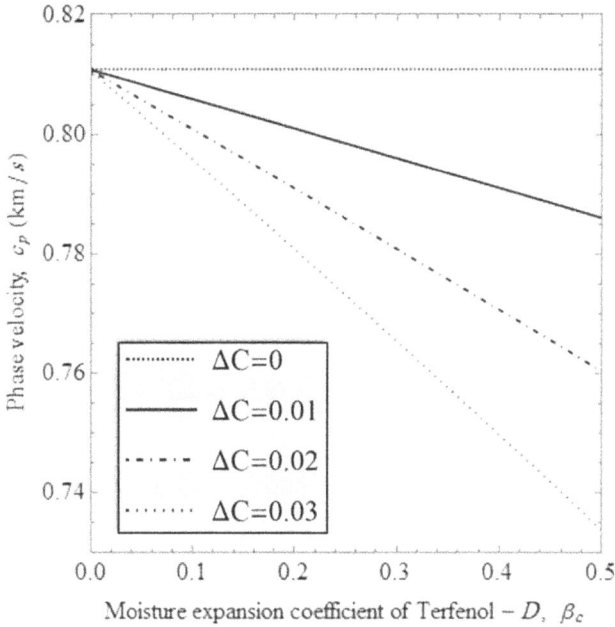

FIGURE 4.11 The phase velocity of MSNPs is affected in different ways at different moisture concentrations due to the moisture expansion coefficient of the magnetostrictive core ($k_w = 0.01$, $k_p = 0.05$, $C_d = 0.1$, $\Delta T = 100$, $\beta = 0.05$ (1/nm), $\lambda = \mu$).

velocity values of MSNPs whenever it is expected that the moisture concentration will be changed. In point of fact, there is a dearth of knowledge about the moisture expansion coefficient of this intelligent substance. Terfenol-D, on the other hand, is indisputably made up of the metallic elements terbium, iron, and dysprosium. This much is without dispute. As a result, the moisture expansion coefficient of Terfenol-D ought to fall within the typical range for this coefficient when applied to metals. This coefficient is regarded to be altered in this figure from 0 to 0.5, and the effects of progressive changes in this coefficient on the wave dispersion speed are noticed. A linear influence with a negative slope can be noticed for the moisture expansion coefficient based on the diagram whenever a nonzero quantity is used for moisture concentration ($\Delta C \neq 0$). This may be seen anytime a moisture concentration is more than zero. In other words, the moisture expansion coefficient of Terfenol-D causes phase velocity to drop constantly in surroundings that are humid, but it does not change in habitats that are dry. This phenomenon occurs in environments that are humid. Beginning right now, the ideal value of 0.3 will be linked to the moisture expansion coefficient of Terfenol-D ($\beta_c = 0.3$ (*wt.* % $H_2O)^{-1}$), which can be found in all of the schematics that will come after this one.

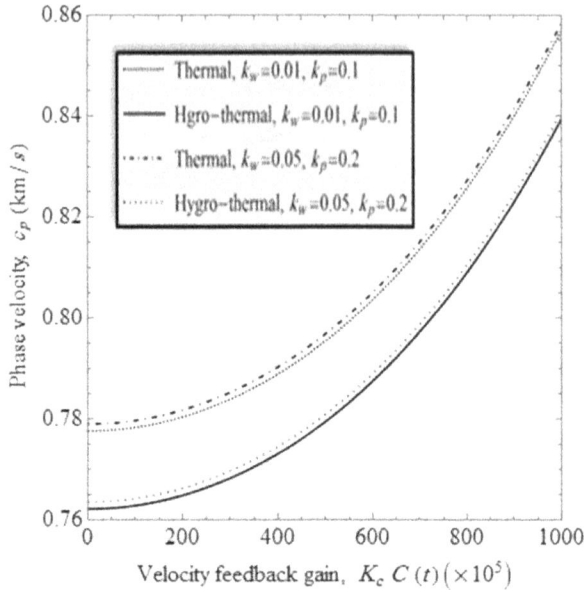

FIGURE 4.12 Studying the influence of velocity feedback gain on the phase velocity of MSNPs in response to varying hygrothermal loadings and medium conditions ($\beta = 0.03$, $\lambda = \mu$, $C_d = 0.2$).

In addition, the effect of the velocity feedback gain on the phase velocity of MSNPs is examined in Figure 4.12, which takes into account a variety of situations as well as a range of values for the foundation parameters. Changing the amount of velocity feedback gain can clearly be shown to have an effect on the phase velocity, which enables the phase velocity to be adjusted. In point of fact, once this control gain is included, a higher increase in phase velocity of MSNPs is seen. On the other hand, it is observable that the wave speed increases in the circumstances with greater Winkler and Pasternak foundations. This is something that can be experienced. As compared to the circumstance in which the structure is put in a hygro-thermal environment, it is interesting to note that the values of phase velocity are greater when the structure is subjected to thermal loading. So, the diminishing impact of moisture concentration may be readily observed again in this picture as it is in Figure 4.11, which is the same as this figure.

The impacts of different surroundings on the phase velocity of MSNPs are depicted in Figure 4.13 with regard to the medium's damping coefficient for a variety of wave numbers. To begin, it is evident that the wave dispersion responses follow a path that is continuously lowering as the damping coefficient rises. On the other hand, the value of the critical damping coefficient that results in the zero value for phase velocity is not the same for all wave numbers. In other words, when a large wave number is used, you will need to use a bigger amount of damping coefficients in order to completely damp the phase velocity of the scattered waves.

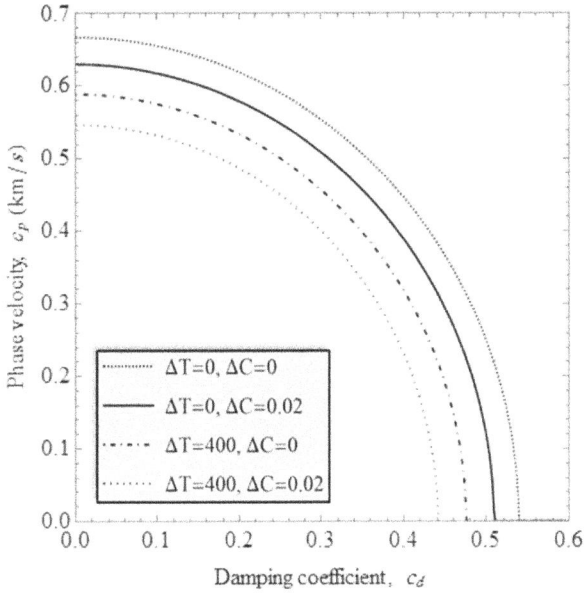

(a) $\beta = 0.04$

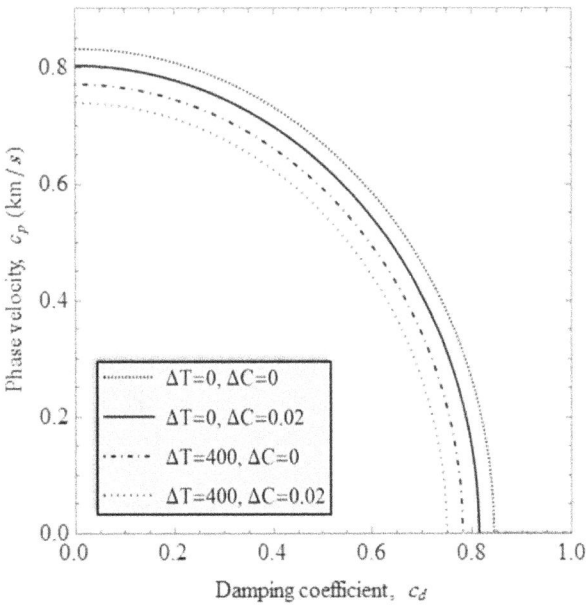

(b) $\beta = 0.05$

FIGURE 4.13 Phase velocity versus damping coefficient as a function of temperature increase and moisture content. ($k_w = 0.01$, $k_p = 0.05$, $\lambda = \mu$). (a) $\beta = 0.04$; (b) $\beta = 0.05$; (c) $\beta = 0.06$.

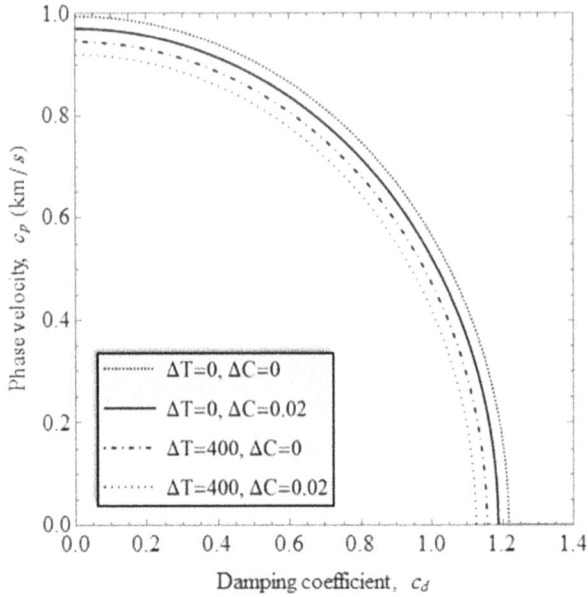

(c) $\beta = 0.06$

FIGURE 4.13   *(Continued)*

As a result, dampers with larger viscosities should be employed in the design methods if high wave numbers are implemented. On the other hand, according to the figure, any increase in temperature rise or moisture concentration leads to reduced wave velocities. This is the case regardless of whatever factor you look at first. So, lesser damping coefficients may be employed once the system is put in hygro-thermal and thermal settings as opposed to the condition of non-hygro-thermal environments. This is due to the fact that humidity and temperature both play a role in the environment.

Last but not least, the combined impacts of moisture concentration, temperature increase, and foundation parameters are explored in Figure 4.14, which depicts the fluctuations of phase velocity versus moisture concentration. When larger foundation parameter values are selected, it is immediately apparent that the phase velocity values will increase as a result. Also, it is easy to comprehend that a reduction in the value of phase velocity follows an increase in the amount of moisture that is present in the substance. Additionally, the same influence may be recorded for any shifts in the temperature increase. This indicates that the greatest amount of phase velocity that can be attained in the event of $T = 0$ is the required moisture concentration. Hence, bigger temperature increases may be employed instead of higher moisture concentrations in the circumstances when there is a constraint for moisture concentration.

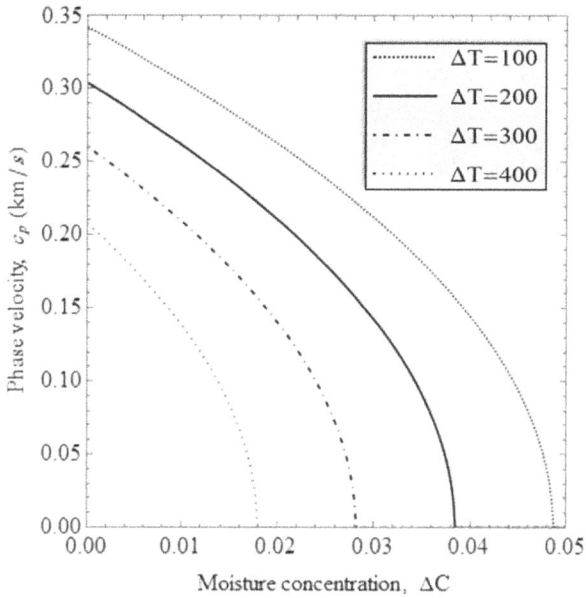

(a) $k_w = 0.01$, $k_p = 0.05$

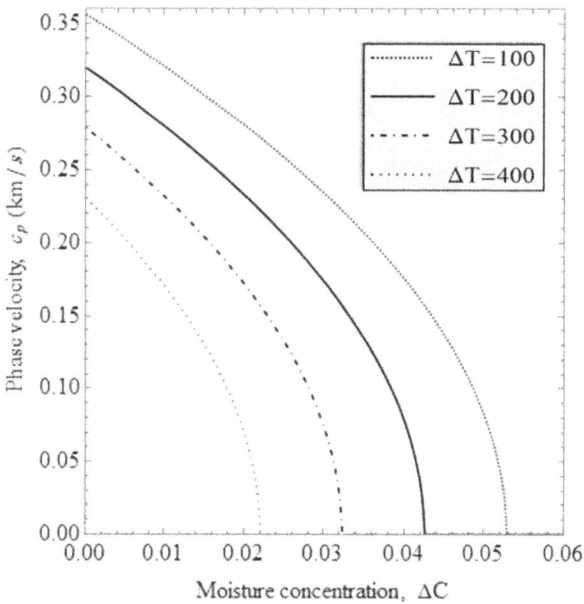

(b) $k_w = 0.1$, $k_p = 0.5$

FIGURE 4.14 The relationship between phase velocity and moisture content as a function of temperature increase and soil factors ($\beta = 0.04$, $\lambda = \mu$, $C_d = 0.2$). (a) $k_w = 0.01$, $k_p = 0.05$; (b) $k_w = 0.1$, $k_p = 0.5$.

In addition, the amount of work done by external forces may be broken down into two distinct categories: the first category is the work done by the elastic medium, and the second category is the work done by the Lorentz force that is caused by the magnetic field. In point of fact, it is speculated that a longitudinally stable magnetic field with an intensity equal to H_0 is acting onto the SLGS. As a result, the formula for the exerted body force that is generated by this field may be stated as follows:

$$f_{Lz} = \eta \left[\underbrace{\underbrace{\nabla \times \left(\nabla \times \left(u \times H_0 \right) \right)}_{h}}_{J} \right] \times H_0 \tag{4.27}$$

where η, ∇, h, and J represent, respectively, the magnetic permeability of MSPs, the gradient operator, a minor disturbance of an applied magnetic field, and the current density vector. The magnetic field may be described as follows within this context:

$$H_0 = H_x \delta_x \hat{\imath} \tag{4.28}$$

where the symbol δ denotes the Kronecker delta tensor. By plugging the equations for Eqs. (4.1) through (4.3) into Eq. (4.27), one may write the applied Lorentz forces per unit volume as:

$$f_{Lz} = \eta H_x^2 \frac{\partial^2 w}{\partial x^2} \tag{4.29}$$

In addition, the resultant force of Lorentz may be expressed in the following form:

$$F_{Lz} = b \int_{-\frac{h}{2}}^{\frac{h}{2}} f_{Lz} dz = \eta b h H_x^2 \frac{\partial^2 w}{\partial x^2} \tag{4.30}$$

Now, we can write down the variation of the external work done by external forces in the following form:

$$\delta V = \int_0^L \left(\begin{array}{c} -k_w \delta w + k_p \left(\dfrac{\partial^2 \delta w}{\partial x^2} + \dfrac{\partial^2 \delta w}{\partial y^2} \right) - c_d \dfrac{\partial w}{\partial t} \\ -N^T \left(\dfrac{\partial^2 \delta w}{\partial x^2} + \dfrac{\partial^2 \delta w}{\partial y^2} \right) - \eta b h H_x^2 \dfrac{\partial^2 w}{\partial x^2} \end{array} \right) dx \tag{4.31}$$

where k_w is the Winkler coefficient, k_p is the Pasternak coefficient, and C_d is the damping coefficient of the elastic medium. Moreover, the thermal loading that is being applied to the system is denoted by the letter N^T. The Euler-Lagrange equations of MSCPs may be stated as follows after Eqs. (2.33), (2.34), (2.42) and (4.31) have been replaced in Eq. (4.8) and the coefficients of u, v, and w have been set to zero:

$$\frac{\partial N_{xx}}{\partial x} + \frac{\partial N_{xy}}{\partial y} = I_0 \frac{\partial^2 u}{\partial t^2} + \frac{e_{31} K_c C(t) h_c}{E_c} \frac{\partial^2 w}{\partial x \partial t} \tag{4.32}$$

$$\frac{\partial N_{xy}}{\partial x} + \frac{\partial N_{yy}}{\partial y} = I_0 \frac{\partial^2 v}{\partial t^2} + \frac{e_{32} K_c C(t) h_c}{E_c} \frac{\partial^2 w}{\partial y \partial t} \tag{4.33}$$

$$\frac{\partial^2 M_{xx}}{\partial x^2} + 2 \frac{\partial^2 M_{xy}}{\partial x \partial y} + \frac{\partial^2 M_{yy}}{\partial y^2} - N^T \nabla^2 w = I_0 \frac{\partial^2 w}{\partial t^2} - I_2 \nabla^2 \frac{\partial^2 w}{\partial t^2}$$

$$+ k_w w - k_p \nabla^2 w + c_d \frac{\partial w}{\partial t} + \frac{e_{31} K_c C(t) h_c}{E_c} \frac{\partial^2 u}{\partial x \partial t} \tag{4.34}$$

$$+ \frac{e_{32} K_c C(t) h_c}{E_c} \frac{\partial^2 v}{\partial y \partial t} - \eta b h H_x^2 \frac{\partial^2 w}{\partial x^2}$$

In the current investigation, the magnetic permeability is thought to be equal to $\eta = 4\pi * 10^7$. In Figure 4.15, the variation of wave frequency versus wave number for various nonlocal and length scale parameters is depicted to put emphasis on the crucial impact of these coefficients on the dispersion responses of MSNPs. This was done so that the reader could see how these coefficients affect the dispersion responses of MSNPs. It is possible to work out that the wave frequency amounts may be lowered whenever the nonlocality is raised from zero to a suitable nonzero value in the case of NE ($\lambda = 0$). This is something that can be found out by working backward from the frequency amounts. As a result, the stiffness-softening impact can be clearly identified in this figure, and it manifests itself as a decrease in wave frequency when a nonlocal parameter is introduced. On the other hand, in a certain quantity of nonlocal parameter that is not zero, an increase in the magnitude of wave frequency might be reported as a length scale parameter that is intended to be intensified. This is because the length scale parameter is expected to be intensified. So, the length scale parameter has the potential to enhance the value of the dispersion replies provided by MSNPs after it has been stimulated to a greater extent.

In order to illustrate the combined effects of magnetic field strength and velocity feedback gain on the phase velocity of MSNPs, Figure 4.16 has been allotted some space. The graphic indicates that an increase in the strength of the magnetic field results in an increase in the phase velocity. In addition, the influence of the magnetic field's strength may be observed more clearly when tiny wave numbers are considered. To put it another way, while the magnetic field strength is held constant, the phase velocity of a signal may be changed more by the wave number of a tiny signal than by a signal with a larger wave number. Changes in the velocity feedback gain, which is an impact that is rising in magnitude, are responsible

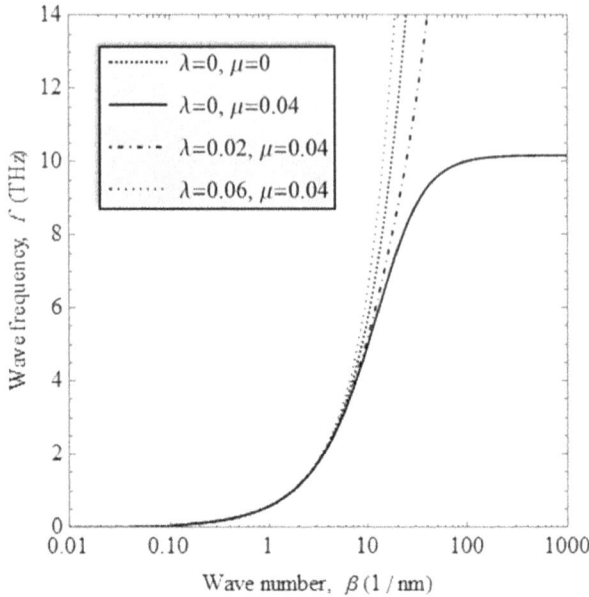

FIGURE 4.15 Variation of wave frequency versus wave number for various nonlocal and length scale parameters ($k_w = 0.05$, $k_p = 0.5$, $C_d = 0.1$, $\Delta T = 100$, $H_x = 0$).

for a further influence. With the addition of velocity feedback gain, amplification of the phase velocity of MSNPs is really possible. So, the result may be summed up in terms of the amplifying influences of both the strength of the magnetic field and the increase in velocity feedback.

In addition, a diagram depicting the fluctuation of phase velocity versus damping coefficient for a variety of temperature increases and foundation parameters is shown in Figure 4.17. As the damping coefficient is increased, the phase velocity of MSNPs decreases, as was previously anticipated. It is important to point out that the damping coefficients show a declining tendency toward lower values if a bigger temperature increase is chosen. In other words, lower phase velocities can be achieved whenever a larger temperature increase is applied, and as a result, the damping operation can be witnessed employing lesser damping coefficients. This is because the higher temperature rise causes the phase velocity to decrease. On the other side, the size of the foundation parameters' influence on the wave propagation responses increases as that influence is increased. To summarize the information presented in this figure, lower phase velocities are attained whenever the temperature rise is increased to its maximum and the foundation parameters are reduced to their minimum.

Figure 4.18 is presented as the concluding part of the presentation. Its purpose is to illustrate the associated impacts of foundation parameters, temperature increase, and magnetic field intensity on the dispersion answers of MSNPs by graphing the changes of phase velocity versus temperature rise. When the temperature rises, it can be seen that the phase velocity follows a path that decreases until it ultimately

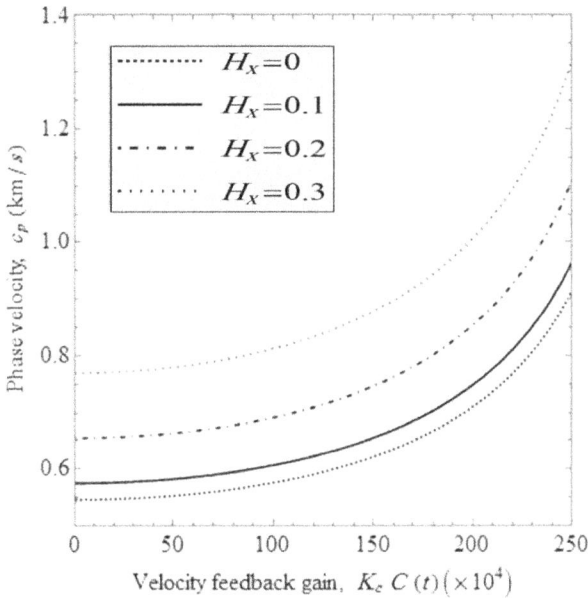

(a) $\beta = 0.01 \times 10^9$

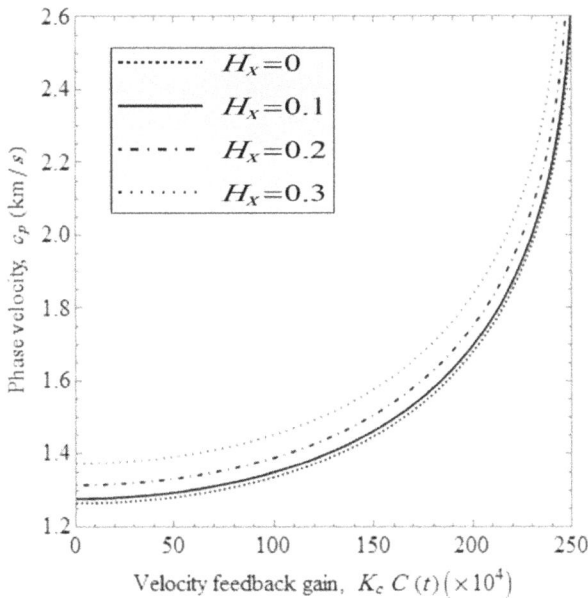

(b) $\beta = 0.05 \times 10^9$

FIGURE 4.16 Variation of phase velocity versus velocity feedback gain for different amounts of magnetic field intensity ($\Delta T = 100$, $k_w = 0.01$, $k_p = 0.1$, $C_d = 0.1$, $\lambda = \mu$). (a) $\beta = 0.01 \times 10^9$; (b) $\beta = 0.05 \times 10^9$.

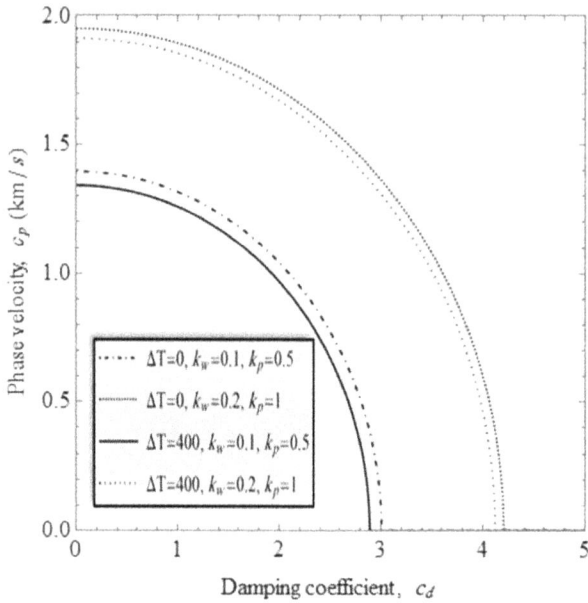

FIGURE 4.17 Variation of phase velocity versus damping coefficient for various Winkler and Pasternak parameters and temperature rises $(\beta = 0.01 \times 10^9, H_x = 0.1, K_c C(t) = 10)$.

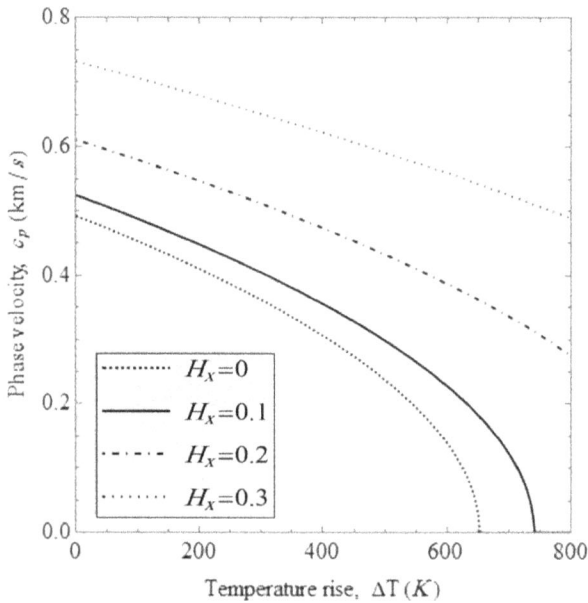

(a) $k_w = 0.01, k_p = 0.05$

FIGURE 4.18 Combined effects of foundation parameters, temperature rise and magnetic field intensity on the phase velocity of MSNPs $(\beta = 0.01 \times 10^9, c_d = 0.1, K_c C(t) = 10)$. (a) $k_w = 0.01, k_p = 0.05$; (b) $k_w = 0.05, k_p = 0.1$.

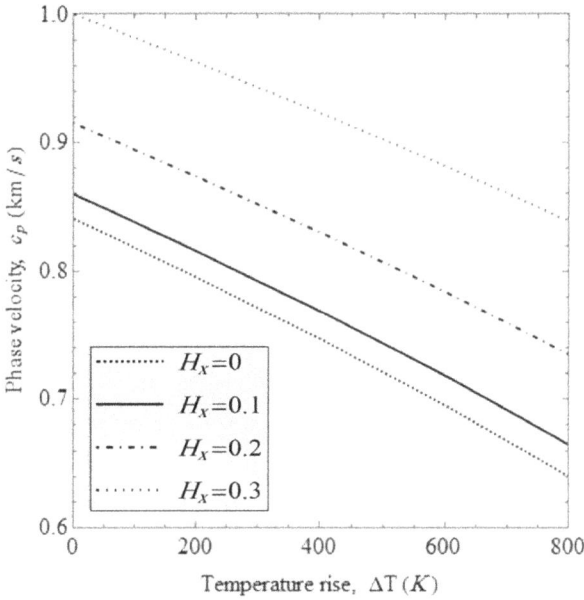

(b) $k_w = 0.05$, $k_p = 0.1$

FIGURE 4.18   (*Continued*)

hits zero. This is a clear observation that can be made. The exact same thing happens here as it does in Figure 4.16: an increase in the magnitude of the magnetic field intensity causes an increase in the phase velocity values whenever the temperature increases. On the other hand, as shown in the picture, wave dispersion responses may be amplified by using larger values for the Winkler and Pasternak coefficients. This can be accomplished by increasing their respective values. It is possible to draw the conclusion that in order to totally damp the phase velocity of MSNPs, a lower temperature increase is necessary after it is determined that the foundation parameters and magnetic field strength may be reduced to their original levels.

4.1.4 CONCLUDING REMARKS

The approach presented in this study is an analytical one, and it is used to analyze the wave propagation of MSNPs in warm conditions. A NSG-based plate model is presented in this model with the purpose of investigating the size effects in more detail. On the basis of a feedback control system, the magnetostriction effects are taken into consideration. A CPT is used to simulate the nanoplate while it is resting on the viscoelastic substrate. On the other hand, governing equations are constructed using Hamilton's principle as a guide. The following is a synopsis of the outcomes that are considered to be the most significant:

- Wave dispersion replies obtained from MSNPs may be increased either by raising the length scale parameter or by reducing the nonlocal parameter. Both of these adjustments may be made independently.
- Higher phase velocities are produced whenever there is an increase in the total amount of velocity feedback gain.
- Including higher Winkler or Pasternak coefficients allows for the addition of phase velocity data.
- When the temperature rise is increased or a greater damping coefficient is applied, the phase velocity of the system will drop, which will result in a reduction in tiny wave numbers.

REFERENCES

1. Ebrahimi, F., A. Dabbagh, and T. Rabczuk, *On wave dispersion characteristics of magnetostrictive sandwich nanoplates in thermal environments.* European Journal of Mechanics-A/Solids, 2021. **85**: p. 104130.
2. Rao, S.S., *Vibration of continuous systems.* 2019: John Wiley & Sons.
3. Ebrahimi, F. and A. Dabbagh, *Wave propagation analysis of magnetostrictive sandwich composite nanoplates via nonlocal strain gradient theory.* Proceedings of the Institution of Mechanical Engineers, Part C: Journal of Mechanical Engineering Science, 2018. **232**(22): p. 4180–4192.
4. Natarajan, S., et al., *Size-dependent free flexural vibration behavior of functionally graded nanoplates.* Computational Materials Science, 2012. **65**: p. 74–80.

5 Dynamic Analysis of Magnetostrictive Materials and Structures

Background

The study of the behavior of a structure when it is exposed to dynamic (activities with high acceleration) loading is covered by the field of structural dynamics, which is a subfield of structural analysis. People, wind, waves, automobile traffic, earthquakes, and explosions are all examples of dynamic loads. Dynamic loading is something that may be applied to any structure. It is possible to identify dynamic displacements, time histories, and modal analyses via the application of dynamic analysis.

Finding out how a physical structure reacts when it is exposed to a force is the primary focus of structural analysis. This action may take the form of load owing to the weight of items such as people, furniture, wind, snow, and so forth, or it can take the shape of some other sort of excitation such as an earthquake, shaking of the ground due to an explosion nearby, and so forth. This action can take place in a variety of ways. Since at one point in time these loads were not present, the essential nature of all of these loads is that they are dynamic. This includes the self-weight of the structure. The difference between dynamic and static analysis is based on whether or not the applied action has sufficient acceleration in relation to the structure's inherent frequency. This is how the dynamic analysis differs from the static analysis. If a load is applied slowly enough, it is possible to disregard the inertia forces, which allows one to simplify the analysis as a static analysis. This is because Newton's first rule of motion allows for this.

Dynamic Analysis of Magnetostrictive Materials and Structures

The last chapter discusses the dynamic behavior of magnetostrictive structures, including magnetostrictive structures that have been exposed to a micro mass spring-damper system with a moving load. A condensed explanation of mathematical connections is provided first as part of an effort to make the process more transparent. Following that was a comparison that was offered in order to verify the mathematical connections. In Section 5.3, numerical examples are given so that the material may be understood more easily.

DOI: 10.1201/9781003355427-5

5.1 DYNAMIC BEHAVIOR OF A MAGNETOSTRICTIVE NANOPLATE SUBJECTED TO A NANO-MASS-SPRING-DAMPER STIMULATOR

As was said before, this will begin with a quick presentation of certain mathematical connections linked to the analysis of the geometry that was discussed. Given that the inverse Laplace transform is used in order to get the governing equation, the solution that was described will be discussed in more detail in this chapter.

5.1.1 LAPLACE TRANSFORMS

In recognition of Pierre Simon De Laplace, a prominent French mathematician, the Laplace transform was given his name (1749–1827). The Laplace transform, like any other transform, takes one signal and converts it into another signal according to a predetermined set of rules or equations. Using the Laplace transformation is by far the most effective strategy for converting differential equations into algebraic equations.

In the field of control system engineering, the Laplace transformation is an extremely important concept. In order to conduct an analysis of the control system, Laplace transforms of a variety of functions need to be performed. In order to do an analysis of the dynamic control system, it is necessary to make use of both the characteristics of the Laplace transform and the inverse Laplace transformation. In this post, we will talk about the Laplace transform in great depth, including its definition, its formula, its characteristics, the Laplace transform table, and the applications of the Laplace transform.

5.1.2 FORMULATION

The present investigation takes into account a sandwich composite that has the following dimensions in the X, Y, and Z axes, respectively: length (a), width (b), and thickness (h). In addition, the magnetostrictive layer contributes to the nano-core plate's layer, and the top and bottom facesheets are made from FG. A mass-spring-damper system, shown in Figure 5.1, is applied to the system that is now under consideration. Winkler and visco-Pasternak have been used so that the foundation medium may be held in place.

We will proceed under the assumption that the nanoplate with linear elastic properties has an orthotropic core (magnetostrictive layer) and isotropic FG facesheets as explained in Eq. (2.21) [1], and variable constants may be defined as Eq. (2.22) [2]. In addition, if the thickness stretching is not zero $(\varepsilon_{zz} \neq 0)$, the three-dimensional elastic constants will have the shape of a bellows when considering the situation [3]:

$$C_{11}^c = C_{22}^c = C_{33}^c = \frac{(1-\vartheta)}{\vartheta} \lambda, C_{44}^c = C_{55}^c = C_{66}^c = \mu, C_{12}^c = C_{13}^c = C_{23}^c = \lambda \quad (5.1)$$

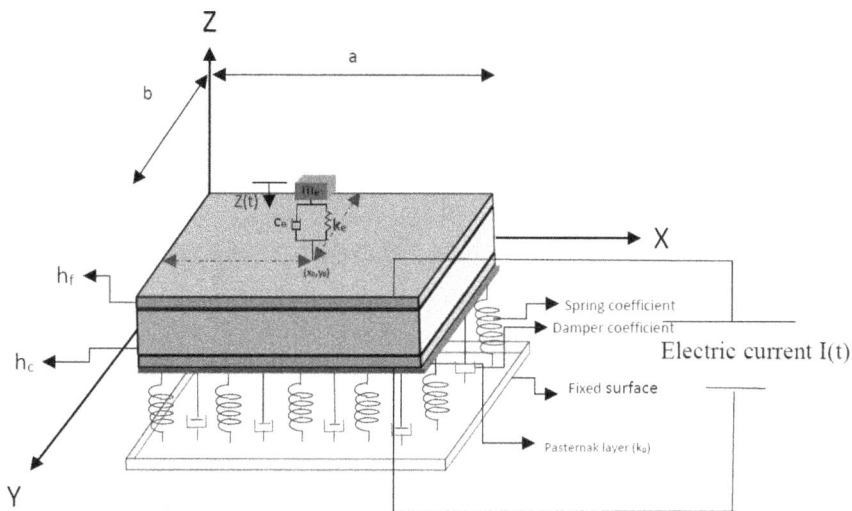

FIGURE 5.1   Sandwich composite configuration.

TABLE 5.1
Characteristics of the FG Facesheets

Material	Al	ZrO2
E (GPa)	70	200
$\rho \left(\dfrac{kg}{m^3} \right)$	2,702	5,700
ϑ	0.3	0.3

Where μ and λ are the coefficients of Lame's method. The following is a definition of Lame's coefficients:

$$\lambda = \frac{\vartheta E}{(1 - 2\vartheta)(1 + \vartheta)}, \mu = \frac{E}{2(1 + \vartheta)} \; . \tag{5.2}$$

In addition, the modulus of elasticity and Poisson's ratio of the magnetostrictive layer are represented by the symbols E_c and v_c, respectively. In the constitutive equations, the symbol for the magnetic field is denoted by H_{zz}, and the converted magnetostrictive moduli may be obtained by solving for $\overline{e}_{ij}$ as explained in Eq. (2.23).

The magnetostrictive core layer's nonzero stresses are able to be defined as Eqs. (2.26 and 2.27). By acknowledging velocity feedback control, the magnetic field intensity (H_{zz}) can be obtained as a function of coil current from Eq. (2.24) [4]. The relation of the facesheets can be expressed as Eq. (2.28) [5].

The nonzero strains of the FG facesheets can be characterized as Eq. (2.30).

5.1.2.1   Governing Equation of the Nanoplate

Within the display part, based on the guideline of minimum potential energy, Hamilton's principle will be actualized to get the set of the governing equations. To this end, this principle can be expressed as [6]:

$$\int_0^t \delta (U - T + V) dt = 0 \tag{5.3}$$

U, T, V represent, respectively, the strain energy, kinetic energy, and work done by external non-conservative forces. The primary variation of the strain energy of the magnetostrictive layer and FG facesheets can be gotten as Eqs. (2.33–2.43).

The proposed system is supposed to be subjected to two kinds of external work.

- Nano mass-spring-damper system
- Visco-Pasternak foundation

In spite of the fact that the Winkler or one-parameter demonstrate is the only shape of a foundation that brings up fair the normal stresses, the Pasternak or two-parameter demonstrate. Apart from that, keep in mind that the vis-co-Pasternak demonstration is formed by adding a damper to the Pasternak demonstration. Since the damping coefficient has such a significant impact on the dynamic response of the material, it is essential that it be taken into consideration throughout the dynamic analysis. As a consequence of this, for the sake of the current study, it is assumed that the nanoplate is supported by the visco-Pasternak elastic medium.

The nano mass-spring-damper framework and the visco-Pasternak foundation are the two types of external work that are anticipated to be carried out on the suggested system [7, 8].

$$\delta V = k_w \cdot \delta w_0 - k_p \left(\frac{\partial^2 \delta w_0}{\partial x^2} + \frac{\partial^2 \delta w_0}{\partial y^2} \right) + c_d \frac{\partial \delta w_0}{\partial t} + q(x, y, t) \tag{5.4}$$

where k_w, k_p, and c_d are the spring, shear and damper constants, respectively.

The first kinetic energy variation can be defined as Eq. (2.42). In the present work, the following simply supported boundary conditions are used as follows:

at:

$x = 0, a$

$$v_0(0,y,t) = v_0(a,y,t) = 0, \; w_0(0,y,t) = w_0(a,y,t)$$
$$= 0, \; N_{xx}(0,y,t) = N_{xx}(a,y,t) = 0,$$
$$M_{xx}(0,y,t) = M_{xx}(a,y,t) = 0, \; F_{xx}(0,y,t) = F_{xx}(a,y,t)$$
$$= 0, \; \phi_y(0,y,t) = \phi_y(a,y,t) = 0$$

(5.5)

at:

$y = 0, b$

$$u_0(x,0,t) = u_0(x,b,t) = 0, \; w_0(x,0,t) = w_0(x,b,t) = 0,$$
$$N_{yy}(x,0,t) = N_{yy}(x,b,t) = 0$$
$$M_{yy}(x,0,t) = M_{yy}(x,b,t) = 0, \; F_{yy}(x,0,t) = F_{yy}(x,b,t) = 0,$$
$$\phi_x(x,0,t) = \phi_x(x,b,t) = 0$$

(5.6)

Imagine a nanoplate with a core layer composed of a magnetostrictive material and top and lower facesheets composed of ferroelectric material. By using Eringen's nonlocal theory and sinusoidal shear deformation theory, we will be able to get the governing equations of the nanoplate, which are as follows:

$$\delta u_0 : \frac{\partial N_{xx}}{\partial x} + \frac{\partial N_{xy}}{\partial y} - e_{31}h_c K_c C(t)\frac{\partial^2 w_0}{\partial x \partial t} = I_0\left(\frac{\partial^2 u_0}{\partial t^2}\right) + I_3\left(\frac{\partial^2 \phi_x}{\partial t^2}\right) - I_1\left(\frac{\partial^3 w_0}{\partial x \partial t^2}\right)$$ (5.7)

$$\delta v_0 : \frac{\partial N_{xx}}{\partial y} + \frac{\partial N_{xy}}{\partial x} - e_{32}h_c K_c C(t)\frac{\partial^2 w_0}{\partial y \partial t} = I_0\left(\frac{\partial^2 v_0}{\partial t^2}\right) + I_3\left(\frac{\partial^2 \phi_y}{\partial t^2}\right) - I_1\left(\frac{\partial^3 w_0}{\partial y \partial t^2}\right)$$ (5.8)

$$\delta w_0 : \frac{\partial^2 M_{xx}}{\partial x^2} + \frac{\partial^2 M_{yy}}{\partial y^2} + 2\frac{\partial^2 M_{xy}}{\partial x \partial y} - K_w \cdot w_0$$

$$+ K_p\left(\frac{\partial^2 w_0}{\partial x^2} + \frac{\partial^2 w_0}{\partial y^2}\right) - c_d\frac{\partial w_0}{\partial t} - e_{31}h_c K_c C(t)\frac{\partial^2 U_x}{\partial x \partial t}$$

$$- e_{32}h_c K_c C(t)\frac{\partial^2 V_y}{\partial y \partial t} + q(x,y,t) = I_0\left(\frac{\partial^2 w_0}{\partial t^2}\right) + I_1\left(\frac{\partial^3 u_0}{\partial x \partial t^2}\right)$$

$$+ I_1\left(\frac{\partial^3 v_0}{\partial y \partial t^2}\right) - I_2\left(\frac{\partial^4 w_0}{\partial x^2 \partial t^2}\right)$$

(5.9)

$$- I_2\left(\frac{\partial^4 w_0}{\partial y^2 \partial t^2}\right) + I_4\left(\frac{\partial^3 \phi_x}{\partial x \partial t^2}\right) + I_4\left(\frac{\partial^3 \phi_y}{\partial y \partial t^2}\right)$$

$$\delta\phi_x : \frac{\partial F_{xx}}{\partial x} + \frac{\partial F_{xy}}{\partial y} - G_{xz} = I_3 \left(\frac{\partial^2 u_0}{\partial t^2} \right) + I_5 \left(\frac{\partial^2 \phi_x}{\partial t^2} \right) - I_4 \left(\frac{\partial^3 w_0}{\partial x \partial t^2} \right) \quad (5.10)$$

$$\delta\phi_y : \frac{\partial F_{yy}}{\partial y} + \frac{\partial F_{xy}}{\partial x} - G_{yz} = I_3 \left(\frac{\partial^2 v_0}{\partial t^2} \right) + I_5 \left(\frac{\partial^2 \phi_y}{\partial t^2} \right) - I_4 \left(\frac{\partial^3 w_0}{\partial y \partial t^2} \right) \quad (5.11)$$

5.1.2.2 Mass-Spring-Damper System's Governing Equation

Let's say that a mass-spring-damper framework has a mass (Me), a spring constant (Ke), and a spring constant (Ce), and that this framework is attached to (x_0, y_0) on the nanoplate. If we use the second law of Newton [9], For the mass-spring-damper structure, the formula for determining the distributed transverse load is as follows:

$$\left(1 - \mu^2 \nabla^2\right) q(x, y, t) = J(x, y, t) \quad (5.12)$$

$$q(x, y, t) = m_e (g - \ddot{z}) \delta (x - x_0) \delta (y - y_0) \quad (5.13)$$

In addition, the equation that controls the linked system may be described as follows: [10]:

$$m_e \ddot{z} + c_e \dot{z} + k_e z = k_e w_0 (x_0, y_0, t) + c_e \dot{w}_0 (x_0, y_0, t) \quad t > 0 \quad (5.14)$$

where z(t) represents the linked mass's location in the vertical plane at time t.

The following is a definition of the relationship between dimensional and non-dimensional quantities:

$$K_e = \frac{k_e a^2}{D_{11}}, \quad M_e = \frac{m_e}{I_0 ab}, \quad \forall = \frac{\sqrt{D_{11}}}{a^2 \sqrt{I_0}}, \quad C_e = \frac{c_e a^2 \forall}{D_{11}} \quad (5.15)$$

In addition, for the sake of ease, the dimensionless foundation characteristics that are listed here are classified as:

$$K_w = \frac{k_w a^4}{D_{11}}, \quad K_p = \frac{k_p a^2}{D_{11}}, \quad C_d = \frac{c_d a^2}{\sqrt{\rho h D_{11}}} \quad (5.16)$$

5.1.3 SOLUTION

Two coupled governing equations of the nanoplate, whereas carrying a damper-spring-mass framework were solved systematically by applying closed-form of the Navier-type arrangement and Laplace transform. It ought to be noted that in Navier's arrangement, the generalized displacement components are communicated in terms of twofold Fourier arrangement as an item of obscure coefficients and known trigonometric functions that fulfill the boundary conditions of the issue. To continue, the equations of the motion (5.7–5.11) of the nanoplate can be expressed in terms of displacements $(u_0, v_0, w_0, \phi_x, \phi_y)$.

Some features of Navier's method are as follow:

- High accuracy, convergence, and performance for different geometries.
- Hold various types of forces such as harmonic and hygro-thermal loads.

By utilizing the Laplace transform, a modern framework of conditions in which time reliance is disposed of will be gotten [11].

According to Navier's solution, the mid-plane displacement and rotation components can be defined by the Fourier series as stated in Eq. (3.45) [12–14].

By substituting Eq. (3.45) for Eqs. (5.7–5.11) and separating the variations of the $u_0, v_0, w_0, \phi_x, \phi_y$ matrix of K, M, and C, the matrix of K, M, and C will be obtained. Appendix 5A indicates the stiffness, mass, and damping matrices K, M, and C, respectively.

In addition, when taking into consideration the double-Fourier sine series, the expression for $q(x,y,t)$ may be written as:

$$q(x,y,t) = \sum_{m=1}^{\infty} \sum_{n=1}^{\infty} Q_{mn}(t) \sin(\alpha x) \sin\left(\frac{n\pi y}{b}\right) \tag{5.17}$$

where with the aid of the orthogonality criterion, the expression $Q_{mn}(t)$ may be represented in the following form:

$$Q_{mn}(t) = \frac{4}{ab} \int_0^a \int_0^b q(x,y,t) \sin\left(\frac{m\pi x}{a}\right) \sin(\beta y) \, dxdy \tag{5.18}$$

The solution to this problem may be found by combining Eqs. (5.17) and (5.18):

$$Q_{mn}(t) = \frac{4}{ab}\left[m_e(g-\ddot{z})\sin(\alpha x_0)\sin(\beta y_0)\right] \tag{5.19}$$

Eq. (5.20) may be derived by substituting Eqs. (3.45) and (5.19) into the motion equations and then sorting the different kinds of components of displacement into their respective categories, as follows:

$$[K]\begin{Bmatrix} U_{mn} \\ V_{mn} \\ W_{mn} \\ \phi_{xmn} \\ \phi_{ymn} \end{Bmatrix} + [M]\begin{Bmatrix} \ddot{U}_{mn} \\ \ddot{V}_{mn} \\ \ddot{W}_{mn} \\ \ddot{\phi}_{xmn} \\ \ddot{\phi}_{ymn} \end{Bmatrix} + [C]\begin{Bmatrix} \dot{U}_{mn} \\ \dot{V}_{mn} \\ \dot{W}_{mn} \\ \dot{\phi}_{xmn} \\ \dot{\phi}_{ymn} \end{Bmatrix} = \begin{Bmatrix} 0 \\ 0 \\ \frac{4}{ab}\left[M_e(g-\ddot{z})\sin(\alpha x_0)\sin(\beta y_0)\right] \\ 0 \\ 0 \end{Bmatrix} \tag{5.20}$$

The linked governing Eq. (5.14) and Eq. (5.20) of the system need to be solved concurrently in order to determine how the system would behave given the following beginning circumstances. [10].

$$U_{mn}(0) = \dot{U}_{mn}(0) = V_{mn}(0) = \dot{V}_{mn}(0) = W_{mn}(0) = \dot{W}_{mn}(0) = \phi_{xmn}(0) =$$
$$\phi_{xmn}(0) = \phi_{ymn}(0) = \dot{\phi}_{ymn}(0) = 0, z(0) = z_0, \dot{z}(0) = 0 \tag{5.21}$$

When the Laplace transform is finally applied to the equations that regulate the system, the following equation will be obtained as a result:

$$\{[K] + s^2[M] + s[C]\}\{\bar{\nabla}\} = \left\{ \begin{array}{c} 0 \\ 0 \\ \dfrac{4}{ab}\left[M_e\left(\dfrac{g}{s} - s^2 Z + s z_0 \right) \sin(\alpha x_0)\sin(\beta y_0) \right] \\ 0 \\ 0 \end{array} \right\} \tag{5.22}$$

where the converted version of the displacement components may be produced by following the steps:

$$\{\bar{\nabla}\} = \{\bar{U}_{mn}, \bar{V}_{mn}, \bar{W}_{mn}, \bar{\phi}_{xmn}, \bar{\phi}_{ymn}\}^T . \tag{5.23}$$

Using the Laplace transform, the governing equation of the interconnected framework may be expressed as follows:

$$\left(m_e s^2 + c_e s + k_e \right) Z = \left(c_e s + k_e \right) W_0\left(x_0, y_0, s \right) + \left(m_e s + c_e \right) z_0 \tag{5.24}$$

To continue, if we take Z from Eq. (5.24) and plug it into Eq. (5.22), we get the following result in Eq. (5.25):

$$\{[K] + s^2[M] + s[C]\}\{\bar{\nabla}\} = \left\{ \begin{array}{c} 0 \\ 0 \\ \dfrac{4}{ab}\left[H\left(x_0, y_0, s \right)\sin(\alpha x_0)\sin(\beta y_0) \right] \\ 0 \\ 0 \end{array} \right\} \tag{5.25}$$

where $H(x_0, y_0, s)$ is defined as:

$$H(x_0, y_0, s) = A - BW_0(x_0, y_0, s) \tag{5.26}$$

where A and B are defined as the following:

$$A = m_e \left(\frac{g}{s} + \frac{K_e s z_0}{M_e s^2 + C_e s + K_e} \right) \quad B = \frac{m_e s^2 (c_e s + k_e)}{m_e s^2 + c_e s + k_e} \tag{5.27}$$

In addition, since $[\aleph_{ii}] = ([K] + s^2[M] + s[C])^{-1}$, the Laplace transformed version of the unknown displacement components may be expressed as:

$$\begin{cases} \bar{U}_{mn} = \frac{4\aleph_{13}}{ab} \left[H(x_0, y_0, s) \sin(\alpha x_0) \sin(\beta y_0) \right] \\[2mm] \bar{V}_{mn} = \frac{4\aleph_{23}}{ab} \left[H(x_0, y_0, s) \sin(\alpha x_0) \sin(\beta y_0) \right] \\[2mm] \bar{W}_{mn} = \frac{4\aleph_{33}}{ab} \left[H(x_0, y_0, s) \sin(\alpha x_0) \sin(\beta y_0) \right] \\[2mm] \bar{\phi}_{xmn} = \frac{4\aleph_{43}}{ab} \left[H(x_0, y_0, s) \sin(\alpha x_0) \sin(\beta y_0) \right] \\[2mm] \bar{\phi}_{ymn} = \frac{4\aleph_{53}}{ab} \left[H(x_0, y_0, s) \sin(\alpha x_0) \sin(\beta y_0) \right] \end{cases} \tag{5.28}$$

In order to determine the displacement component along the Z-axis, by developing the third equation, the value of $\bar{W}_0(x, y, s)$ may be reached as:

$$\bar{W}_0(x, y, s) = \frac{4}{ab} \sum_{m=1}^{\infty} \sum_{n=1}^{\infty} \aleph_{33} \left[(A - BW_0(x_0, y_0, s)) \sin(\alpha x_0) \sin(\beta y_0) \right] \\ \sin(\alpha x) \sin(\beta y) \tag{5.29}$$

So, by substituting (x, y) with (x_0, y_0) on both sides of Eq. (5.28), we get the following result:

$$\bar{W}_0(x_0, y_0, s) = \frac{F_1(x_0, y_0)}{1 + F_2(x_0, y_0)} \tag{5.30}$$

where $F_1(x_0, y_0)$ and $F_1(x_0, y_0)$ are defined as:

$$\begin{cases} F_1(x_0, y_0) = \frac{4}{ab} \sum_{m=1}^{\infty} \sum_{n=1}^{\infty} \aleph_{33} \left[A \sin^2(\alpha x_0) \sin^2(\beta y_0) \right] \\[2mm] F_2(x_0, y_0) = \frac{4}{ab} \sum_{m=1}^{\infty} \sum_{n=1}^{\infty} \aleph_{33} B \left[\sin^2(\alpha x_0) \sin^2(\beta y_0) \right] \end{cases} \tag{5.31}$$

The framework's natural frequencies may be obtained by setting the denominator of Eq. (5.30) to zero and solving the resulting equation. For example, the poles of Eq. (5.30), which represent the natural frequencies of the coupled system, are shown next. The system's inherent frequencies are represented by the imaginary components of the denominator, which represents Eq. (5.30). When Eq. (5.30), which represents the displacement, is substituted into Eq. (5.25), and the Laplace transformed version of the Naiver solution is used, the items representing the displacement will be obtained in the time domain.

Combining Eq. (5.30) with Eq. (5.24) and using the inverse Laplace transform results in the following vertical location of the suspended mass in the Laplace transform domain:

$$Z = \frac{(c_e s + k_e)}{(m_e s^2 + c_e s + k_e)} \frac{F_1(x_0, y_0)}{1 + F_2(x_0, y_0)} + \frac{(m_e s + c_e) z_0}{(m_e s^2 + c_e s + k_e)} \qquad (5.32)$$

In this section, the suggested framework is first verified by comparing the obtained results with the literature results. This is done in order to ensure that the proposed framework is accurate. Subsequently, a number of parametric studies are carried out in order to highlight the effect that a variety of factors have on the frequency response and the dynamic response of the system.

As a preliminary illustration of this concept, comparative research has been carried out for the particular circumstance of a single-layer isotropic FG plate based on the higher-order SSDT. According to the findings that were published in the past by Neves et al. [15], utilizing 13^2 Chebyshev points for $((\varepsilon_{zz} = 0)$ and $((\varepsilon_{zz} \neq 0))$, as well as Vel and Batra [16] and Qian et al. [17]. The magnetostrictive layer was not included in the analysis since its exclusion would have made the comparison less reliable. Materials on the plate range from zirconium dioxide (on the plate's bottom surface) to aluminum (on the plate's top surface). It is possible to verify that there is a wonderful arrangement between the outcomes, with the current results, and the references from the past.

The natural frequency is noticeably lower in theories that do not take into account the stretching of thickness, as compared to theories that do take into account the stretching of thickness. A careful examination of Table 5.2 for the sandwich composite nanoplate indicates that the current model is capable of reaching very trustworthy predictions for all types of mechanical loading issues. In addition, the Error is characterized by the formula Error% = |Present-Reference|/Reference100. Comparing dimensionless natural frequencies with reliable references [18] is what Table 5.3 would be about for a single-layered square magnetostrictive nanoplate.

Analytical solution for the non-dimensional natural frequencies of a magnetostrictive nanoplate is compared with those found in references [19, 20]. The results of this comparison are shown in Table 5.4.

As can be seen in Tables 5.3 and 5.4, the results indicate a significant degree of concordance regardless of the value chosen for the feedback gain parameter.

TABLE 5.2

Four Variations of the Gradient Index Were Used to Compare the Frequencies of an SSSS Isotropic Functionally Graded Plate (Al/ZrO$_2$)

$(a/h = 5, \ \Omega = \dfrac{\omega a \sqrt{\rho_c}}{\sqrt{E_c}})$

Gradient index	Ref. [17]	Ref. [16]	Ref. [15] $(\varepsilon_{zz} \neq 0)$	Ref. [15] $(\varepsilon_{zz} \neq 0)$	Present	Error%
k = 1	0.2152	0.2192	0.2193	0.2184	0.216257	0.98
k = 2	0.2153	0.2197	0.2198	0.2189	0.217556	0.61
k = 3	0.2172	0.2211	0.2212	0.2202	0.219041	0.52
k = 5	0.2194	0.2225	0.2225	0.2215	0.221662	0.07

TABLE 5.3

Natural Frequency Comparison ($K_c C(t) = 10^4$, SSSS, $\gamma = 1$, $h_c = 0.1a$,

$\left(\mu^2 = 0.06 \times a^2, \Omega = \dfrac{\omega a \sqrt{\rho_c}}{\sqrt{E_c}} \right)$

Dimensionless frequency	Ref. [18]	Present	Error %
Ω_{11}	0.3841	0.381638	0.64

TABLE 5.4

Natural Frequency Comparison Natural Frequencies Without Dimensions

$\left(\gamma = 1, \alpha_m = 0.2, \Omega = \dfrac{\omega h_c \sqrt{\rho_c}}{\sqrt{E_c}} \right)$

hc = 2	$K_c C(t) = 10^3$		
	Ref. [20]	Ref. [19]	Present
	1.046	1.000	1.0067

5.1.4 RESULTS

In this part of the chapter, numerical findings are carried out for the sandwich composite that consists of a coupled nano mass-spring-damper system. It is hypothesized that the core layer of the three-layered composite is made up of magnetostrictive material, and that the two facesheet layers are made up of FG material.

In each and every one of the outcomes that are shown here, the recommended quantities have been assumed, unless otherwise specified.

$a = b = 20$ nm, $h_c = 1$ nm, $h_f = 0.2$ nm, $K_w = K_p = 0$,

$$(x_0, y_0) = \left(\frac{a}{2}, \frac{b}{2}\right) Y_n(y), \text{ Kc C(t)} = 1\times10^5$$

The effect of the damper (the visco-Pasternak setup) on the center deflection of the nanoplate is shown as a plot in Figure 5.2, which may be seen here. According to what is shown in this figure, increasing the damper's coefficient results in a reduction in the capability of the nanoplate's central deflection to meet its requirements. This is because viscose damping has a favorable impact

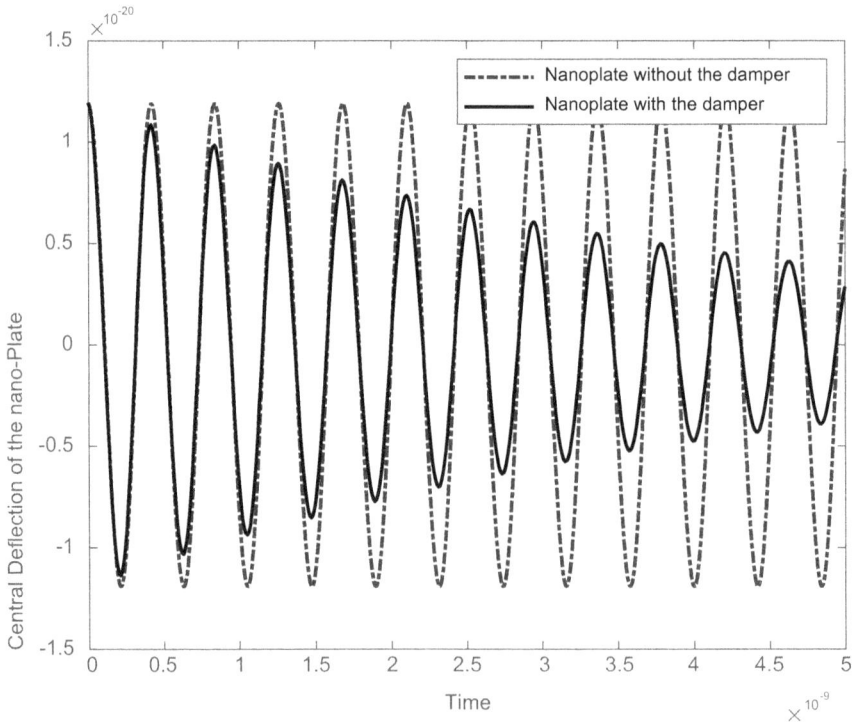

FIGURE 5.2 Nanoplate central deflection (W (a/2, b/2, t)) versus time with varying damping coefficients.

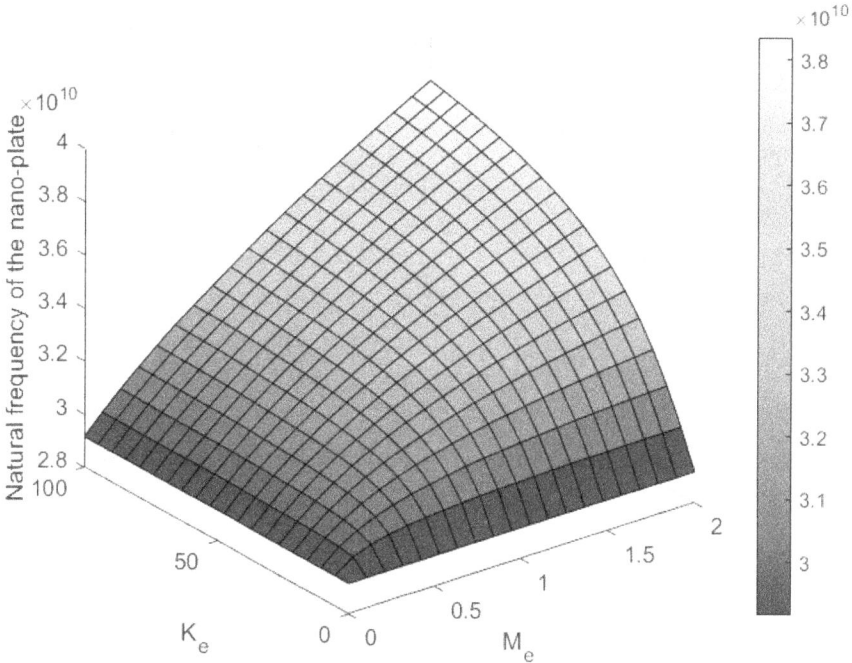

FIGURE 5.3 The effect of the mass-spring structure on the nano-fundamental plate's frequency ($K_w = 10$, $K_p = 3$, $k = 0$, $\mu^2 = 3nm^2$).

on the reduction of the amplitude of the oscillations, which has led to this result.

Figure 5.3 illustrates the most important effects that the suspended mass and the spring coefficient have on the vibration of the nanoplate. According to what is shown in Figure 5.3, increasing the value of the suspended mass and the spring's coefficient causes an increase in the natural frequency of the nanoplate.

In order to better comprehend how the thickness of the magnetostrictive layer (core layer) influences the nano-natural plate's frequency, Figure 5.4 has been constructed. As expected, the frequency of the nano-natural plate rises as its core layer thickness grows. The natural frequency likewise decreases when the nonlocal parameter is given a bigger value. This occurs because nonlocality softens the rigidity of the nanoplates.

Figure 5.5 illustrates, in a straightforward and understandable manner, the primary effects that the thickness of the core layer and the FG facesheets have on the vibration of the nanoplate. As was to be predicted, raising the value of the magnetostrictive facesheets and the FG facesheets leads to an increase in the natural frequency of the nanoplate.

An in-depth analysis of Figure 5.6 reveals a few distinguishing highlights of the impact of the velocity feedback gain in sandwich nanoplate's vibrational

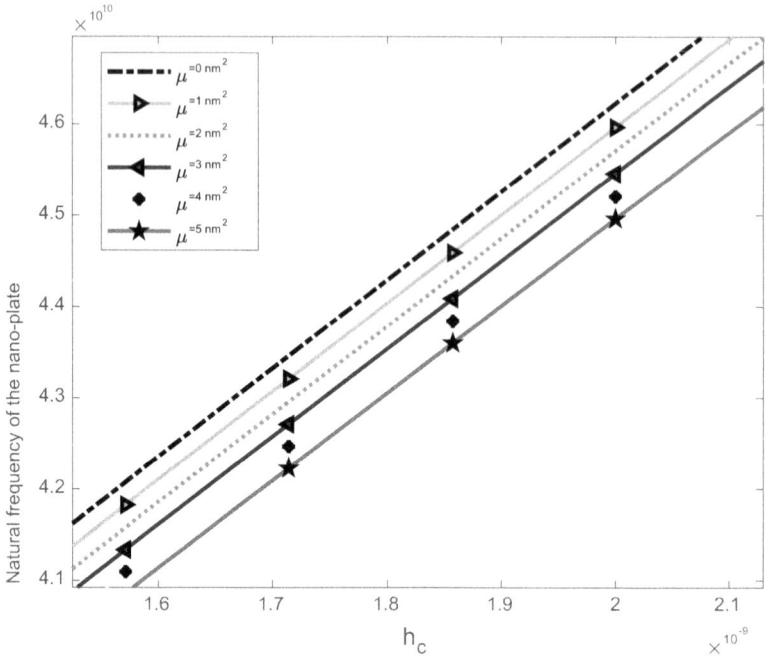

FIGURE 5.4 Relationship between the nanoplate's natural frequency and the magnetostrictive layer's thickness.

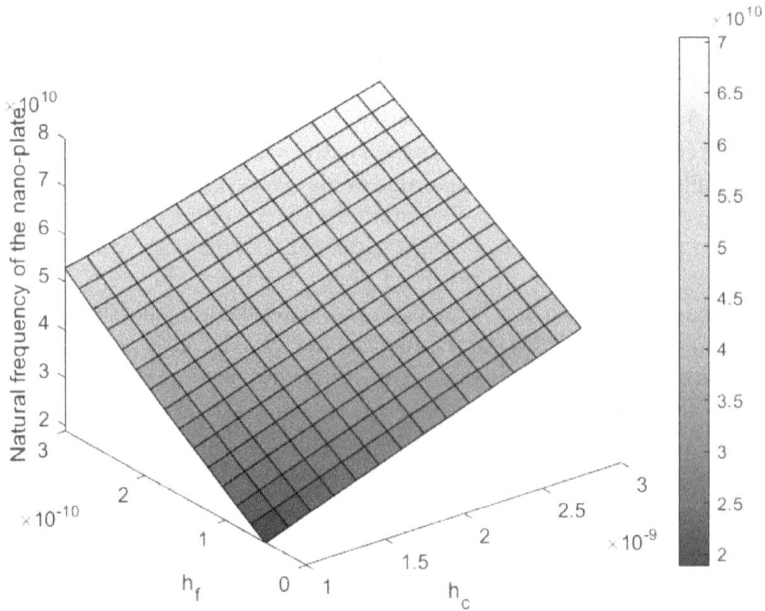

FIGURE 5.5 Relationship between nanoplate frequency, magnetostrictive layer thickness, and (FG) facesheet thickness ($K_w = 10$, $K_p = 3$, $k = 0$, $\mu^2 = 2\,nm^2$).

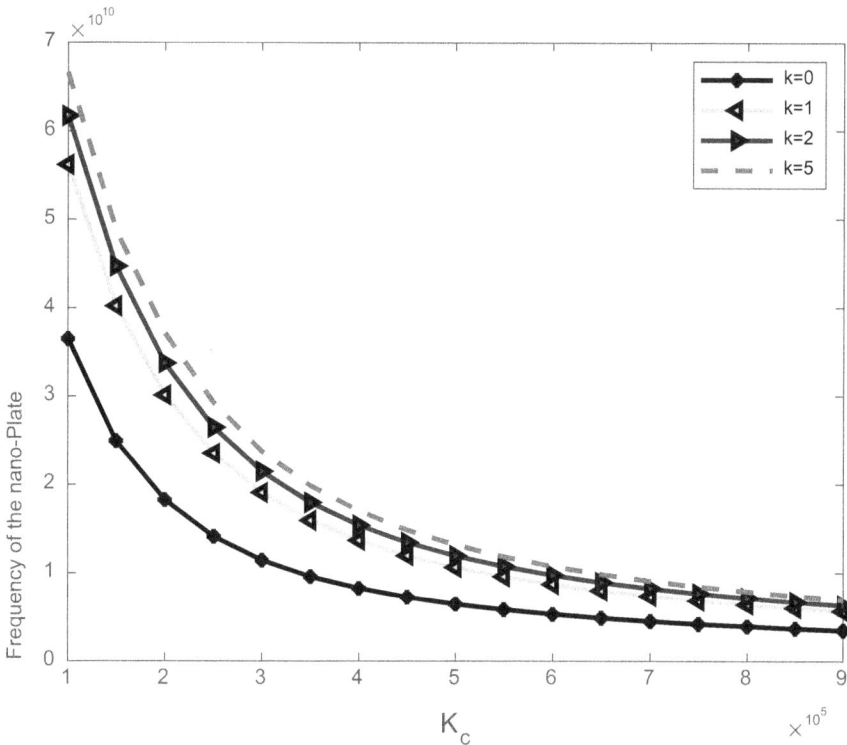

FIGURE 5.6   Nanoplate frequency versus velocity-gain feedback ($\mu^2 = 1\,nm^2$).

behavior for the four different gradient index values in general. These highlights are revealed as a result of the impact of the velocity feedback gain. The dampening impact that the magnetostriction phenomenon has on the behavior of the system causes the natural frequencies to decrease as the feedback gain parameter is increased. This is owing to the fact that raising the feedback gain parameter affects the behavior of the system. To put it another way, the velocity feedback gain parameter plays a controlling role in the framework's overall stability.

The reaction of the natural frequency in relation to the elastic foundation and the nonlocal parameter is seen in Figure 5.7. When the nonlocal parameter experiences an increase, the natural frequency experiences a decrease when it is in close proximity to an elastic institution. Figure 5.7 is an illustration of the influence that the visco-Pasternak medium has on the natural frequency of the nanoplate when the boundary condition is simply sustained. As seen in Figure 5.7, an increase in the value of the Winkler foundation results in an increase in the natural frequency. In addition, increasing the value of the Pasternak coefficient causes a proportional improvement in the natural frequencies of composite magnetostrictive nanoplates that include FGM facesheets.

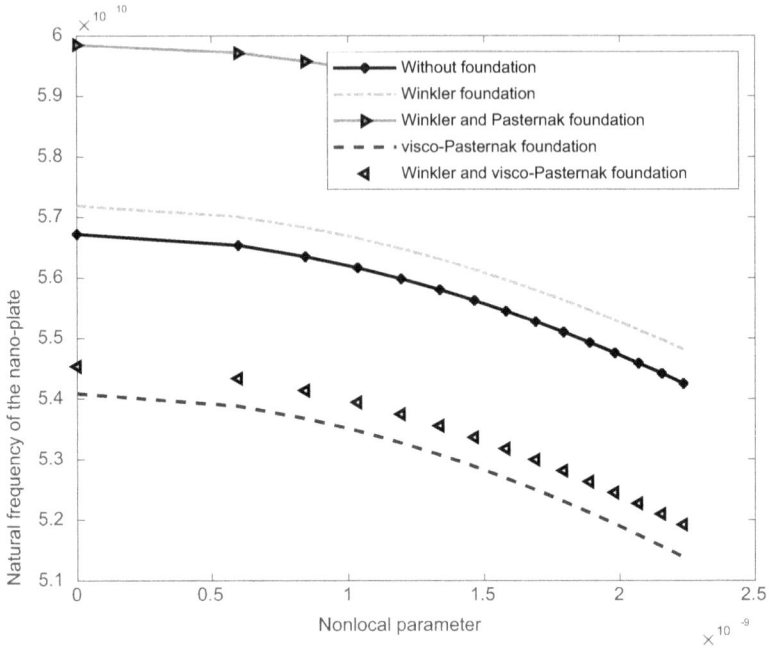

FIGURE 5.7　Nanoplate frequency against elastic substrate.

Figure 5.8 presents a scheme for the five various values of the nonlocal parameter of the Eringen in order to analyze how the parameter of the suspended spring affects the frequency of the nanoplate. This analysis will help determine how the parameter influences the frequency of the nanoplate. As can be seen in Figure 5.8, raising the weight of the suspended spring's coefficient leads to an increase in frequency. This is because the overall stiffness of the nanoplate decreases as a consequence of the increase in the frequency.

The description of the suggested nanoplate behavior when it was exposed to a system consisting of a mass-spring-damper is then brought to a close by the final Figure 5.9. As can be seen in Figure 5.9, a rise in the value of the suspended mass causes an increase in the natural frequency of the nanoplate.

Table 5.5 displays the data in a manner that indicates the impact of the spring's coefficient on the frequency of the mass-spring-damper system and the nanoplate. The first four occurrences of a certain frequency are listed in Table 5.5. When the spring's coefficient is increased, so too are the nanoplate's inherent frequencies. Based on the data presented here, it seems that a modest rise in natural frequencies is possible when the spring's coefficient is small.

Also, the effect of the velocity feedback gain on the natural frequency is shown in Table 5.5 for each of the possible value combinations. As was to be predicted, raising the amount of the velocity feedback gain has the effect of lowering the natural frequency, which manifests itself most obviously in the frequency of the nanoplate.

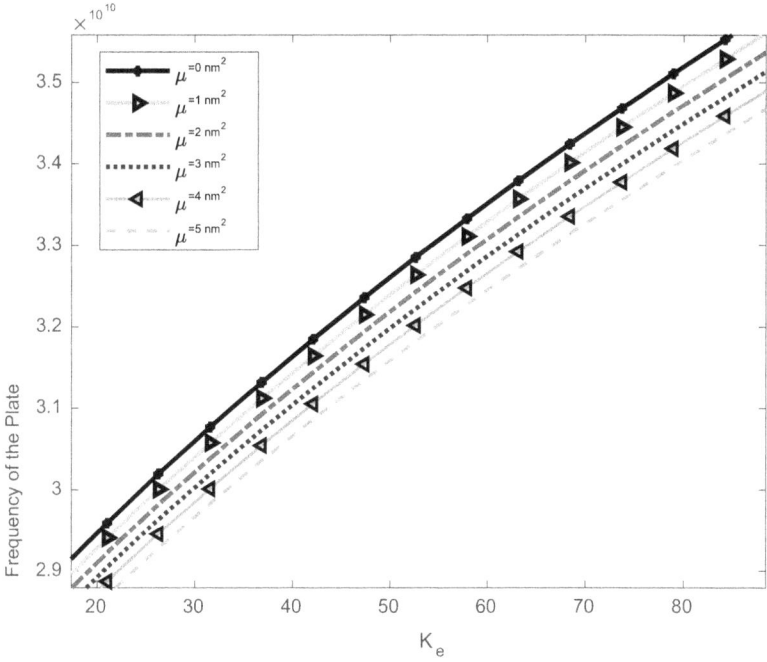

FIGURE 5.8 Influence of the coefficient of the suspended spring on the nanoplate's frequency for varying values of the nonlocal parameter.

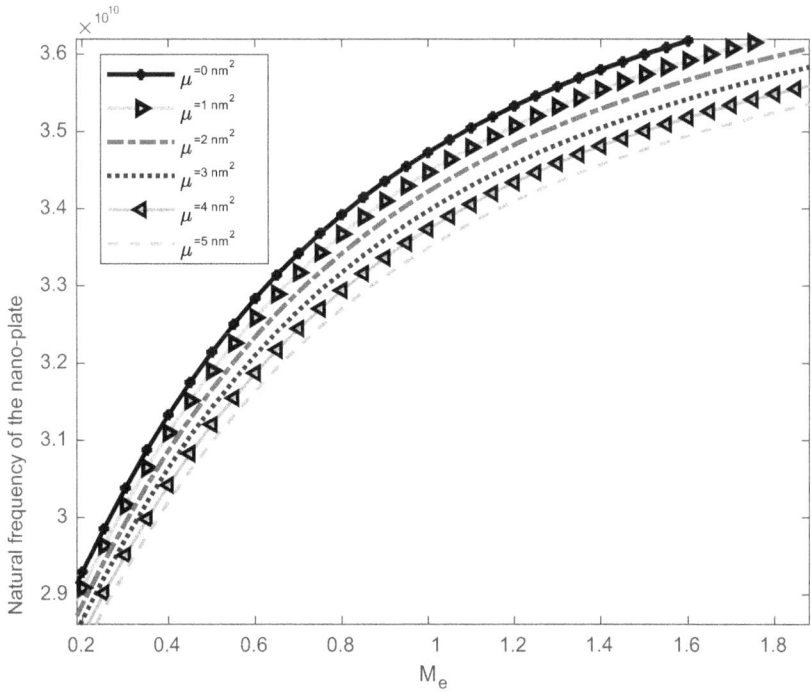

FIGURE 5.9 Quantifying the impact of a nonlocal parameter on the frequency of a nanoplate with a suspended mass ($k = 0$).

TABLE 5.5

Frequency (in GHz) of a Nanoplate for Various Amounts of Spring's Coefficient ($k = 1$, $C_d = 0$, $C_e = 0$, $Me = 2$, $\mu^2 = 0$)

Spring's coefficient	Velocity feedback gain	Mass-spring-damper Mechanism Frequency		Nanoplates' natural frequency			
		Separated	Attached (ϖ)	ω_{11}	ω_{12}	ω_{22}	ω_{13}
$M_e = 0$	1×10^5	–	–	40.42	97.3263	145.753	182.776
$K_e = 1\times10^{-3}$	1×10^5	0.085166	0.085165	40.4202	97.3263	145.753	182.776
$K_e = 1\times10^{-3}$	5×10^5	–	0.085165	9.85363	23.7847	35.6789	44.7982
$K_e = 5\times10^{-3}$	1×10^5	0.190437	0.190432	40.4212	97.3263	145.753	182.776
$K_e = 5\times10^{-3}$	5×10^5	–	0.190432	9.85387	23.7847	35.6789	44.7982
$K_e = 9\times10^{-3}$	1×10^5	0.255498	0.255485	40.4222	97.3263	145.753	182.776
$K_e = 9\times10^{-?}$	5×10^5	–	0.255485	9.85410	23.7847	35.6789	44.7983

TABLE 5.6

A Nano-Resonant Plate's Frequency (in GHz) as a Function of the Spring's Coefficient ($k = 1$, $C_d = 0$, $C_e = 0$, $Me = 2$, $\mu^2 = 0$)

	The mass spring and damper system's fundamental frequency		Nanoplates' natural frequency			
	Separated	Attached (ϖ)	ω_{11}	ω_{12}	ω_{22}	ω_{13}
$M_e = 0$	–	–	40.42	97.3263	145.753	182.776
$K_e = 1$	2.69319	2.67732	40.6589	97.3263	145.753	182.828
$K_e = 10$	8.51661	8.06609	42.6709	97.3263	145.753	183.296
$K_e = 100$	26.9319	19.1621	56.7204	97.3263	145.753	187.815

When there is a sufficient quantity of Ke present, the effect of the spring is more important and has a greater impact on the frequency, as can be seen by comparing Tables 5.5 and 5.6.

Table 5.7 illustrates the natural frequencies for the linked system, taking into consideration, in particular, the value of the suspended mass. In doing so, it is possible to provide a more comprehensive explanation for the sandwich composite nano-response plate's to a mass-spring-damper system. It's not surprising that the natural frequency of the nanoplate rises as the value of the suspended mass rises. This is mainly because an increase in the overall amount of mass that is

TABLE 5.7

Calculating the Nano-Fundamental Plate's Natural Frequency (in GHz) for a Range of Spring Coefficients ($k = 1$, $C_d = 0$, $C_e = 0$, $Ke = 200$, $\mu^2 = 0$, $hc = 3$ nm, $h_f = 0.3$ nm)

	Frequency of mass-spring-damper System		Frequency of nanoplate			
	Separated	Attached (ϖ)	ω_{11}	ω_{12}	ω_{22}	ω_{13}
$M_e = 0$	–	–	63.6948	134.854	181.025	239.977
$M_e = 1 \times 10^3$	3.64687	1.79881	127.106	134.854	181.025	249.421
$M_e = 5 \times 10^3$	1.63093	0.80426	127.136	134.854	181.025	249.425
$M_e = 9 \times 10^3$	1.21562	0.599444	127.139	134.854	181.025	249.426

connected will cause the system's weight to grow. Moreover, the natural frequencies of the suggested system's mass-spring-damper combination decrease when the overall mass of the system is raised. Specifically, the proportional relationship between the frequency of the mass-spring-damper system and the suspended mass, as illustrated by the equation $e = K/M$, which can be seen in the preceding phrase, gives birth to this phenomenon. Indeed, the frequency will decrease when the value of the suspended mass is increased since a larger denominator means a lower frequency.

Table 5.8 summarizes, with a particular emphasis on the nonlocal parameter, the natural frequencies that correspond to the various form modes. As can be gathered from Table 5.8, changes in the natural frequencies of the nanoplate in terms of the nonlocal parameter have an effect on the frequency of the nanoplate when it is coupled to a connected mass-spring-damper system. Due to the fact that the stiffness of nanostructures lowers with an increase in nonlocal parameter, growing the Eringen's nonlocal parameter causes reducing the frequencies, which is caused by the fact that the frequencies drop. This clearly demonstrates the impact of softening.

Table 5.9 illustrates the impacts of the location of the mass-spring-damper system as well as the gradient index parameter on the frequency of a nanoplate that is rectangular (a = b = 20 nm) and has a constant total thickness (H = $hc + 2h_f$). It has been determined that as the mass-spring-damper system is brought closer to the center of the nanoplate, frequencies begin to have a greater amount of influence. A rise in the value of the gradient-index parameter in this instance unequivocally indicates an increase in the natural frequency, and the significance of this increase is amplified if the mass-spring-damper locates in the middle of the nanoplate. In addition, as can be shown in Table 5.9, the frequency of the

TABLE 5.8

The Nano-Coupled Plate's Frequency (GHz) for Different Values of Eringen's Nonlocal Parameter (k = 2, C_d = 0, Ce = 0, Ke = 0.5, Me = 2×10⁻⁴)

Natural Frequency	Feedback gain	$\mu^2 = 0\,\mathrm{nm}^2$	$\mu^2 = 1\,\mathrm{nm}^2$	$\mu^2 = 2\,\mathrm{nm}^2$	$\mu^2 = 3\,\mathrm{nm}^2$	$\mu^2 = 4\,\mathrm{nm}^2$	$\mu^2 = 5\,\mathrm{nm}^2$
ω_{11}	$k_cC(t) = 1\times10^5$	44.1647	43.7707	43.3872	43.0135	42.6495	42.2946
ω_{11}	$k_cC(t) = 5\times10^5$	10.9768	10.9706	10.9645	10.9584	10.9522	10.9461
ω_{11}	$k_cC(t) = 9\times10^5$	6.14682	6.14574	6.14465	6.14357	6.14249	6.14141
ω_{22}	$k_cC(t) = 1\times10^5$	157.713	152.256	147.3337	142.872	138.795	135.052
ω_{22}	$k_cC(t) = 5\times10^5$	39.37	39.2789	39.1884	39.0986	39.0095	38.9209
ω_{22}	$k_cC(t) = 9\times10^5$	22.0525	22.0364	22.0204	22.0044	21.9884	21.9725
ω_{33}	$k_cC(t) = 1\times10^5$	305.349	282.916	264.883	249.959	237.332	226.463
ω_{33}	$k_cC(t) = 5\times10^5$	76.472	76.0654	75.6656	75.2724	74.8857	74.5054
ω_{33}	$k_cC(t) = 9\times10^5$	42.8438	427717	42.700	42.6287	42.5578	42.4873

TABLE 5.9

Nanoplate Mass-Spring-Damper Coupled Frequency (in GHz) (C_d = 0, C_e = 0, Ke = 100, Me = 2, μ^2 = 0)

Gradient index	$(x_0, y_0) = (0,0)$	$(x_0, y_0) = \left(\dfrac{a}{2}, \dfrac{b}{2}\right)$	$(x_0, y_0) = \left(\dfrac{a}{4}, \dfrac{3b}{4}\right)$	$(x_0, y_0) = (a, b)$
$k = 0$	26.7789	36.729	29.2627	26.7789
$k = 0.5$	36.2875	50.599	39.9363	36.2875
$k = 1$	40.42	56.7204	44.6095	40.42
$k = 2$	44.1587	62.3019	48.8541	44.1587
$k = 5$	47.4921	67.3072	52.651	47.4921

TABLE 5.10

For a Range of Elastic Media Densities, the Coupled Frequency (GHz) of a Nanoplate ($k = 0$, $\mu^2 = 2\text{nm}^2$, $Me = 2$, $Ke = 100$)

Winkler foundation	Spring's coefficient	Damper's coefficient	Natural frequency
$K_w = 0$	$K_p = 0$	$C_d = 3$	34.9455
		$C_d = 8$	32.8738
	$K_p = 3$	$C_d = 3$	36.9375
		$C_d = 8$	34.9075
	$K_p = 5$	$C_d = 3$	38.1892
		$C_d = 8$	36.1806
$K_w = 10$	$K_p = 0$	$C_d = 3$	35.2927
		$C_d = 8$	33.2291
	$K_p = 3$	$C_d = 3$	37.2598
		$C_d = 8$	35.2356
	$K_p = 5$	$C_d = 3$	38.4979
		$C_d = 8$	36.4941
$K_w = 20$	$K_p = 0$	$C_d = 3$	35.6353
		$C_d = 8$	33.5793
	$K_p = 3$	$C_d = 3$	37.5785
		$C_d = 8$	35.5599
	$K_p = 5$	$C_d = 3$	38.8033
		$C_d = 8$	36.8042

nanoplate rises as the value of the gradient index does as well. This is because zirconium dioxide has a larger Young's modulus than aluminum, which is the reason for the previous statement.

Table 5.10 illustrates how the natural frequency of a nanoplate that is simply supported (SS) fluctuates as a result of changes to the elastic basis. The numerical findings show that an increase in frequency occurs as a direct consequence of the inclusion of the Winkler medium in the system. A similar phenomenon occurs when one adds a spring to an existing foundation; the consequence is an increase in the natural frequency. If the damper is aligned such that it is parallel to the spring, then an increase in the damper's coefficient will result in a reduction in the frequency of the nanoplate.

Table 5.11 provides a retrospective look at the thickness parameter versus the natural frequency in order to investigate the thickness impact. It cannot be disputed that the thickness of the core layer and the FG facesheets has a direct correlation to a substantial rise in the sandwich nanoplate's level of stiffness. The

TABLE 5.11

Nanoplate's Coupled Frequency (GHz) for Different Magnetostrictive Core Layer Thicknesses ($k = 1$, $K_w = 20$, $Kp = 5$, $C_d = 3$, $\mu^2 = 1nm^2$, $Ke = 0.5$, $Me = 2 \times 10^{-4}$)

	$h_c = 0.5$ nm	$h_c = 0.75$ nm	$h_c = 1$ nm	$h_c = 1.5$ nm
$h_f = 0.2$ nm	41.5071	43.1398	44.999	48.9294
$h_f = 0.1$ nm	23.0234	25.3543	27.7476	32.4946

values of the thickness of the core layer were evaluated in Table 5.11, together with two other possibilities for the thickness of the facesheets.

5.1.5 CONCLUDING REMARKS

In light of the increasing demand for the analysis of the vibrational behavior of sandwich composite nanoplates integrated with functionally graded facesheets and a connected nano mass-spring-damper system, this research was carried out. The research was based on the higher-order sinusoidal shear deformation theory. Analytical solutions are found for the two coupled governing equations of the system by generating a solution of the Navier-type and using the Laplace transform technique. The influence of a number of factors, such as a nonlocal parameter, the value of the suspended mass, the spring's coefficient, and the mass's coefficient, was investigated using a parametric approach. The theoretical research and numerical findings that have been provided here may be used in the process of modeling and assessing nanostructures that include magnetostrictive layers. In addition, the numerical findings of this study may serve as benchmark values for other researchers to verify their various analyses, which are used in the process of constructing these sorts of nanostructures.

The following is a list of some of the most astounding findings that have emerged from this work:

1- The natural frequency may be made lower by avoiding the nanoplate's center and moving farther away from it.
2- Raising the feedback gain parameter has a considerable influence on both the frequency of the nanoplate and the frequency of the mass-spring-damper system. This is because both of these frequencies are proportional to the feedback gain parameter.
3- The fluctuation in the frequency is more perceptible when there is a greater significance in the spring's coefficient.
4- The frequency of a connected mass-spring-damper system is lower than that of a mass-spring-damper system that is detached.
5- Raising the value of the suspended mass causes a large increase in the frequency of the nanoplate, notably for the first shape mode.

Appendix 5A

$$K_{11} = -A_{11}\left(\alpha^2\right) - A_{66}\left(\beta^2\right)$$

$$K_{12} = -A_{12}\left(\alpha\beta\right) - A_{66}\left(\alpha\beta\right)$$

$$K_{13} = B_{11}\left(\alpha^3\right) + B_{12}\left(\alpha\beta^2\right) + 2B_{66}\left(\alpha\beta^2\right)$$

$$K_{14} = -S_{11}\left(\alpha^2\right) - S_{66}\left(\beta^2\right)$$

$$K_{15} = -S_{12}\left(\alpha\beta\right) - S_{66}\left(\alpha\beta\right)$$

$$K_{21} = -A_{21}\left(\alpha\beta\right) - A_{66}\left(\alpha\beta\right)$$

$$K_{22} = -A_{22}\left(\beta^2\right) - A_{66}\left(\alpha^2\right)$$

$$K_{23} = B_{21}\left(\beta\alpha^2\right) + B_{22}\left(\beta^3\right) + 2B_{66}\left(\beta\alpha^2\right)$$

$$K_{24} = -S_{21}\left(\alpha\beta\right) - S_{66}\left(\alpha\beta\right)$$

$$K_{25} = -S_{22}\left(\alpha^2\right) - S_{66}\left(\beta^2\right)$$

$$K_{31} = B_{11}\left(\alpha^3\right) + B_{12}\left(\alpha\beta^2\right) + 2B_{66}\left(\alpha\beta^2\right)$$

$$K_{32} = B_{21}\left(\beta\alpha^2\right) + B_{22}\left(\beta^3\right) + 2B_{66}\left(\beta\alpha^2\right)$$

$$K_{33} = -D_{11}\left(\alpha^4\right) - D_{12}\left(\alpha^2\beta^2\right) - D_{22}\left(\beta^4\right) - 4D_{66}\left(\alpha^2\beta^2\right)$$
$$-K_w - K_g\left(\alpha^2\right) - K_g\left(\beta^2\right) - K_w\left(\mu^2\alpha^2\right) - K_w\left(\mu^2\beta^2\right)$$
$$-K_g\left(\mu^2\alpha^4\right) - K_g\left(\mu^2\beta^4\right) - 2K_g\left(\mu^2\alpha^2\beta^2\right)K_{12} = -A_{12}\left(\alpha\beta\right) - A_{66}\left(\alpha\beta\right)$$

$$K_{13} = B_{11}\left(\alpha^3\right) + B_{12}\left(\alpha\beta^2\right) + 2B_{66}\left(\alpha\beta^2\right)$$

$$K_{14} = -S_{11}\left(\alpha^2\right) - S_{66}\left(\beta^2\right)$$

$$K_{15} = -S_{12}\left(\alpha\beta\right) - S_{66}\left(\alpha\beta\right)$$

$$K_{21} = -A_{21}\left(\alpha\beta\right) - A_{66}\left(\alpha\beta\right)$$

$$K_{22} = -A_{22}\left(\beta^2\right) - A_{66}\left(\alpha^2\right)$$

$$K_{23} = B_{21}\left(\beta\alpha^2\right) + B_{22}\left(\beta^3\right) + 2B_{66}\left(\beta\alpha^2\right)$$

$$K_{24} = -S_{21}\left(\alpha\beta\right) - S_{66}\left(\alpha\beta\right)$$

$$K_{25} = -S_{22}\left(\alpha^2\right) - S_{66}\left(\beta^2\right)$$

$$K_{31} = B_{11}\left(\alpha^3\right) + B_{12}\left(\alpha\beta^2\right) + 2B_{66}\left(\alpha\beta^2\right)$$

$$K_{32} = B_{21}\left(\beta\alpha^2\right) + B_{22}\left(\beta^3\right) + 2B_{66}\left(\beta\alpha^2\right)$$

$$K_{33} = -D_{11}\left(\alpha^4\right) - D_{12}\left(\alpha^2\beta^2\right) - D_{22}\left(\beta^4\right) - 4D_{66}\left(\alpha^2\beta^2\right) - K_w$$
$$-K_g\left(\alpha^2\right) - K_g\left(\beta^2\right) - K_w\left(\mu^2\alpha^2\right) - K_w\left(\mu^2\beta^2\right) - K_g\left(\mu^2\beta^4\right) - 2K_g\left(\mu^2\alpha^2\beta^2\right)$$

$$K_{34} = M_{11}\left(\alpha^3\right) + M_{12}\left(\alpha\beta^2\right) + 2M_{66}\left(\alpha\beta^2\right)$$

$$K_{35} = M_{12}\left(\alpha^2\beta\right) + M_{22}\left(\beta^3\right) + 2M_{66}\left(\alpha^2\beta\right)$$

$$K_{41} = -S_{11}\left(\alpha^2\right) - S_{66}\left(\beta^2\right) \qquad K_{42} = -S_{21}\left(\alpha\beta\right) - S_{66}\left(\alpha\beta\right)$$

$$K_{43} = M_{11}\left(\alpha^3\right) + M_{12}\left(\alpha\beta^2\right) + 2M_{66}\left(\alpha\beta^2\right)$$

$$K_{44} = -O_{11}\left(\alpha^2\right) - O_{66}\left(\beta^2\right) - L_{55}$$

$$K_{45} = -O_{12}(\alpha\beta) - O_{66}(\alpha\beta)$$

$$K_{51} = -S_{12}(\alpha\beta) - S_{66}(\alpha\beta) \qquad K_{52} = -S_{22}(\alpha^2) - S_{66}(\beta^2)$$

$$K_{53} = M_{12}(\alpha^2\beta) + M_{22}(\beta^3) + 2M_{66}(\alpha^2\beta)$$

$$K_{54} = -O_{12}(\alpha\beta) - O_{66}(\alpha\beta)$$

$$K_{55} = -O_{22}(\beta^2) - O_{66}(\alpha^2) - L_{44}$$

$$M_{11} = I_0\left(1 + \mu^2\alpha^2 + \mu^2\beta^2\right)$$

$$M_{11} = I_0\left(1 + \mu^2\alpha^2 + \mu^2\beta^2\right)$$

$$M_{13} = -I_1\left(\alpha + \mu^2\alpha^3 + \mu^2\alpha\beta^2\right)$$

$$M_{14} = +I_3\left(1 + \mu^2\alpha^2 + \mu^2\beta^2\right)$$

$$M_{22} = +I_0\left(1 + \mu^2\alpha^2 + \mu^2\beta^2\right)$$

$$M_{23} = -I_1\left(\beta + \mu^2\beta\alpha^2 + \mu^2\beta^3\right)$$

$$M_{25} = +I_3\left(1 + \mu^2\alpha^2 + \mu^2\beta^2\right)$$

$$M_{31} = -I_1\left(\alpha + \mu^2\alpha^3 + \mu^2\alpha\beta^2\right)$$

$$M_{32} = -I_1\left(\beta + \mu^2\beta\alpha^2 + \mu^2\beta^3\right)$$

$$M_{33} = I_0\left(1 + \mu^2\alpha^2 + \mu^2\beta^2\right) + I_2\left(\alpha + \mu^2\alpha^3 + \mu^2\alpha\beta^2\right) + I_2\left(\beta + \mu^2\beta^3 + \mu^2\beta\alpha^2\right)$$

$$M_{34} = -I_4\left(\alpha + \mu^2\alpha^3 + \mu^2\alpha\beta^2\right)$$

$$M_{35} = -I_4 \left(\beta + \mu^2 \beta \alpha^2 + \mu^2 \beta^3 \right)$$

$$M_{41} = I_3 \left(1 + \mu^2 \alpha^2 + \mu^2 \beta^2 \right)$$

$$M_{34} = -I_4 \left(\alpha + \mu^2 \alpha^3 + \mu^2 \alpha \beta^2 \right)$$

$$M_{44} = I_5 \left(1 + \mu^2 \alpha^2 + \mu^2 \beta^2 \right) \qquad M_{52} = I_1 \left(1 + \mu^2 \alpha^2 + \mu^2 \beta^2 \right)$$

$$M_{55} = I_2 \left(1 + \mu^2 \alpha^2 + \mu^2 \beta^2 \right)$$

$$C_{11} = C_{12} = C_{14} = C_{15} = 0 \qquad C_{13} = \int_{-\frac{h_c}{2}}^{\frac{h_c}{2}} e_{31} \left(\alpha \right) K_c C(t) dz$$

$$C_{21} = C_{22} = C_{24} = C_{25} = 0 \qquad C_{23} = \int_{-\frac{h_c}{2}}^{\frac{h_c}{2}} e_{32} \left(\beta \right) K_c C(t) dz$$

$$C_{31} = \int_{-\frac{h_c}{2}}^{\frac{h_c}{2}} e_{31} \left(\alpha \right) K_c C(t) dz \qquad C_{32} = \int_{-\frac{h_c}{2}}^{\frac{h_c}{2}} e_{31} \left(\beta \right) K_c C(t) dz$$

$$C_{33} = c_d \left(1 + \mu^2 \alpha^2 + \mu^2 \beta^2 \right) \qquad C_{34} = \int_{-\frac{h_c}{2}}^{\frac{h_c}{2}} z e_{31} \left(\alpha \right) K_c C(t) dz$$

$$C_{35} = \int_{-\frac{h_c}{2}}^{\frac{h_c}{2}} z e_{32} \left(\beta \right) K_c C(t) dz$$

$$C_{41} = C_{42} = C_{44} = C_{45} = 0 \qquad C_{43} = \int_{-\frac{h_c}{2}}^{\frac{h_c}{2}} z e_{31} \left(\alpha \right) K_c C(t) dz$$

$$C_{51} = C_{52} = C_{54} = C_{55} = 0 \qquad C_{53} = \int_{-\frac{h_c}{2}}^{\frac{h_c}{2}} z e_{32} \left(\beta \right) K_c C(t) dz$$

$$A_{11} = \int_{-h_c/2}^{h_c/2} C_{11}^c dz + \int_{-\frac{(hc+h_f)}{2}}^{\frac{-hc}{2}} C_{11}^f dz + \int_{\frac{h_c}{2}}^{\frac{(h_c+h_f)}{2}} C_{11}^f dz$$

$$A_{12} = \int_{-h_c/2}^{h_c/2} C_{12}^c dz + \int_{-\frac{(hc+h_f)}{2}}^{\frac{-hc}{2}} C_{12}^f dz + \int_{\frac{h_c}{2}}^{\frac{(h_c+h_f)}{2}} C_{12}^f dz$$

$$A_{22} = \int_{-h_c/2}^{h_c/2} C_{22}^c dz + \int_{-\frac{(hc+h_f)}{2}}^{\frac{-hc}{2}} C_{22}^f dz + \int_{\frac{h_c}{2}}^{\frac{(h_c+h_f)}{2}} C_{22}^f dz$$

$$A_{44} = \int_{-h_c/2}^{h_c/2} C_{44}^c dz + \int_{-\frac{(hc+h_f)}{2}}^{\frac{-hc}{2}} C_{44}^f dz + \int_{\frac{h_c}{2}}^{\frac{(h_c+h_f)}{2}} C_{44}^f dz$$

$$A_{55} = \int_{-h_c/2}^{h_c/2} C_{55}^c dz + \int_{-\frac{(hc+h_f)}{2}}^{\frac{-hc}{2}} C_{55}^f dz + \int_{\frac{h_c}{2}}^{\frac{(h_c+h_f)}{2}} C_{55}^f dz$$

$$A_{66} = \int_{-h_c/2}^{h_c/2} C_{66}^c dz + \int_{-\frac{(hc+h_f)}{2}}^{\frac{-hc}{2}} C_{66}^f dz + \int_{\frac{h_c}{2}}^{\frac{(h_c+h_f)}{2}} C_{66}^f dz$$

$$B_{11} = \int_{-h_c/2}^{h_c/2} zC_{11}^c dz + \int_{-\frac{(hc+h_f)}{2}}^{\frac{-hc}{2}} zC_{11}^f dz + \int_{\frac{h_c}{2}}^{\frac{(h_c+h_f)}{2}} zC_{11}^f dz$$

$$B_{12} = \int_{-h_c/2}^{h_c/2} zC_{12}^c dz + \int_{-\frac{(hc+h_f)}{2}}^{\frac{-hc}{2}} zC_{12}^f dz + \int_{\frac{h_c}{2}}^{\frac{(h_c+h_f)}{2}} zC_{12}^f dz$$

$$B_{22} = \int_{-h_c/2}^{h_c/2} zC_{22}^c dz + \int_{-\frac{(hc+h_f)}{2}}^{\frac{-hc}{2}} zC_{22}^f dz + \int_{\frac{h_c}{2}}^{\frac{(h_c+h_f)}{2}} zC_{22}^f dz$$

$$B_{44} = \int_{-h_c/2}^{h_c/2} zC_{44}^c dz + \int_{-\frac{(hc+h_f)}{2}}^{\frac{-hc}{2}} zC_{44}^f dz + \int_{\frac{h_c}{2}}^{\frac{(h_c+h_f)}{2}} zC_{44}^f dz$$

$$B_{55} = \int_{-h_c/2}^{h_c/2} zC_{55}^c dz + \int_{-\frac{(hc+h_f)}{2}}^{\frac{-hc}{2}} zC_{22}^f dz + \int_{\frac{h_c}{2}}^{\frac{(h_c+h_f)}{2}} zC_{55}^f dz$$

$$B_{66} = \int_{-h_c/2}^{h_c/2} zC_{66}^c dz + \int_{-\frac{(hc+h_f)}{2}}^{\frac{-hc}{2}} zC_{66}^f dz + \int_{\frac{h_c}{2}}^{\frac{(h_c+h_f)}{2}} zC_{66}^f dz$$

$$D_{11} = \int_{-h_c/2}^{h_c/2} z^2 C_{11}^c dz + \int_{-\frac{(hc+h_f)}{2}}^{\frac{-hc}{2}} z^2 C_{11}^f dz + \int_{\frac{h_c}{2}}^{\frac{(h_c+h_f)}{2}} z^2 C_{11}^f dz$$

$$D_{12} = \int_{-h_c/2}^{h_c/2} z^2 C_{12}^c dz + \int_{-\frac{(hc+h_f)}{2}}^{\frac{-hc}{2}} z^2 C_{12}^f dz + \int_{\frac{h_c}{2}}^{\frac{(h_c+h_f)}{2}} z^2 C_{12}^f dz$$

$$D_{22} = \int_{-h_c/2}^{h_c/2} z^2 C_{22}^c dz + \int_{-\frac{(hc+h_f)}{2}}^{\frac{-hc}{2}} z^2 C_{22}^f dz + \int_{\frac{h_c}{2}}^{\frac{(h_c+h_f)}{2}} z^2 C_{22}^f dz$$

$$D_{44} = \int_{-h_c/2}^{h_c/2} z^2 C_{44}^c dz + \int_{-\frac{(hc+h_f)}{2}}^{\frac{-hc}{2}} z^2 C_{44}^f dz + \int_{\frac{h_c}{2}}^{\frac{(h_c+h_f)}{2}} z^2 C_{44}^f dz$$

$$D_{55} = \int_{-h_c/2}^{h_c/2} z^2 C_{55}^c dz + \int_{-\frac{(hc+h_f)}{2}}^{\frac{-hc}{2}} z^2 C_{55}^f dz + \int_{\frac{h_c}{2}}^{\frac{(h_c+h_f)}{2}} z^2 C_{55}^f dz$$

$$D_{66} = \int_{-h_c/2}^{h_c/2} z^2 C_{66}^c dz + \int_{-\frac{(hc+h_f)}{2}}^{\frac{-hc}{2}} z^2 C_{66}^f dz + \int_{\frac{h_c}{2}}^{\frac{(h_c+h_f)}{2}} z^2 C_{66}^f dz$$

$$S_{11} = \int_{-h_c/2}^{h_c/2} f(z) C_{11}^c dz + \int_{-\frac{(hc+h_f)}{2}}^{\frac{-hc}{2}} f(z) C_{11}^f dz + \int_{\frac{h_c}{2}}^{\frac{(h_c+h_f)}{2}} f(z) C_{11}^f dz$$

$$S_{12} = \int_{-h_c/2}^{h_c/2} f(z) C_{12}^c dz + \int_{-\frac{(hc+h_f)}{2}}^{\frac{-hc}{2}} f(z) C_{12}^f dz + \int_{\frac{h_c}{2}}^{\frac{(h_c+h_f)}{2}} f(z) C_{12}^f dz$$

$$S_{22} = \int_{-h_c/2}^{h_c/2} f(z)C_{22}^c dz + \int_{-\frac{(hc+h_f)}{2}}^{\frac{-hc}{2}} f(z)C_{22}^f dz + \int_{\frac{h_c}{2}}^{\frac{(h_c+h_f)}{2}} f(z)C_{22}^f dz$$

$$S_{44} = \int_{-h_c/2}^{h_c/2} f(z)C_{44}^c dz + \int_{-\frac{(hc+h_f)}{2}}^{\frac{-hc}{2}} f(z)C_{44}^f dz + \int_{\frac{h_c}{2}}^{\frac{(h_c+h_f)}{2}} f(z)C_{44}^f dz$$

$$S_{55} = \int_{-h_c/2}^{h_c/2} f(z)C_{55}^c dz + \int_{-\frac{(hc+h_f)}{2}}^{\frac{-hc}{2}} f(z)C_{55}^f dz + \int_{\frac{h_c}{2}}^{\frac{(h_c+h_f)}{2}} f(z)C_{55}^f dz$$

$$S_{66} = \int_{-h_c/2}^{h_c/2} f(z)C_{66}^c dz + \int_{-\frac{(hc+h_f)}{2}}^{\frac{-hc}{2}} f(z)C_{66}^f dz + \int_{\frac{h_c}{2}}^{\frac{(h_c+h_f)}{2}} f(z)C_{66}^f dz$$

$$M_{11} = \int_{-h_c/2}^{h_c/2} f(z)zC_{11}^c dz + \int_{-\frac{(hc+h_f)}{2}}^{\frac{-hc}{2}} f(z)zC_{11}^f dz + \int_{\frac{h_c}{2}}^{\frac{(h_c+h_f)}{2}} f(z)zC_{11}^f dz$$

$$M_{12} = \int_{-h_c/2}^{h_c/2} f(z)zC_{12}^c dz + \int_{-\frac{(hc+h_f)}{2}}^{\frac{-hc}{2}} f(z)zC_{12}^f dz + \int_{\frac{h_c}{2}}^{\frac{(h_c+h_f)}{2}} f(z)zC_{12}^f dz$$

$$M_{22} = \int_{-h_c/2}^{h_c/2} f(z)zC_{22}^c dz + \int_{-\frac{(hc+h_f)}{2}}^{\frac{-hc}{2}} f(z)zC_{22}^f dz + \int_{\frac{h_c}{2}}^{\frac{(h_c+h_f)}{2}} f(z)zC_{22}^f dz$$

$$M_{44} = \int_{-h_c/2}^{h_c/2} f(z)zC_{44}^c dz + \int_{-\frac{(hc+h_f)}{2}}^{\frac{-hc}{2}} f(z)zC_{44}^f dz + \int_{\frac{h_c}{2}}^{\frac{(h_c+h_f)}{2}} f(z)zC_{44}^f dz$$

$$M_{55} = \int_{-h_c/2}^{h_c/2} f(z)zC_{55}^c dz + \int_{-\frac{(hc+h_f)}{2}}^{\frac{-hc}{2}} f(z)zC_{22}^f dz + \int_{\frac{h_c}{2}}^{\frac{(h_c+h_f)}{2}} f(z)zC_{55}^f dz$$

$$M_{66} = \int_{-h_c/2}^{h_c/2} f(z)zC_{66}^c dz + \int_{-\frac{(hc+h_f)}{2}}^{\frac{-hc}{2}} f(z)zC_{66}^f dz + \int_{\frac{h_c}{2}}^{\frac{(h_c+h_f)}{2}} f(z)zC_{66}^f dz$$

$$O_{11} = \int_{-h_c/2}^{h_c/2} f(z)^2 C_{11}^c dz + \int_{-\frac{(hc+h_f)}{2}}^{\frac{-hc}{2}} f(z)^2 C_{11}^f dz + \int_{\frac{h_c}{2}}^{\frac{(h_c+h_f)}{2}} f(z)^2 C_{11}^f dz$$

$$O_{12} = \int_{-h_c/2}^{h_c/2} f(z)^2 C_{12}^c dz + \int_{-\frac{(hc+h_f)}{2}}^{\frac{-hc}{2}} f(z)^2 C_{12}^f dz + \int_{\frac{h_c}{2}}^{\frac{(h_c+h_f)}{2}} f(z)^2 C_{12}^f dz$$

$$O_{22} = \int_{-h_c/2}^{h_c/2} f(z)^2 C_{22}^c dz + \int_{-\frac{(hc+h_f)}{2}}^{\frac{-hc}{2}} f(z)^2 C_{22}^f dz + \int_{\frac{h_c}{2}}^{\frac{(h_c+h_f)}{2}} f(z)^2 C_{22}^f dz$$

$$O_{44} = \int_{-h_c/2}^{h_c/2} f(z)^2 C_{44}^c dz + \int_{-\frac{(hc+h_f)}{2}}^{\frac{-hc}{2}} f(z)^2 C_{44}^f dz + \int_{\frac{h_c}{2}}^{\frac{(h_c+h_f)}{2}} f(z)^2 C_{44}^f dz$$

$$O_{55} = \int_{-h_c/2}^{h_c/2} f(z)^2 C_{55}^c dz + \int_{-\frac{(hc+h_f)}{2}}^{\frac{-hc}{2}} f(z)^2 C_{55}^f dz + \int_{\frac{h_c}{2}}^{\frac{(h_c+h_f)}{2}} f(z)^2 C_{55}^f dz$$

$$O_{66} = \int_{-h_c/2}^{h_c/2} f(z)^2 C_{66}^c dz + \int_{-\frac{(hc+h_f)}{2}}^{\frac{-hc}{2}} f(z)^2 C_{66}^f dz + \int_{\frac{h_c}{2}}^{\frac{(h_c+h_f)}{2}} f(z)^2 C_{66}^f dz$$

$$L_{44} = \int_{-h_c/2}^{h_c/2} f(z)'^2 C_{44}^c dz + \int_{-\frac{(hc+h_f)}{2}}^{\frac{-hc}{2}} f(z)'^2 C_{44}^f dz + \int_{\frac{h_c}{2}}^{\frac{(h_c+h_f)}{2}} f(z)'^2 C_{44}^f dz$$

$$L_{55} = \int_{-h_c/2}^{h_c/2} f(z)'^2 C_{55}^c dz + \int_{-\frac{(hc+h_f)}{2}}^{\frac{-hc}{2}} f(z)'^2 C_{55}^f dz + \int_{\frac{h_c}{2}}^{\frac{(h_c+h_f)}{2}} f(z)'^2 C_{55}^f dz$$

REFERENCES

1. Ebrahimi, F., A. Dabbagh, and T. Rabczuk, *On wave dispersion characteristics of magnetostrictive sandwich nanoplates in thermal environments.* European Journal of Mechanics-A/Solids, 2021. **85**: p. 104130.
2. Ghorbani, K., et al., *Investigation of surface effects on the natural frequency of a functionally graded cylindrical nanoshell based on nonlocal strain gradient theory.* The European Physical Journal Plus, 2020. **135**(9): p. 1–23.
3. Hebali, H., et al., *New quasi-3D hyperbolic shear deformation theory for the static and free vibration analysis of functionally graded plates.* Journal of Engineering Mechanics, 2014. **140**(2): p. 374–383.

4. Ebrahimi, F., et al., *Hygro-thermal effects on wave dispersion responses of magneto-strictive sandwich nanoplates.* Advances in Nano Research, 2019. **7**(3): p. 157.

5. Mahinzare, M., H. Akhavan, and M. Ghadiri, *A nonlocal strain gradient theory for rotating thermo-mechanical characteristics on magnetically actuated viscoelastic functionally graded nanoshell.* Journal of Intelligent Material Systems and Structures, 2020. **31**(12): p. 1511–1523.

6. Rao, S.S., *Vibration of continuous systems.* Vol. 464. 2007: Wiley Online Library.

7. Ebrahimi, F., R. Nopour, and A. Dabbagh, *Smart laminates with an auxetic ply rested on visco-Pasternak medium: Active control of the system's oscillation.* Engineering with Computers, 2021: p. 1–11.

8. Ebrahimi, F. and S.B. Sedighi, *Wave dispersion characteristics of a rectangular sandwich composite plate with tunable magneto-rheological fluid core rested on a visco-Pasternak foundation.* Mechanics Based Design of Structures and Machines, 2022. **50**(1): p. 170–183.

9. Singiresu, S.R., *Mechanical vibrations.* 1995: Addison Wesley Boston, MA.

10. Zhou, D. and T. Ji, *Free vibration of rectangular plates with continuously distributed spring-mass.* International Journal of Solids and Structures, 2006. **43**(21): p. 6502–6520.

11. Williams, J., *Laplace transforms.* 1973: Allen & Unwin.

12. Oktem, A.S. and R.A. Chaudhuri, *Higher-order theory based boundary-discontinuous Fourier analysis of simply supported thick cross-ply doubly curved panels.* Composite Structures, 2009. **89**(3): p. 448–458.

13. Oktem, A.S. and R.A. Chaudhuri, *Sensitivity of the response of thick cross-ply doubly curved panels to edge clamping.* Composite Structures, 2009. **87**(4): p. 293–306.

14. Oktem, A.S. and C.G. Soares, *Boundary discontinuous Fourier solution for plates and doubly curved panels using a higher order theory.* Composites Part B: Engineering, 2011. **42**(4): p. 842–850.

15. Neves, A., et al., *A quasi-3D sinusoidal shear deformation theory for the static and free vibration analysis of functionally graded plates.* Composites Part B: Engineering, 2012. **43**(2): p. 711–725.

16. Vel, S.S. and R. Batra, *Three-dimensional exact solution for the vibration of functionally graded rectangular plates.* Journal of Sound and Vibration, 2004. **272**(3–5): p. 703–730.

17. Qian, L., R. Batra, and L. Chen, *Static and dynamic deformations of thick functionally graded elastic plates by using higher-order shear and normal deformable plate theory and meshless local Petrov-Galerkin method.* Composites Part B: Engineering, 2004. **35**(6–8): p. 685–697.

18. Khani Arani, H., M. Shariyat, and A. Mohammadian, *Vibration analysis of magnetostrictive nano-plate by using modified couple stress and nonlocal elasticity theories.* International Journal of Materials and Metallurgical Engineering, 2020. **14**(9): p. 229–234.

19. Ghorbanpour Arani, A., Z. Khoddami Maraghi, and H. Khani Arani, *Orthotropic patterns of Pasternak foundation in smart vibration analysis of magnetostrictive nanoplate.* Proceedings of the Institution of Mechanical Engineers, Part C: Journal of Mechanical Engineering Science, 2016. **230**(4): p. 559–572.

20. Ghorbanpour Arani, A., H. Khani Arani, and Z. Khoddami Maraghi, *Size-dependent in vibration analysis of magnetostrictive sandwich composite micro-plate in magnetic field using modified couple stress theory.* Journal of Sandwich Structures & Materials, 2019. **21**(2): p. 580–603.

Index

Note: Page numbers in *italics* indicate a figure and page numbers in **bold** indicate a table on the corresponding page.

For Product Safety Concerns and Information please contact our EU
representative GPSR@taylorandfrancis.com
Taylor & Francis Verlag GmbH, Kaufingerstraße 24, 80331 München, Germany